Infrared Vapour Spectra

Infrared Vapour Spectra

Group frequency correlations, sample handling and the examination of gas chromatographic fractions

by D. Welti

Unilever Research Laboratory, Sharnbrook, Bedford

PUBLISHED BY	HEYDEN & SON LTD
IN CO-OPERATION WITH	SADTLER RESEARCH LABORATORIES INC.

Heyden & Son Ltd., Spectrum House, Alderton Crescent, London, NW4
Heyden & Son Inc., 1220 Lexington Avenue, New York, N.Y. 10028, USA
Heyden & Son GmbH, Steinfurter Str. 45, 4440 Rheine/Westf., Germany

Co-published for exclusive distribution
in the USA by
Sadtler Research Laboratories Inc.,
3314–20 Spring Garden Street,
Philadelphia Pa. 19104

Library of Congress Catalog Card No. 75-124536
ISBN 0 85501 019 3

Printed by photo-lithography and made in Great Britain at
the Pitman Press, Bath

Contents

Foreword

Much of the early work in infrared spectroscopy was concerned with the examination of simple gases. Their spectra are ideal subjects for theoretical study due to the simplicity, and frequently the high symmetry, of these small molecules. However the widespread growth of interest in infrared spectroscopy followed a quite separate development, that of the examination of complex molecules in the liquid or solid phase, and the empirical correlation of the appearance of absorption bands at particular frequencies with the presence of particular functional groups in the molecule. Thus infrared spectroscopy has become a significant tool in the elucidation of molecular structure. It has subsequently been developed as a rapid method for the identification of unknown substances. This depends upon comparing the spectrum of an unknown with a library of spectra of known compounds, recognising the infrared spectrum as a 'molecular fingerprint'. As the number of library spectra has increased, so rapid methods of comparing spectra have been developed. It is now possible to compare an unknown spectrum with many thousands of reference spectra in a few seconds by computer methods, and thus a simple and rapid identification of an unknown sample may be achieved.

The use of such methods has however revealed difficulties, of which the most formidable is the inability to obtain useful results from mixtures. This had led to a much increased interest in the combination of infrared spectroscopy with efficient separation procedures, and the present work stems originally from Mr. Welti's interest in the combination of infrared spectroscopy with vapour phase chromatography. The direct collection and examination of chromatographic fractions in the vapour phase, usually in a heated gas cell, is a simple and efficient operation. The only substantial problem which has remained in its application as a method of identification of unknowns is the lack of good quality vapour phase spectra in the literature, and a relative lack of knowledge of the differences to be expected between the spectrum of the vapour and the condensed phase. Thus, after a period of relative neglect, interest in the spectroscopy of substances in the vapour phase has been renewed. Apart from the extensive compilation of vapour phase spectra which Mr. Welti has prepared, the examination of relatively complex materials in the vapour phase, particularly those in which intermolecular hydrogen bonding can occur, is shown to be a most interesting study in its own right, which is both of theoretical interest and of practical importance to the industrial spectroscopist.

H. A. Willis

Preface

The spectra of gases have been examined since the earliest days of infrared spectroscopy. However, the greater number of vapour spectra that are found in the literature or in the commercial reference spectra are of simple compounds of low molecular weight and high volatility, and have been determined at room temperature. Although the use of high temperature gas cells increases the range of compounds that can be examined in the vapour state, so far, comparatively few vapour spectra of less volatile compounds have been published.

The object of this publication is to emphasise that vapour spectra can be used for structural determinations as easily as solution liquid or solid spectra. This is especially true for the examination of compounds which are already in the vapour state after separation by gas chromatography.

It is important, therefore, to convince both spectroscopists, and gas chromatographers who do their own spectroscopic determinations, that vapour spectra can be just as useful as, and complementary to, the spectra obtained from other states.

The book is intended to be of use both as a reference to analytical spectroscopists carrying out routine structural identifications, and as a stimulant to those people who are developing the techniques for directly combining gas chromatography and infrared spectroscopy. The vapour spectra of larger molecules are not exceptional. They are generally similar to normal liquid or solution spectra and as a rule they do not have the well resolved rotational lines observed in the vibrational spectra of such small molecules as hydrogen chloride, ammonia or water.

Hydrogen bonding molecules are usually monomeric in the gas phase and have different spectra to the dimers or polymers of the condensed states or solutions. Non-bonding molecules produce vapour spectra very similar to those of the condensed states or solutions, but their individual bands shift in frequency due to the reduced intermolecular interactions in the vapour state.

In addition to the collection of spectra, sections are included on band contours, sample handling and group frequency correlations. There is also a section on the role of vapour spectra in the development of gas chromatographic/infrared combination techniques.

Acknowledgements

I am grateful to Unilever Ltd. for permission to publish this work, and to Mr. H. A. Willis of ICI Plastics Division for the opportunity and encouragement to write it. In addition thanks are due to Drs. R. P. W. Scott and W. Kelly for helpful advice, Miss C. Browning and Messrs. I. A. Fowlis and D. Hughes for assistance, and to Mr. H. Deakin for some photography.

The publishers of the Sadtler Spectra, the DMS Spectra, the ASTM collection of Wyandotte/ASTM infrared punched cards and Drs. Erley and Blake of the Dow Chemical Company must be thanked for giving permission to use their spectra or cards in the accumulation of the general infrared vapour spectra index.

In addition, I am grateful to the following organisations or people for permission to reproduce some of the illustrations in the text:

The Research and Industrial Instrument Co. for Figures 1.1, 1.2, 1.5
Perkin-Elmer Ltd. for Figures 1.4 and 1.9
Wilks Scientific for Figures 1.6, 1.7 and 4.4
The Editor of the *Journal of Gas Chromatography* for Figure 1.8
The Elsevier Publishing Co. for Table 3.1, Figures 3.1 and 4.2
Varian Aerograph for Table 4.1 and Figure 4.1
The Editor of *Analytical Chemistry* for Figures 4.3 and 4.4
Beckman Instruments Inc. for Figure 4.5
The Institute of Petroleum for Figure 4.6
The Pergamon Press for Figure 4.7
Block Engineering Inc. for Figure 4.8 and its legend
The Warner and Swasey Co. for Figure 4.9

1
Sample handling

a. Criteria for efficient gas cells

In common with all types of spectroscopy where the sample size is small, it is necessary to obtain the maximum sensitivity from the cells that are available. Although for instrumental reasons the experimental absorbance value of an infrared band is not strictly proportional to the concentration of the solution or the pathlength of the cell, it is still dependent to a large degree on these factors. In the case of a vapour, the concentration or partial vapour pressure is dependent on the sample size and the volume of the cell. For a fixed sample size, the absorbance is controlled by the pathlength and volume of the cell. For the greatest cell efficiency, the pathlength must be high and, to keep the concentration high, the volume must be low. One criterion of the cell efficiency is therefore the ratio of pathlength to volume, the P/V ratio. The unit is a reciprocal area and in terms of centimetres is cm^{-2}. When there is a choice of cells of the same P/V ratio and different pathlength, it is better to use the cell with the higher percentage light transmission.

Unfortunately, with a reduced cell volume the normal collimated radiation is heavily attenuated. This means either that a low cell transmission has to be tolerated, or that some form of light condensation and recollimation has to be used. The former is simpler if the available light is sufficient to give a power output large enough to drive the servo system efficiently. Otherwise the cell must be built into an external optical system such as those used with reflecting beam condensers or multiple attenuated total reflectance attachments.

Modern spectrometers, including the small prism and grating models, can work down to very low light transmission values. A transmission of 5% in the sample cell gives a spectrum with a good signal-to-noise value. Transmissions of down to 1% will give workable spectra, but the noise level will increase, and the spectrometer comb attenuator will not be working efficiently. A fixed additional attenuator in the reference beam will lower the recorder base-line, but in so doing it will remove more energy from the detector and therefore, the servo system.

It must be remembered that the design of the cell must be compatible with the optics of the spectrometer. Every spectrometer has different optics, so ideally, every spectrometer should have its own especially designed cell. In practice, a cell can be used in a number of spectrometers, provided that the operator realises that its performance will differ in each spectometer.

b. Vapour pressures and the use of cells at elevated temperatures

The maximum concentration of a vapour is fixed by its saturated vapour pressure at the experimental temperature. Many organic compounds possess too low a vapour pressure at room temperature to produce strong or even recognisable spectra, but on heating, their partial vapour pressures can be increased sufficiently to give strong spectra. The gas cell should therefore be capable of working at elevated temperatures, when it can be used to examine both less volatile samples that arc introduced singly and also fractions that appear in a hot gas chromatographic effluent.

The sample weight, vapour pressure and the molecular weight are all related to each other. Polar molecules give more intense spectra than non-polar molecules, so whereas a 0·1 mg sample might give a strong spectrum of acetone, 0·8 mg might be needed to give a strong spectrum of a paraffin hydrocarbon.

After calculating the sample size necessary for a cell of a particular volume, it is necessary to confirm that the partial vapour pressure is below the saturation vapour pressure of the molecule at the temperature of the cell. If it is not, the sample will condense in the cell (most probably on the windows, which are the coolest part of the cell) and the spectrum could include that of the liquid film. It must be remembered that a decrease in cell volume means an increase in vapour pressure for a fixed weight of sample, so it may be necessary to increase the cell temperature to keep above the saturation vapour pressure. On the other hand the temperature must not be increased to such an extent that the sample decomposes.

Jordan has published a most valuable collection of vapour pressure/temperature data in *Vapour Pressure of Organic Compounds.*[1]

c. Introduction of samples into gas cells

For the determination of standard reference vapour spectra of higher molecular weight compounds, it is not necessary to introduce the sample into an evacuated cell under controlled pressure conditions. It is quite sufficient to inject a calculated quantity of the sample into the cell at atmospheric pressure in the presence of air, or if there is some danger of oxidation, in the presence of an inert gas such as nitrogen or argon. Solids can be inserted directly into the cell in crystalline or powdered form.

In the case of compounds that are gases at the ambient temperature and pressure, it is necessary to use proper gas-handling apparatus for transferring the gas under controlled pressure conditions. As a result the gas cells must be capable of working under vacuum or under positive pressure. The intensity of vapour spectra can be enhanced by increasing the vapour pressure, using pressures above atmospheric if necessary; but when measuring the rotational-vibrational spectra of small molecules, high resolution spectra show a broadening in the bands as the pressure is increased. In addition, the band intensity and broadening are increased as the absolute pressure is increased by the admission of an inert gas into the cell, while keeping the partial vapour pressure of the sample constant.

In gas chromatographic/infrared experiments, it is assumed that all measurements will be carried out at atmospheric pressure in the presence of inert carrier gases.

The cells are generally purged with an inert gas until they are free of the previous sample. In the case of the combination with gas chromatography the cell can either be purged through a separate gas line or by the carrier gas stream that is eluted between well-separated fractions.

d. Choice of cell and window materials

Gas cells have been made from materials such as glass, stainless steel, mild steel, copper, brass and nickel. The steels are probably the best materials for the manufacture of heated cells, as they are less likely to react with the vapours or window materials and can withstand high temperatures. Mild steel is reputed to be the easier to coat with gold, which is commonly used to make the inner surface of the cell more inert and efficient as a reflecting surface.

Silver chloride is one of the most useful of the common window materials that are available. It is sufficiently soft and malleable to form a self-seal between the window flanges, thereby overcoming the dangers of adsorption on to gasket materials; at the same time it is less prone to crack under the pressures that sometimes develop on heating cells. Its transmission properties over the 2–15 micron range are good and when new it will give a straight base line, even in a small 5 ml 'minimum volume' cell, without a reference cell. Silver chloride tends to tarnish in visible light after long exposure and the windows will begin to 'bow' at temperatures above 100°C. Both these effects influence the cell transmission. In the first stages, any discrepancy in the base line can be balanced out with similar pieces of silver chloride in the reference cell. Continued use at temperatures above 150°C exaggerates the 'bowing' effect until the windows adopt lens-like properties which diminish the light transmission of the cell. Otherwise, a cell can remain in use for months without having to change the windows.

It is important to store cells fitted with silver chloride windows in an oven above room temperature, e.g. 60–70°C. Any moisture condensing on the metal promotes an electrolytic action between the silver chloride and the metal and the windows soon dissolve and collapse. This effect is much more pronounced with brass and copper, although it does occur with steel.

No molecules that can form silver complexes should be introduced into the cell. Hot ethylene diamine vapour can ruin a pair of windows very quickly.

All other window materials have drawbacks. All are more brittle than silver chloride and need gaskets, except for KRS-5. This, however, is a toxic material which would be dangerous at high temperatures. The Irtran* series of windows all have some absorption in the fundamental infrared and are very brittle, as are the fluorides. Sodium chloride and potassium bromide are the cheapest and most readily available materials, but in addition to needing gaskets, if they come into contact with carboxylic acids for a prolonged period of time at high temperatures, layers of the acid salt will form on their inner surface.

Some substances will decompose at high temperatures. Saturated carboxylic acids decompose to give an increasing CO_2 spectrum at 190°C. Pyruvic

* Registered trademark of Eastman Kodak Co.

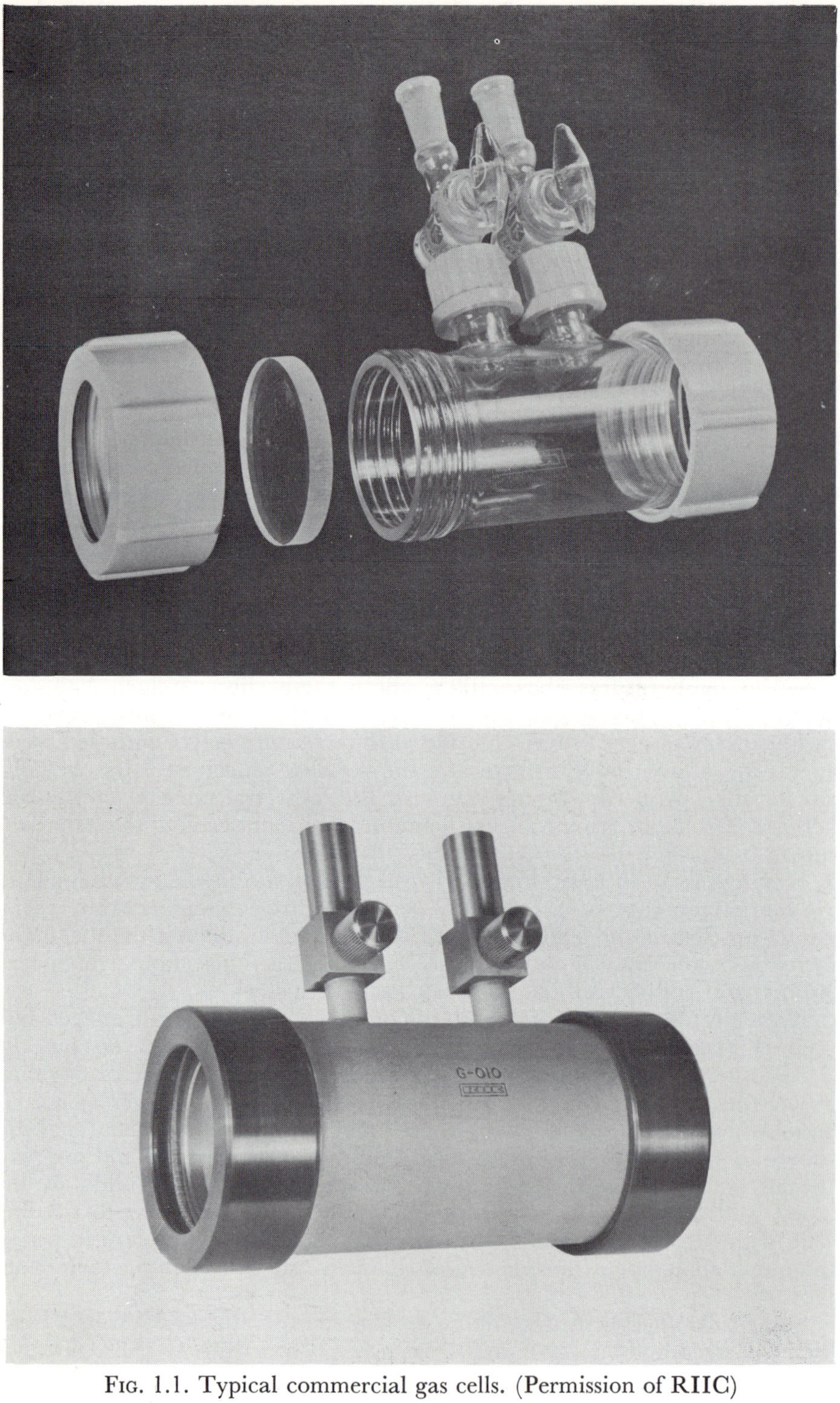

FIG. 1.1. Typical commercial gas cells. (Permission of RIIC)

and propynoic acids begin to decompose above 100°C. It is difficult to say whether this is solely an effect of temperature or whether it is due to contact with the material of the cell after the gold surface has been worn away. Amines decompose to give ammonia more easily as the temperature increases. In all these cases and there are probably more, it is preferable to obtain a spectrum at the lowest temperature at which the vapour pressure is sufficient for a reasonable spectrum to be measured.

High boiling compounds and particularly those which form strong intermolecular bonds, can provide 'memory' troubles which are shown by the reappearance of their spectra after the cell has been purged. This can probably be associated with the problem of adsorption on, or solution in, parts of the gas cell. Some causes of this are the presence of gaskets between the cell windows and the flanges, the presence of O-rings in the unions attached to the cell, and the septa through which samples can be injected. As the cell volume and the sample size decreases, adsorption problems become increasingly important. In one experimental cell with a volume of 0·1 ml, a 10 microgram sample which initially gave a strong spectrum, was completely 'lost' within a minute. Tests showed that the cell was vacuum and pressure tight and that the losses occurred solely due to adsorption on to an O-ring in a union attached to the entry tube. As mentioned previously the need for window gaskets can be overcome by using silver chloride windows. If gaskets have to be used, the lead or aluminium gaskets as used by Sterling and Haines[2] or the PTFE gaskets used by Willis* are quite suitable.

e. Examples of gas cells in use

The most common type of cell is represented in Fig. 1.1, which shows typical (Research and Industrial Instrument Company[3]) glass and metal cells. This type of cell is generally a cylinder about 4 cm in diameter and about 10 cm

FIG. 1.2. Minimum volume gas cell. (Permission of RIIC)

long, the maximum length that will give a direct transmission through the cell compartment of most modern infrared spectrometers. The resultant volume of

* H. A. Willis. Private communication.

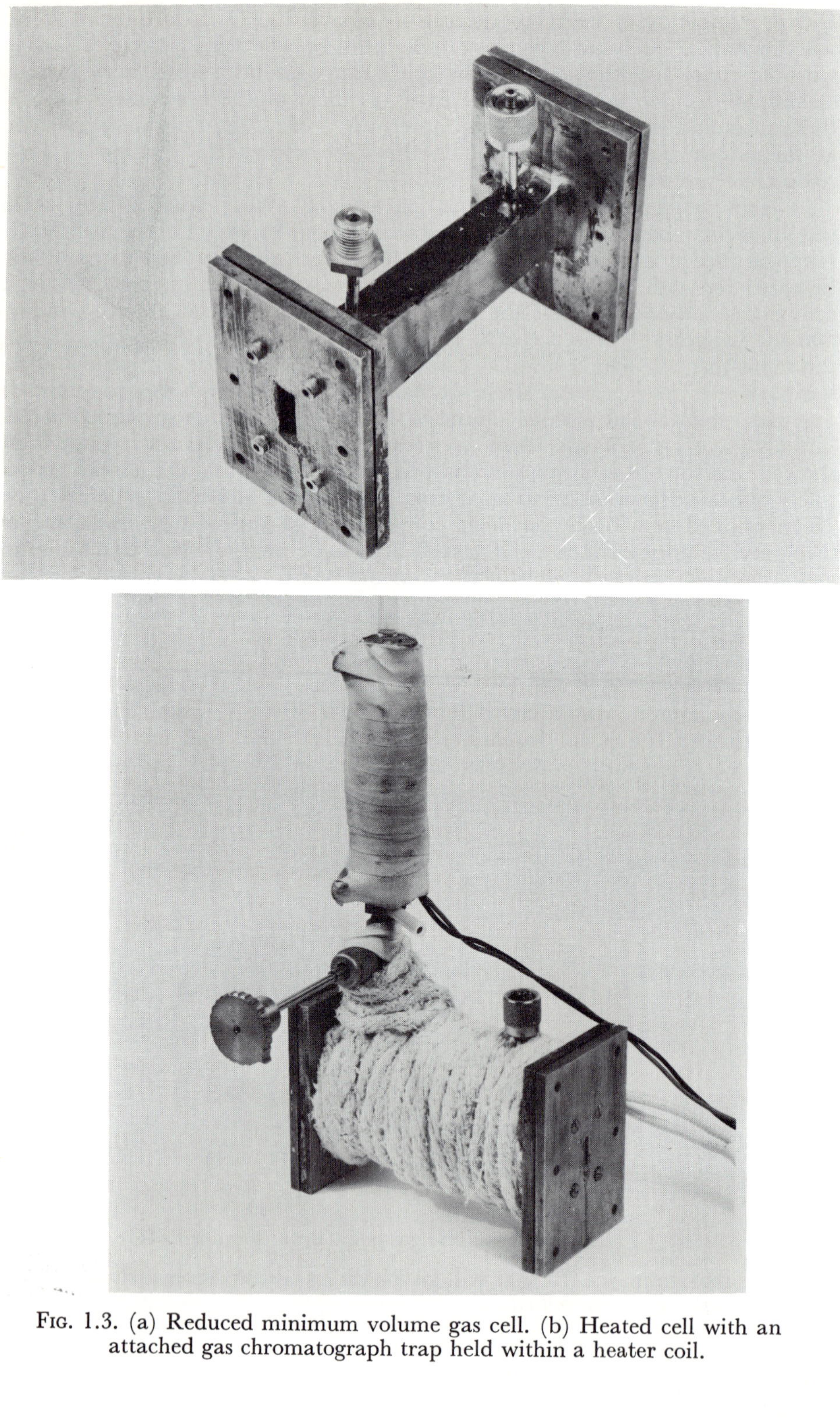

FIG. 1.3. (a) Reduced minimum volume gas cell. (b) Heated cell with an attached gas chromatograph trap held within a heater coil.

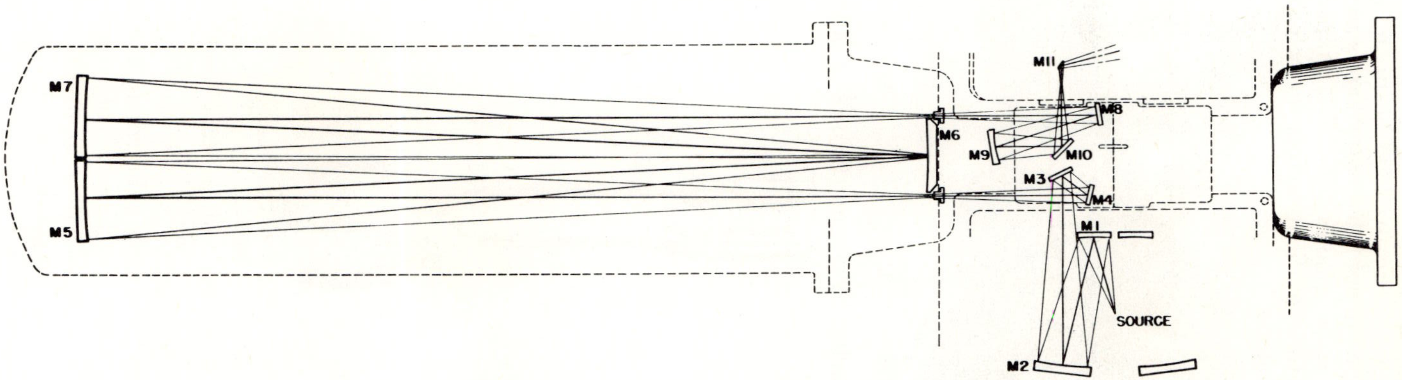

FIG. 1.4. (a) Optical path of a Perkin-Elmer 40 metre multipath cell.

70–120 ml gives a pathlength/volume (P/V) ratio of 0·08 to 0·15 cm^{-2}. A large part of the sample is outside the effective light beam and is unused. Most of the commercial cells of this type are not fitted with heaters and can be used at room temperature only. They can be modified for high temperature work, but they will have greater temperature gradients across the cell and cell windows than cells with smaller cross-sections.

FIG. 1.4. (b) A dual multipath system. (Permission of Perkin-Elmer Ltd.)

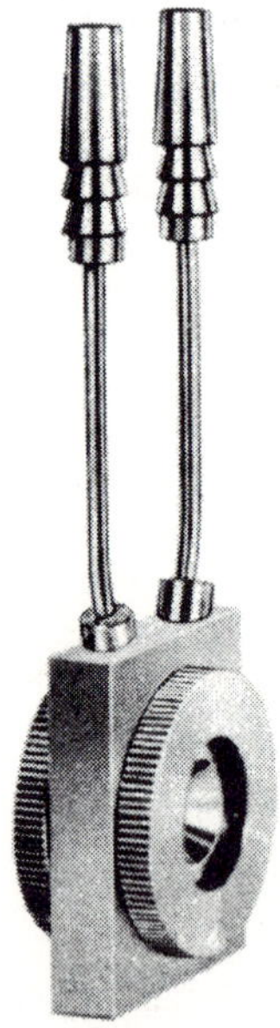

FIG. 1.5. Micro-gas cell for use with a beam condenser. (Permission of RIIC)

To prevent the waste of sample inherent in the large cylindrical cells, 'minimum volume' cells were developed. They have either a conical or pyramidal structure which approximately follows the shape of the effective light beam passing through the cell compartment. An example of a minimum volume cell for use with a light beam focussed at the photometer of the end cell compartment is given in Fig. 1.2. Cells with pathlengths of 9–10 cm, volumes of about 25 ml and a P/V ratio of about 0·4 cm^{-2} can be used in the Unicam[4]

SP200 or Perkin-Elmer[5] 137 series of instruments. This is an improvement over the normal type of cell and all the sample is utilised. The transmission is still close to 100%. The cells can be heated to 200°C and should be gold plated.

The intensity of light which reaches the detector is not the same over the whole effective cross-section of the light beam. In practice the centre of the beam tends to be more intense than its perimeter. The cross section of the beam can therefore be reduced without proportionately reducing the percentage transmission of a cell. A 'reduced minimum volume' cell is shown in Fig. 1.3. It is 10 cm long and with its truncated-pyramid geometry, has a volume of 5 ml, a P/V ratio of about 2·0 cm^{-2} and a transmission of 45–50%. This cell can also be heated to 200°C and is internally gold-plated.

Another cell which has an increased P/V ratio is the multiple path cell, which was first described by White, Alpert, Ward and Gallaway.[6] An example of a large volume Perkin-Elmer[5] multiple path cell is shown in Fig. 1.4. Internal mirrors reflect the light beam many times along the length of the cell and through the sample. The cell used by Flett and Hughes[7] has a pathlength of 240 cm, a volume of 130 ml and a P/V ratio of 1·8 cm^{-2}. A Grubb Parsons[8] cell can be used at elevated temperatures, but the mirrors have to be adjusted to suit the physical dimensions at the temperature to be used. One difficulty in using such a cell is the possibility of the adsorption of the sample on to the mirror mountings, with the allied memory effects and loss of sample. In addition, the mounting of mirrors within the cell means that the space around the mounts and behind the mirrors becomes dead space and it is possible that despite the multi-reflections that half the sample may never see the light beam.

Micro gas cells for use with beam condensers are also available. A Research and Industrial Instrument Co.[3] unheated cell is shown in Fig. 1.5. It has a pathlength of 0·8 cm, a volume of 0·15 ml and a P/V ratio of 5·3 cm^{-2}.

Many of the cells used in the direct combination of gas chromatography and infrared spectroscopy are so called 'light-pipes'. These are metal tubes of circular or rectangular cross-section with highly reflecting, generally gold-plated, surfaces. The light source is moved back so that the focus, which coincides with the entrance to the light-pipe, is transferred to the exit of the light-pipe, but the process is not completely efficient. The light emitted by the pipe

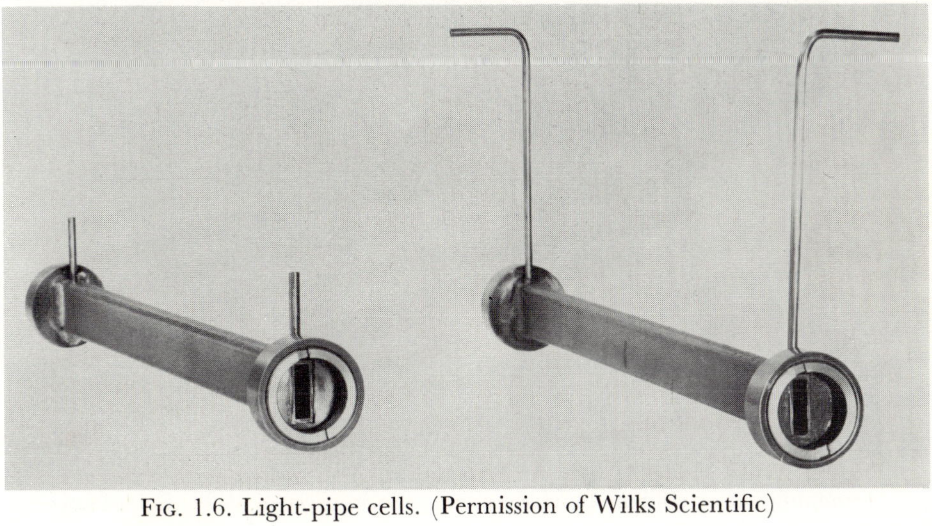

FIG. 1.6. Light-pipe cells. (Permission of Wilks Scientific)

TABLE 1.1

Origin of Light-pipe	Dimensions in cm: Length	Dimensions in cm: Cross-section	Volume in cm^3	P/V Ratio in cm^{-2}	Trans-mission (%)	Maximum Tempera-ture (°C)
Bartz and Ruhl[9]	30	0·32 × 1·27	12	2·5	40	250
Wilks and Brown[10]	30	0·5 × 1·5	22·5	1·3	50	250
Beckman IR 102[11]	10	—	3	3·33	30	350
	20	—	6	3·33	15	
	30	—	9	3·33	7	
Low and Freeman[12]	5	0·4 id	0·6	8·3	—	Used at 150
Carle Instruments[13]	6	—	2·5	2·4	30	250
Wilks[14] Model 41 (see Fig. 1.7)	6	0·1 cm^2	0·6	10	25–35	250

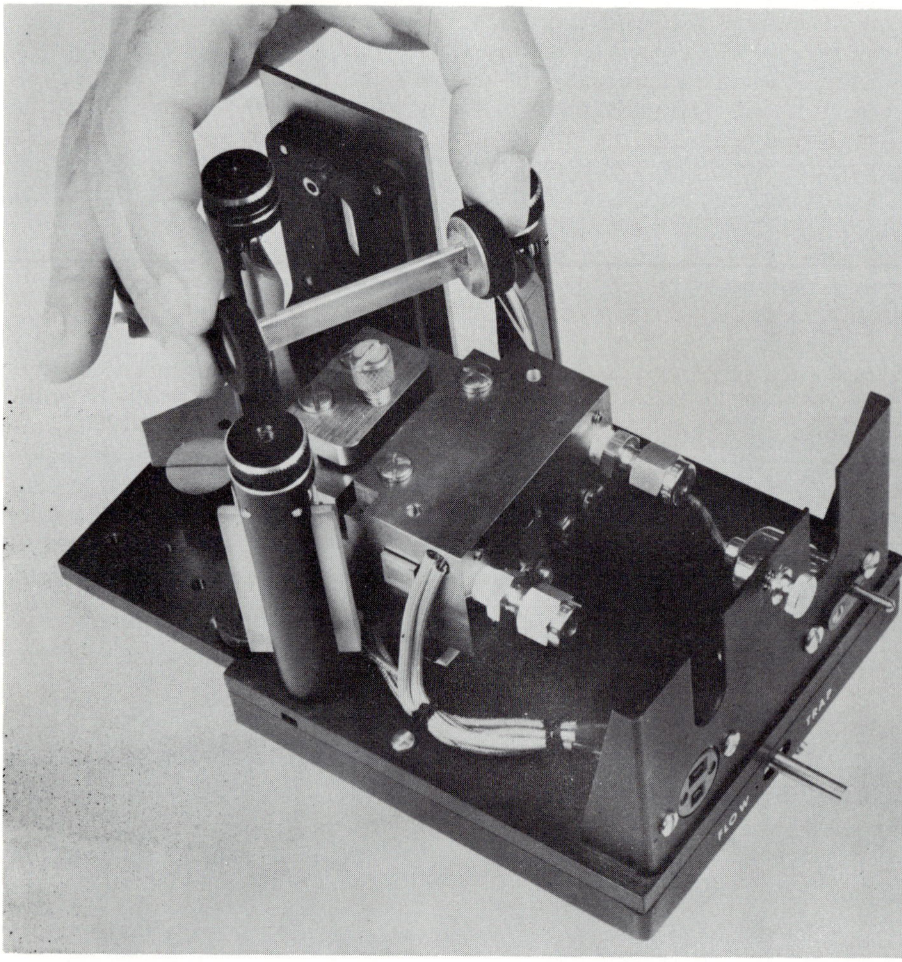

FIG. 1.7. Wilks Model 41 light-pipe, including the beam condenser and gas chromatographic effluent by-pass system. (Permission of Wilks Scientific)

is therefore either allowed directly into the photometer unit without further collimation and with the resultant loss in transmission, or it is recollimated into the photometer as if it had never seen the cell. In particular cases internal reflections within the pipe increase its apparent pathlength up to an additional 33%. Examples of such cells are shown in Fig. 1.6.

The properties of a selection of light-pipes are given in Table 1.1. Those of Wilks and Brown, Carle and Wilks can be used in conventional spectrometers.

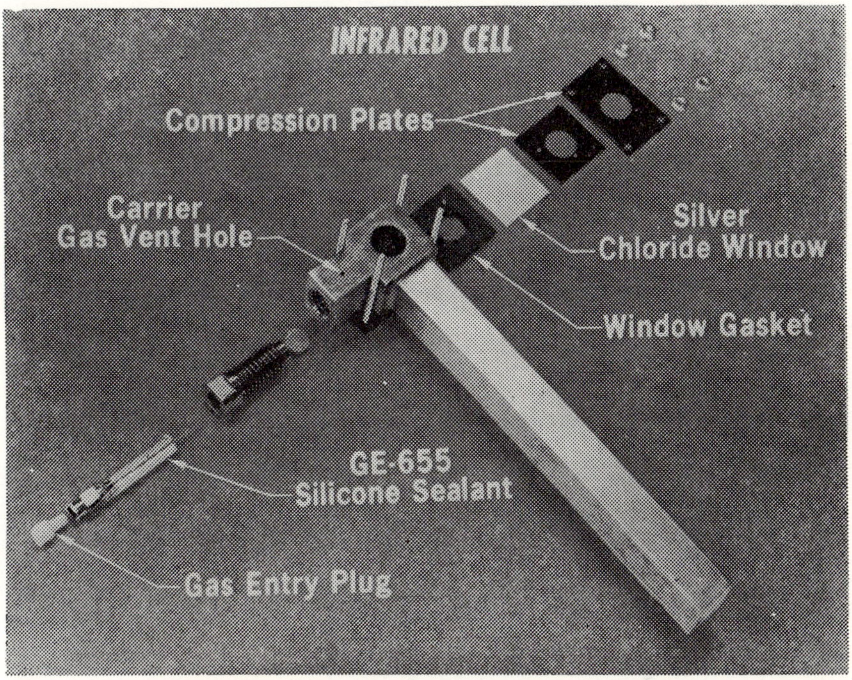

FIG. 1.8. The Ballinger, Bartels and Taylor composite gas chromatographic fraction trap/infrared gas cell. (Permission of the Editor of the *Journal of Chromatography*)

Two other gas cells are of interest. Sterling and Haines[2] have described a cell about 10 cm long and 80 ml volume (P/V ratio = 0·125) which can be heated to higher than 200°C and can hold a vacuum of 10^{-3} mm Hg over 24 hours. It can be evacuated to 10^{-5} mm. They describe the use of pyrolytic and gas chromatographic accessories. Ballinger, Bartels and Taylor[15] have constructed a cell (Fig. 1.8) which is both a gas chromatographic fraction trap and an infrared cell. Its pathlength is 2·54 cm, volume 3·17 ml and P/V ratio 0·8 cm^{-2}. Using silver chloride windows and a beam condenser its light transmission is 70%, without a beam condenser it is 63%. It can be heated to 200°C, but can be cooled with liquid nitrogen or any other coolant for chromatographic trapping purposes.

Large volume, long pathlength cells have been used for examining either the absorption or emission spectra of trace amounts of permanent gases where sample size is no limit and a low partial vapour pressure is the only problem. A Perkin-Elmer[5] 10 metre cell is shown in Fig. 1.9 and a 40 metre multipath cell is shown in Fig. 1.4.

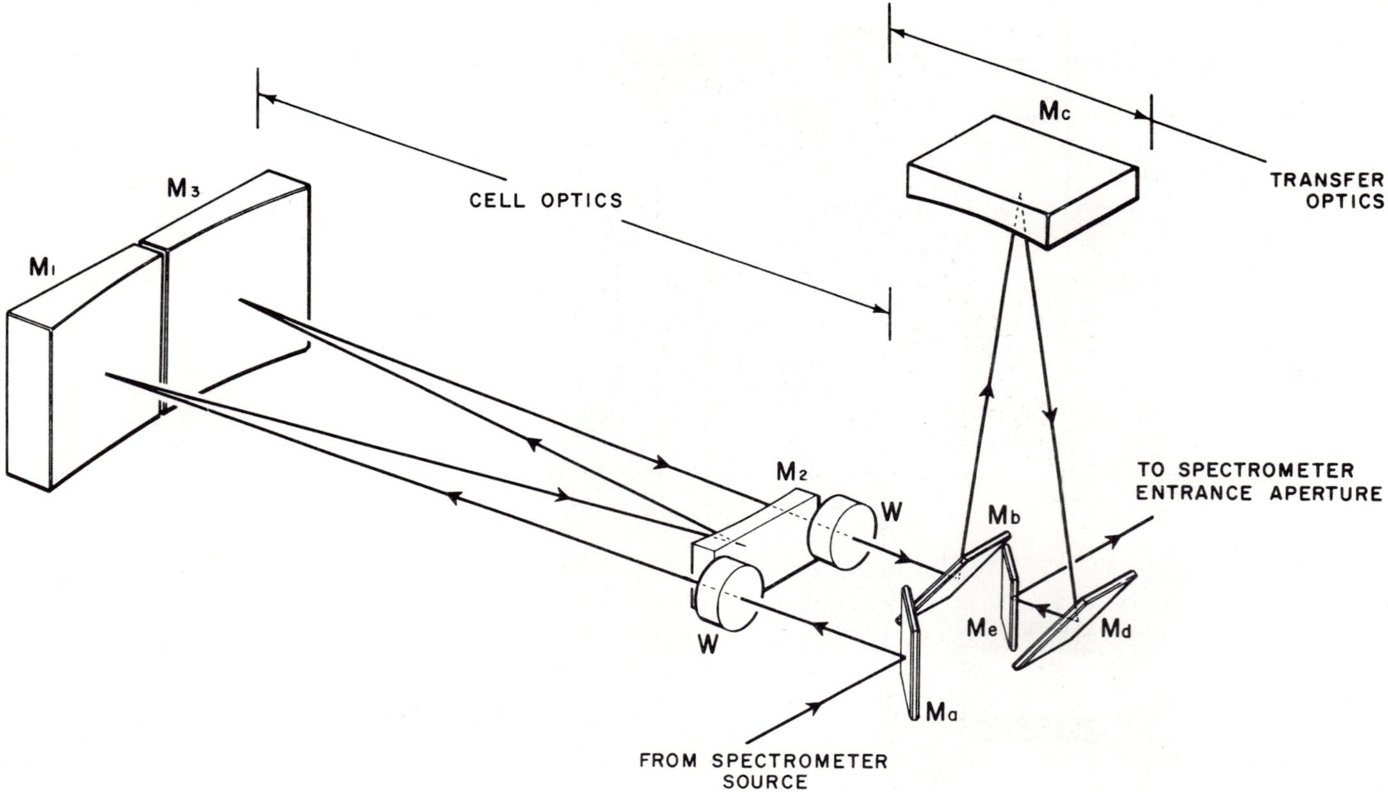

FIG. 1.9. A Perkin-Elmer large volume 10 metre gas cell. (a) Optical path schematic.

FIG. 1.9. (b) Cell. (Permission of Perkin-Elmer Co.)

Although this book is intended to deal primarily with absorption vapour spectra, it is appropriate at this point to emphasise briefly the large amount of work done on emission spectra, principally in optical detection systems. Recently Low and Coleman,[16] have developed the Block Model 200 interference spectrometer[17] for determining emission spectra and Low and Claney[18] have described its use for the remote sensing and characterisation of stack gases. In another type of approach to the subject Robinson, Woodward and Barnes[19] have detected gaseous organic compounds by the infrared emission stimulated by a laser beam.

2
Frequency correlations of vapour spectra

a. General observation

It will be seen from the general index of vapour spectra in Chapter 6 that although the vapour spectra of nearly 2000 compounds are available in the literature, a large number of these are of polyhalogenated and inorganic molecules which form separate specialised fields and will not be considered here. As stated previously, the intention of this book is to show that the data that can be obtained from vapour spectra of ordinary simple organic compounds can easily be included in the normal group frequency correlation tables and, indeed, valuable extra evidence may be obtained in many cases by comparing liquid and vapour spectra.

Although use is made of additional data obtained from the literature or from commercial spectra in considering the frequency range from 650–1500 cm^{-1}, the values of the ν(C=O), ν(O—H) and ν(N—H) bands are taken mainly from the spectra presented in Chapter 5. This is principally due to some doubts on the calibration of the many prism spectra where the higher frequency values disagree randomly with the more internally consistent spectra presented in Chapter 5. Welti and Stephany[20] have quoted the frequency values of some of the interesting bands in a number of these spectra in Chapter 5. The values are not reproduced here, but more generalised sets of group frequency correlations are tabulated in this section.

In solution spectra, the frequency of a band maximum shifts on changing from a more polar to a less polar solvent, due to a decrease in the strength of the intermolecular forces. A vapour can be thought of as the extreme state of dilute non-polar solution where interaction is very weak or negligible. The band positions are shifted to those expected for infinitely dilute solutions.

The bands in the spectra of all molecules shift to a greater or lesser degree on a change of state. Bands associated with stretching vibrations move to higher frequencies in the vapour state and bands associated with deformation vibrations move to lower frequencies. Monomeric ν(OH) bands can show differences of 20–40 cm^{-1} between their positions in carbon tetrachloride solution and vapour spectra; ν(C=O) bands show smaller differences of from 5–20 cm^{-1}. The spectra of molecules that are normally hydrogen bonded show more profound changes. Most of them exist as dimers or polymers in concentrated solutions, liquids or solids; but on diluting the solutions, the probability of intermolecular interaction decreases and the monomeric form is preferred. It is no surprise therefore that most hydrogen bonding compounds exist predominantly as monomers in the vapour state. The most notable exceptions are the carboxylic acids which exist mainly as dimers at low temperatures, but even in these cases the monomeric form predominates at higher temperatures. The band shifts of the ν(OH) and ν(C=O) bands seem to be exceptionally large since the solvent shifts must be added to the dimer → monomer shifts.

It is often overlooked that where a molecule has a vapour pressure of a few millimetres of mercury, it is easier to obtain the vapour spectrum than the solution spectrum. Apart from the ease of sample handling, there is no blanking out of whole regions of the spectrum due to the high absorption of long path-lengths of solvent. This is particularly true in the case of carboxylic acids.

With the very simple molecules, the P and R branches of the bands may be resolved into rotational fine structure. As the molecule becomes more complex, the spacing between rotational lines narrows and the P and R branches appear as broad maxima. The band contours due to this rudimentary rotational branching can be correlated with the geometry of the molecule, and the technique is introduced briefly in the next section.

Due to the increased population of the higher rotational energy levels with increasing temperature, the wings of the vibration bands increase in intensity, making them look broader than their counterparts in solution or condensed state spectra. Shoulders and inflections that are prominent on the sides of bands in other spectra can appear to be missing, or very weak, in vapour spectra.

In recent years a number of papers[21–25] have been published describing the transfer to computer programmes of the information that is available on the Wyandotte/ASTM infrared punched cards.[26] These programmes were written specifically to aid in the rapid and automatic identification of spectra by comparing them with the spectra in the literature or the commercial reference collections. Cross, Shields, Proudlock and Welti[27] describe the routine applications of a programme which was first described by Cross, Haw and Shields[28] and then developed by jointly ICI Ltd. and Unilever Ltd. They show that in the case of non-hydrogen bonded compounds, search requests based on spectra in any physical state can be matched with reference spectra in any physical state and this includes the vapour state. However, in the case of compounds capable of molecular association, particularly hydrogen bonding, spectra may be matched reliably only with reference spectra obtained from molecules in the same physical state. This is almost invariably the case with carboxylic acids and amines. Some alcohols can probably be cross-matched, while phenols, particularly those which are intramolecularly bonded, are much more amenable to cross-matching.

Although it is hoped that the following discussions on the vapour spectra of differing types of molecules are self-supporting, the reader must be assumed to have some knowledge of group frequency correlations, such as those given by Bellamy,[29,30] Jones and Sandorfy,[31] Flett,[32] Beaven, Johnson, Miller and Willis[33] and the Irscot system.[34]

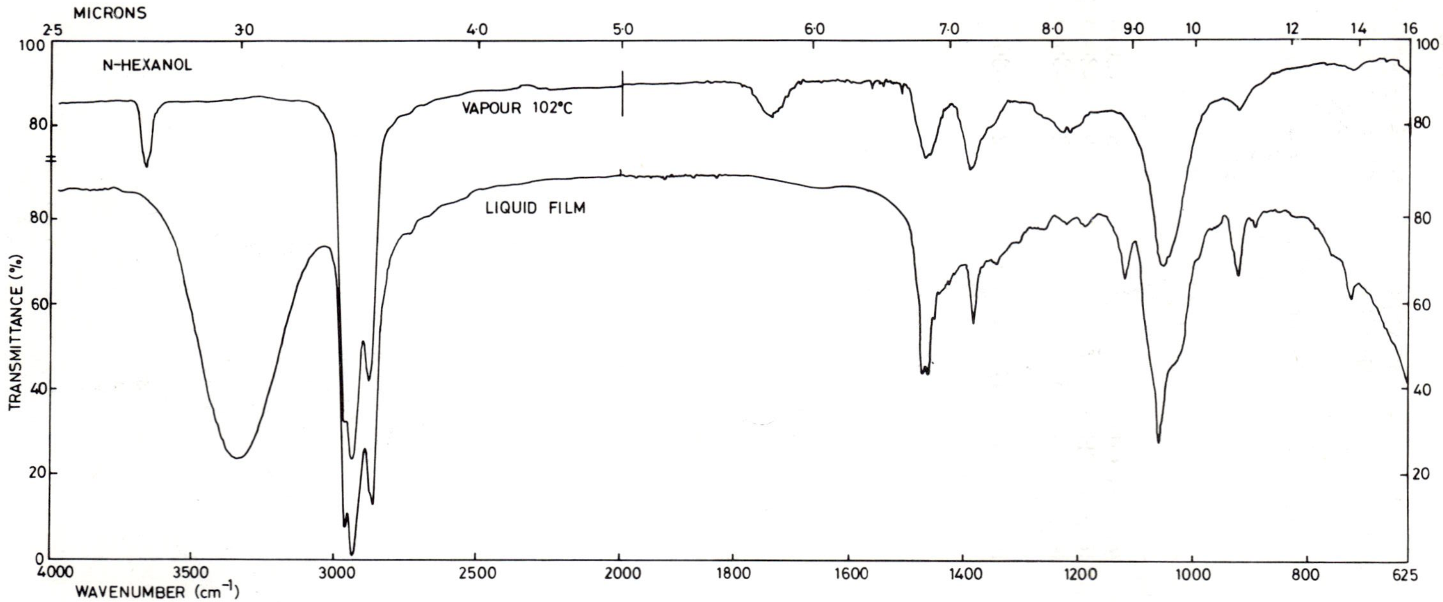

FIG. 2.1. Comparison of the vapour and liquid spectra of n-hexanol.

b. Alcohols

The typical frequencies found in the vapour spectra of alcohols are given in Table 2.1. The vapour and liquid spectra of a simple alcohol, n-hexanol, are compared in Fig. 2.1.

TABLE 2.1. Band maxima in the infrared spectra of alcohols

Class		No. of spectra observed	Frequency (cm^{-1})
v (OH) free or monomeric[a]			
Primary saturated	R—CH_2OH	5	~3670 (w-m) sharp
Secondary saturated	R_1R_2—CHOH	9	~3660 (w-m) sharp
Tertiary saturated	$R_1R_2R_3$—COH	2	~3640 (w-m) sharp
Primary 3-enol	R—CH=$CHCH_2CH_2OH$	2	~3670 + 3625 (w-m) sharp
Secondary '3'-enol	R_1—CH=$CHCH_2CH(R_2)OH$	2	~3660 + 3625 (w-m) sharp
Primary 2-enol	R—CH=$CHCH_2OH$	5	~3660 (w-m) sharp
Secondary '3'-enol	R_1—CH=$CHCH(R_2)OH$	3	~3650 (w-m) sharp
2-hydroxy esters	$(COOR_1)CH(R_2)OH$	4	~3640 + 3575 (w-m) sharp
3-hydroxy esters	$(COOR_1)CH_2CH(R_2)OH$	1	sh + 3590 (w-m) sharp
3-hydroxybutan-2-one	$CH_3C(=O)CH(OH)CH_3$	1	~3660 + 3527 (w-m) sharp
δ(CH), see text			
All alcohols			~1380 (m) broad
δ(OH)			
All, except tertiary alcohols			~1270–1220 (w) broad
Tertiary alcohols		3	~1330–1315 (w) broad
v(C—O)			
All primary alcohols, except 2-enols		9	~1065–1050 (s)
Primary 2-enols		3	~1020 (s)
All secondary alcohols, except '2'-enols		11	~1145–1130 (s)
All secondary alcohols			bands between 1100–1000 (s)

[a] Steric and conformational factors will modify these values, particularly in cyclic structures.

The free monomeric OH stretching bands occur at frequencies that are anything up to 35 cm^{-1} higher than the equivalent bands in the spectra of dilute solutions: the frequency differences between the various types of alcohols are maintained. Steric and conformational effects also cause shifts in the frequency values. The spectra of various geometric and conformational isomers of cyclic hydroxy compounds have been obtained from dilute solutions. The frequency shifts of these isomers should also be apparent in their vapour spectra.

Alcohols capable of intramolecular bonding normally show two ν(OH) bands in the vapour state, as they do in solutions. Generally the bonded OH band is stronger than the free OH band. For example, 1,3-propanediol (Spectrum No. 49) which can form a hydrogen bonded 6-membered ring (**1**)

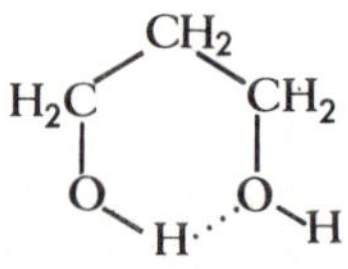

(1)

shows two distinct bands at 3605 and 3660 cm^{-1}. On the other hand, 1,2-ethanediol (Spectrum No. 47) and 1,2-propanediol (Spectrum No. 48) show only complex branched bands around 3650 cm^{-1}. Presumably the 5-membered ring structure (**2**)

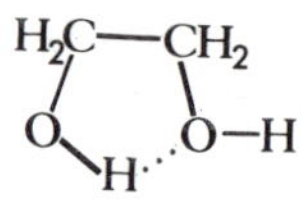

(2)

is less stable than the six-membered ring of the 1,3-diol.

Intramolecularly bonding hydroxy esters and ketones also show two bands (Table 2.1). The 2-hydroxy esters and 3-hydroxybutanone, which form 5-membered rings (**3**, **4** and **5**) show two distinct bands, but the 3-hydroxy ester

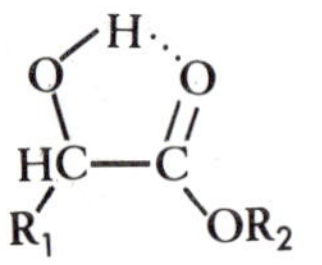

(3)

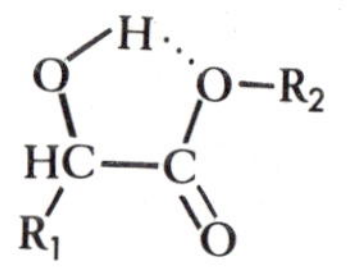

(4)

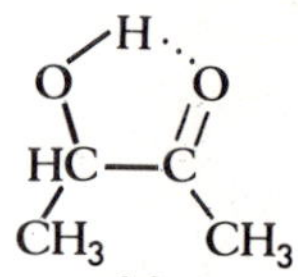

(5)

shows only a shoulder with the intramolecularly bonded hydroxyl band at 3590 cm^{-1}. This could be caused by the hydrogen bonded 6-membered ring (**6** and **7**) being stable enough to preclude the existence of any non-bonded molecules in any quantity.

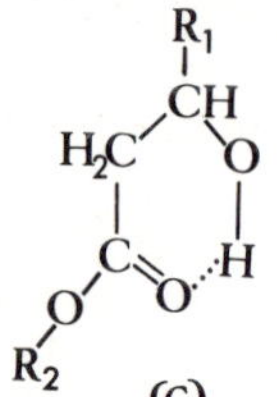

(6)

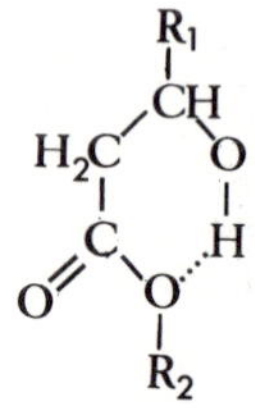

(7)

The OH group hydrogen bonds with the π electrons of $\beta\gamma$ (**8**), but not $\alpha\beta$ (**9**), double bonds in unsaturated alcohols.

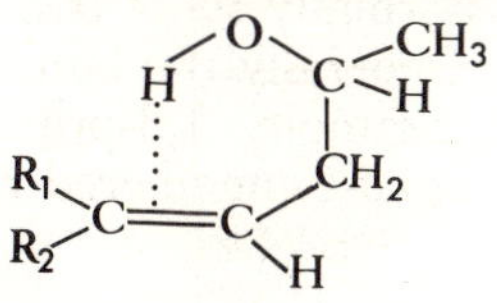

(8)

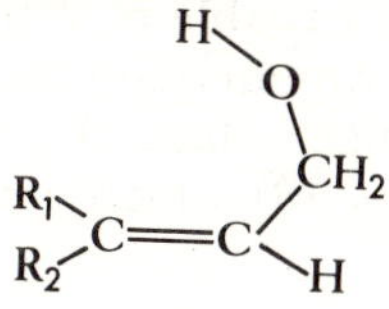

(9)

The band that appears at 1380 cm^{-1} in all alcohols must be considered in conjunction with the δ(OH) and ν(C—O) bands. Although there is confusion in the literature on the location of the δ(OH) and ν(CO) bands, we have followed the assignments of Stuart and Sutherland,[35] which are discussed by Tarte and Deponthiere.[36] While the ν(C—O) bands near 1050 cm^{-1} in primary alcohols, or the group of bands including the 1130–1145 cm^{-1} band in secondary alcohols are easily recognisable, the δ(OH) band is by no means so distinctive. In fact the 1380 cm^{-1} band is more intense than the δ(OH) band, and it appears in all alcohols. It is very noticeable that its intensity is increased in the vapour spectra of most alcohols, including benzyl alcohol, furfuryl alcohol and hex-5-en-1-ol, where it cannot be attributed to a CH_3 deformation vibration. Krimm, Liang and Sutherland[37] have suggested that in the liquid state the δ(OH) mode interacts with a δ(CH) mode, giving two mixed vibrations δ(OH + CH) with bands close to 1330 cm^{-1} and 1410 cm^{-1}. They suggest that with the monomeric molecules found in the vapour state, this interaction disappears leaving a δ(OH) band near 1250 cm^{-1} and a δ(CH) band near 1380 cm^{-1}. Whether this is correct or not, it seems odd that while the δ(CH) and δ(OH) bands interact, the ν(C—O) band remains virtually at the same frequency, with the same comparative intensity. More evidence is also needed to explain why the 1380 cm^{-1} band in the vapour spectrum is so consistent in position and intensity whatever groups (aromatic rings, saturated rings, alkyl groups, double bonds) are attached to the —CH_2OH. Experiments involving the partial exchange of the OH groups in octan-2-ol and octan-3-ol with D_2O, show that while the band near 1250 cm^{-1} changes in intensity, the 1380 cm^{-1} band does not, confirming that the 1250 cm^{-1} band is probably an OH band while the 1380 cm^{-1} band has some other origin.

c. Phenols

The hydroxyl stretching frequencies found in the vapour spectra of phenols are given in Table 2.2. The vapour and solid film spectra of phenol are compared in Fig. 2.2.

Phenols show only one OH band in the vapour phase, unless intramolecular hydrogen bonding is possible, when a second band appears at lower frequency. The free OH bands are always much weaker than the bonded OH bands in these cases. When compared with their positions in dilute solutions, the free OH bands are shifted 35–45 cm^{-1} to higher frequencies, while the intramolecularly bonded OH bands show a shift of only 27–32 cm^{-1}.

The spectra obtained from phenols in different physical states, are generally very similar, but as would be expected, the positions of the δ(OH) and δ(CO) bands differ significantly between the bonded and non-bonded states.

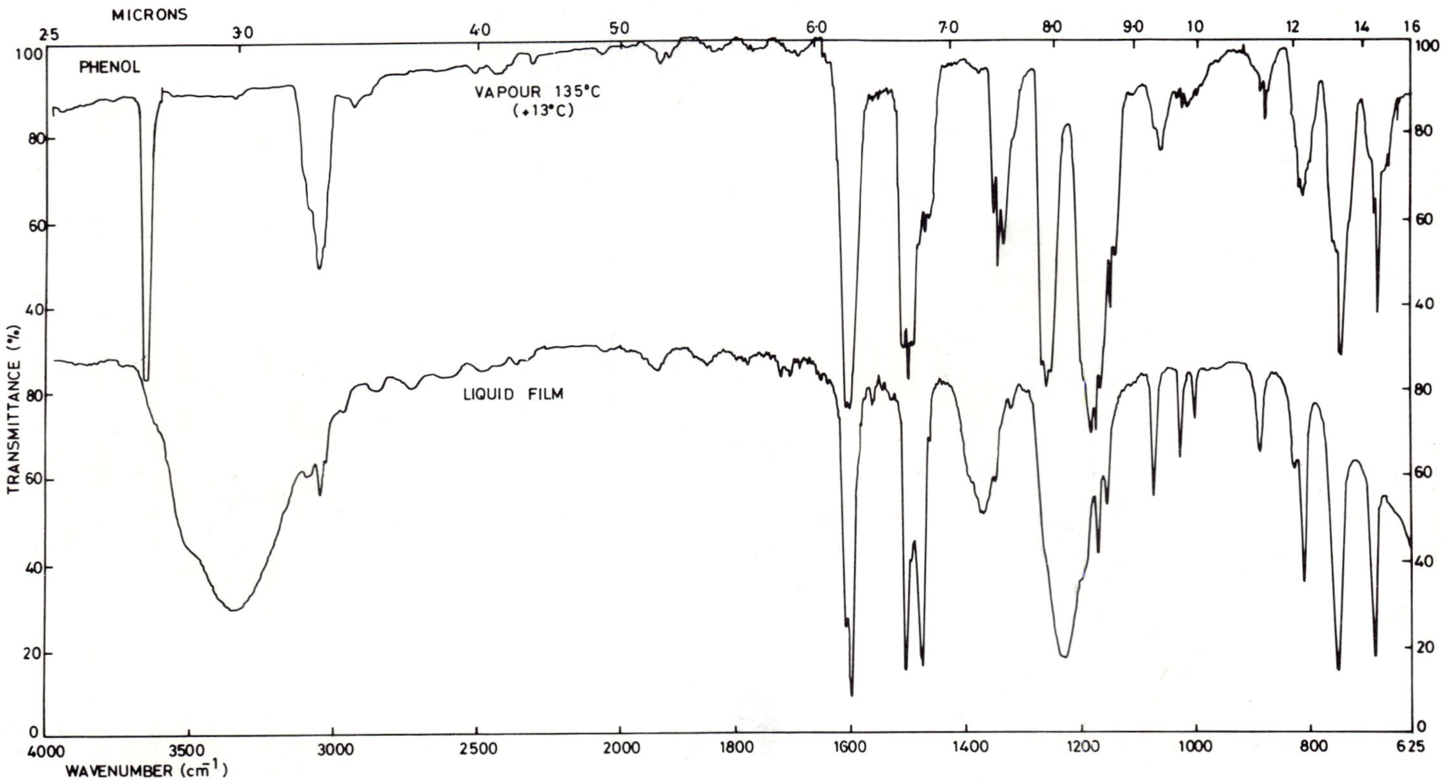

FIG. 2.2. Comparison of the vapour and liquid spectra of phenol.

TABLE 2.2. Band maxima in the infrared spectra of phenols [a]

Compound	No. of spectra observed	Frequency (cm^{-1})
ν (OH) (free or monomeric)		
Alkyl substituted	16	3655–3650 (m) sharp
2-methoxyphenols	3	~3590 (m) sharp
2,6-dimethoxyphenol	1	~3580 (m) sharp
2-hydroxyphenols	3	~3663 (w) + 3603 (m) sharp
2-fluorophenol[38]	1	~3635 (m) sharp
2-chlorophenol[38]	1	~3650 (w) + 3583 (m) sharp
2-bromophenol[38]	1	~3650 (w) 3560 (m) sharp
2-iodophenol[38]	1	~3650 (w) + 3538 (m) sharp
δ(OH) and ν(C—O) See the text		

[a]Additional spectra by Mecke are included in the DMS collection.[39]

Unfortunately the literature is not unanimous over the assignment of the ν(C—O), δ(OH) and ring vibrations.[40–42] The spectra of one series of phenols (phenol, *p*-cresol, *p*-chlorophenol and *p*-bromophenol) are very similar. The more preferable band assignments for phenol have been given by Davies and Jones[41] and those for *p*-cresol have been given by Jakobsen.[42] They produced comparable assignments, which associated the 1250 cm^{-1} band with partial ν(C—O) character and the 1185 cm^{-1} band with the δ(OH) vibration. While these assignments could probably be extended to a number of other phenols, it would be unwise to do so without further proof.

d. Carboxylic Acids

The typical frequencies found in the vapour spectra of carboxylic acids are given in Table 2.3. The vapour and liquid spectra of a simple acid, n-hexanoic acid, are compared in Fig. 2.3.

The spectra of carboxylic acids in the vapour state are greatly superior to those measured in dilute solutions and represent the most clear example of the value of vapour phase spectra in structural work. To obtain a spectrum of the monomeric acid from a dilute solution entails the use of solutions of such dilution that relatively long pathlength cells are needed to produce absorption bands of sufficient intensity. Although the common solvents are transparent in the 3000 cm^{-1} region, they absorb so much energy at frequencies below 1700 cm^{-1} that each solvent allows only comparatively narrow windows for an examination. The vapour spectrum gives the whole spectrum with fewer sample handling problems and it is immeasurably superior to the solution spectrum. Carboxylic acids can exist either as dimers or monomers in the vapour state. The dimer is predominant at lower temperatures, but as the temperature increases so the equilibrium shifts until above about 150°C it is very much in favour of the monomer. Examples of the two forms of formic and acetic acids are shown in Spectra 73 to 76.

The maxima for the free ν(OH) bands are shifted about 60 cm^{-1} to higher frequencies when compared to those obtained from solution spectra[43] and look like normal free ν(OH) bands.

Table 2.3. Band maxima in the infrared spectra of carboxylic acids

Compound		No. of spectra observed	Frequency (cm^{-1})
ν(OH) (free or monomeric)			
Normal saturated	RCOOH	10	~3575 (m) sharp
α-branched	$R_1CH(R_2)COOH$	4	3575–3570 (m) sharp
αβ-unsaturated	RCH=CHCOOH	11	3585–3580 (m) sharp
βγ-unsaturated	$RCH{=}CHCH_2COOH$	3	~3580 (m) sharp
ν(C=O) monomeric			
Normal saturated	RCOOH	10	1780–1777 (s)
α-methyl branched	$RCH(CH_3)COOH$	2	1778–1777 (s)
α-ethyl branched	$RCH(CH_2CH_3)COOH$	2	1772–1770 (s)
αβ-unsaturated	RCH=CHCOOH	10	1764–1760 (s)
αβ-γδ-unsaturated	RCH=CHCH=CHCOOH	1	1755 (s)
βγ-unsaturated	$RCH{=}CHCH_2COOH$	3	1782–1778 (s)
Cf. dimeric normal saturated (vapour)			~1730 (s)
δ(OH)			
Normal saturated and α-branched	RCOOH, $R_1CH(R_2)COOH$	13	1385–1380 (wm)
αβ-unsaturated	RCH=CHCOOH	11	1365–1335 (wm)
βγ-unsaturated	$RCH{=}CHCH_2COOH$	3	1370–1360 (wm)
ν(C—O)			
Normal saturated	RCOOH	10	1160–1130 (m)
α-branched	$R_1CH(R_2)COOH$	4	1180–1130 (v) (sometimes infl. or sh. to 1100 cm^{-1} band)
αβ-unsaturated	RCH=CHCOOH	7	1150–1140 (m)
α-branched αβ-unsaturated; β-branched, αβ-unsaturated	$R_1CH{=}C(R_2)COOH$; $R_1C(R_2){=}CHCOOH$	3	1125–1115 (m)
βγ-unsaturated	$RCH{=}CHCH_2COOH$	3	1140–1125 (m) broad
1100 cm^{-1} bands			
Normal saturated	RCOOH	10	1110–1080 (m)
α-branched	$R_1CH(R_2)COOH$	4	~1100 (m)
αβ-unsaturated	RCH=CHCOOH	7	1110–1100 (m)
α-branched, αβ-unsaturated; β-branched, αβ-unsaturated	$R_1CH{=}C(R_2)COOH$; $R_1C(R_2){=}CHCOOH$	3	1080–1070 (sh)
βγ-unsaturated	$RCH{=}CHCH_2COOH$	3	unresolved from the ν(C—O) band

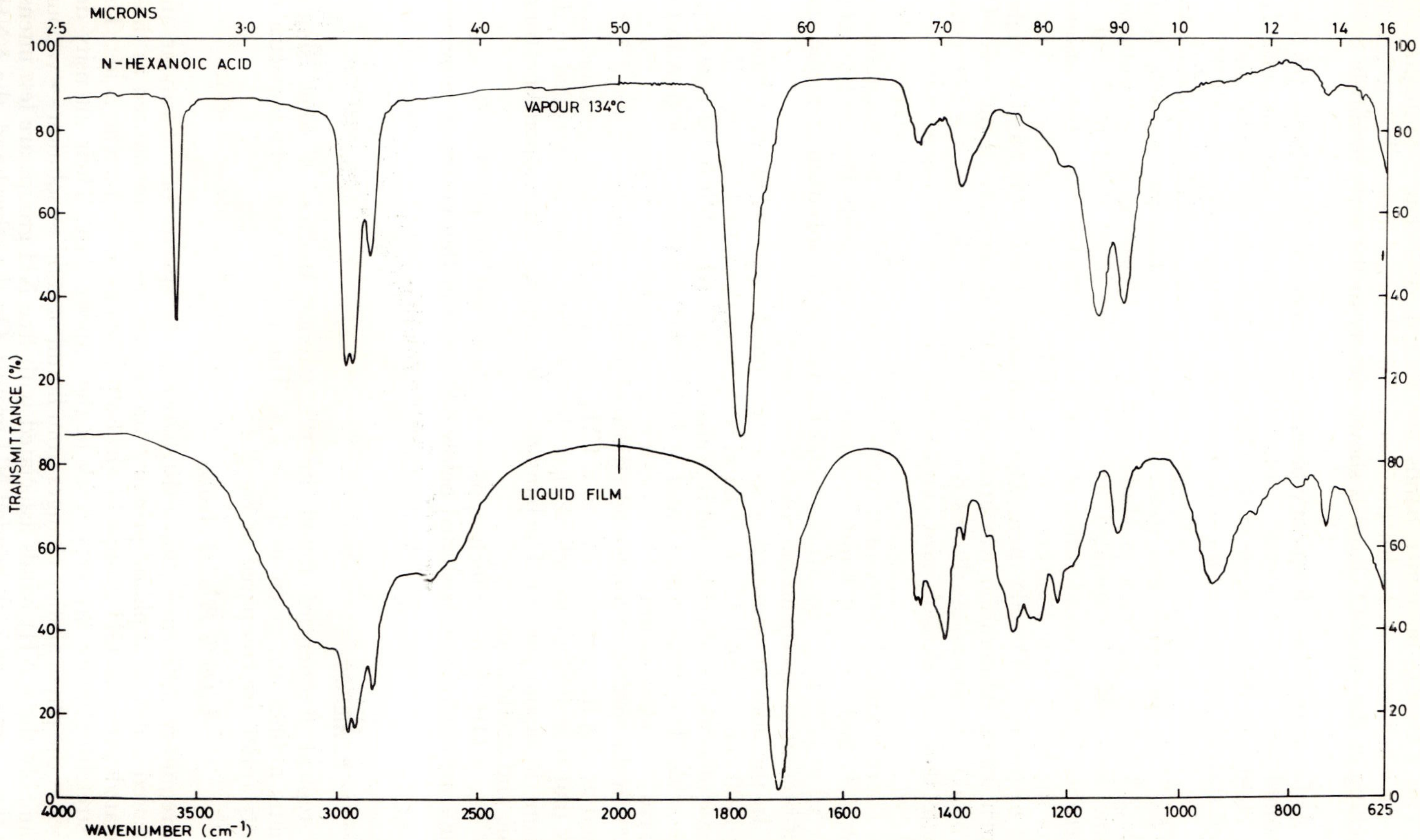

FIG. 2.3. Comparison of the vapour and liquid spectra of n-hexanoic acid.

The ν(C=O) monomer bands are shifted 20–25 cm^{-1} to higher frequencies when compared with the spectra of dilute solutions. The rotational branching of the ν(C=O) band is shown clearly in the early members of the series.

As far as the lower frequency bands are concerned, the typical spectra of the homologous series of normal saturated acids begin with butanoic acid. The large change in the character of the OH group between the vapour and condensed states is reflected in the complete shift in frequencies of the δ(OH) and ν(C—O) bands which give strongly coupled bands around 1420 cm^{-1} and 1300 cm^{-1} in the condensed state. The frequency assignments given in the Table 2.3 follow the work of Hadzi and Pintar,[44] who examined butanoic, benzoic and some halogenated acids. From the band shifts produced by deuteration and the consideration of work found in the literature, they assigned one of the bands near 1380 cm^{-1} to the δ(OH) vibration and the 1150 cm^{-1} band to the ν(C—O) vibration. The small shift of ν(C—O) band to higher frequencies on deuteration was attributed to the decoupling of the ν(C—O) and δ(OD) vibrations. As discussed in the section on alcohols, it is interesting to note that the alcohols also consistently possess an enhanced band near 1380 cm^{-1}.

There is some disagreement in the literature,[43,45–52] on the assignment of the δ(OH) and ν(C—O) bands in formic and acetic acid. The δ(OH) band observed at 1229 cm^{-1} by Millikan and Pitzer[46] in the spectrum of formic acid is not present in Spectrum No. 74 in Chapter 5.

In the normal saturated acids a twin band to the 1150 cm^{-1} band lies near 1100 cm^{-1}. The 1100 cm^{-1} band is stronger than the 1150 cm^{-1} band in the case of α-branched acids, and is probably confused with it in the case of the 3-ene acids. The benzoic acid band at 1180 cm^{-1} is possibly analogous to the 1100 cm^{-1} band in saturated acids as it is strong and does not move on deuteration.[44] In the spectra of condensed saturated acids, it has been suggested[53] that a band at this frequency is due to a CH_3 vibration. However, the band is so weak that it is difficult to believe that the correlation can be extended to the monomeric vapour spectra. The band does not appear in the vapour spectra of formic and acetic acids.

The γ(OH) band, which appears at about 900–950 cm^{-1} in the spectra of the condensed states, is shifted below 600 cm^{-1} in the spectra of monomers and is therefore out of range of the present spectra.

e. Amines

The typical frequencies found in the vapour spectra of some simple aliphatic amines are given in Table 2.4. The vapour and liquid spectra of n-pentylamine are compared in Fig. 2.4.

(i) *Aliphatic Primary Amines*

The frequency differences that exist between the spectra of the vapour and condensed states of aliphatic primary amines are not so pronounced as those associated with the OH groups in alcohols, phenols or carboxylic acids. The NH stretching bands are very weak in the vapour state. Their intensity decreases as the physical state changes from liquid or solid → solution → vapour. Presumably the ν(NH) bands associated with free NH groups are less intense than those associated with bonded NH groups. On the other hand, the δ(NH)

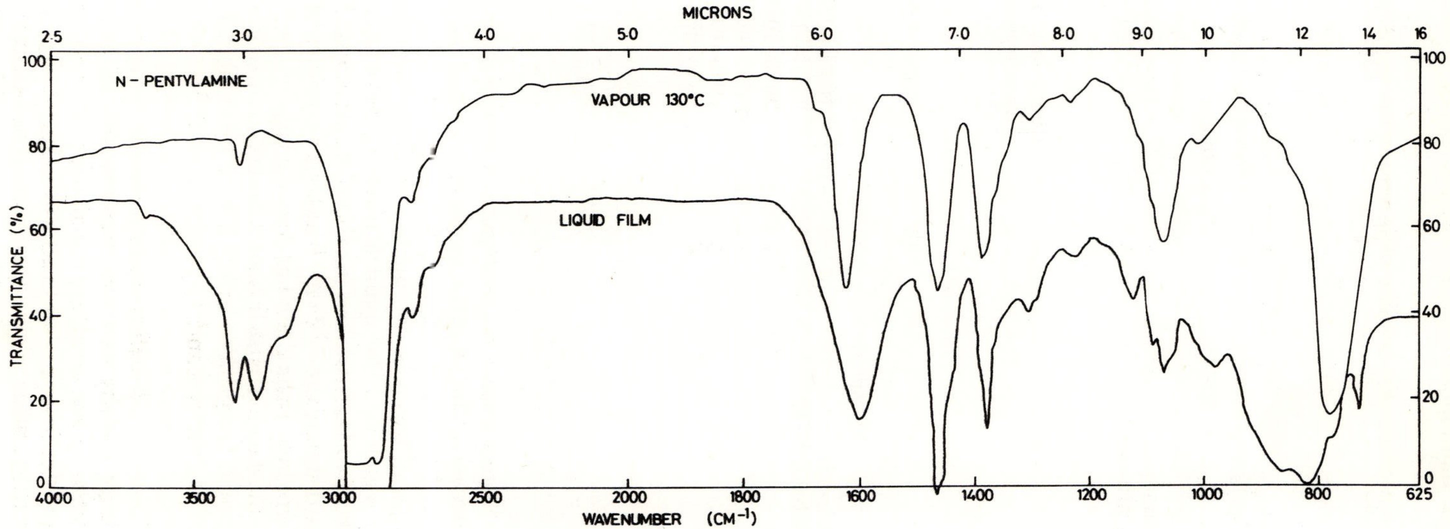

FIG. 2.4. Comparison of the vapour and liquid spectra of n-pentylamine, l-aminopentane.

TABLE 2.4. Band maxima in the infrared spectra of amines

Compound	No. of spectra observed	Frequency (cm^{-1})
ν(NH) antisymmetric		
Normal aliphatic primary	7	3430–3410 (v.v.w.)
α-branched aliphatic primary		too weak to see normally
ν(NH) symmetric		
Normal aliphatic primary	7	3350–3340 (v.w.)
α-branched aliphatic primary	5	~3330 (v.w.)
ν(NH)		
Aliphatic secondary		too weak to see normally (dimethylamine quoted at 3355 [54,55])
δ(NH)		
Aliphatic primary	14	1625–1620 (m)
Bands that include some ν(C—N) character		
Primary aliphatic—normal	9	1090–1070 (w)
Primary aliphatic—α-branched	5	1160–1145 (ms)
	5	1070–1010 (w)
Secondary aliphatic	7	1155–1135 (s)
	6	1025–1015 (w)
	6	905–875 or 965–930 (w)
β(NH)		
Primary aliphatic	14	790–765 (s)
Secondary aliphatic	7	725–710 (s)

band near 1640 cm^{-1} becomes sharper and more intense in the vapour spectrum. These differences between the spectra of amines in the condensed and the vapour phase may lead to a misinterpretation of the spectrum. The apparent absence of a ν(NH) band taken in conjunction with the sharpening of the NH deformation bands, which are the main features of aliphatic amine spectra, could allow the spectra to be mistaken for those of unsaturated hydrocarbons.

In the condensed states, the 780 cm^{-1} band is a complex of bands, centred at a slightly higher frequency than the band in the vapour spectrum. Although sharper in the vapour state, it is still broad and asymmetric enough to be a mixture of superimposed bands, but it is the most prominent feature of simple amine spectra. The CH_2 rocking band appears as an inflection or shoulder near 725 cm^{-1}.

The 1620–1625 cm^{-1} NH deformation band is very consistent in its position, besides having a slightly narrower bandwidth in the vapour state.

The C—N stretching bands of the normal aliphatic amines are too weak to be of much use in identifying a spectrum as being that of an amine. They probably appear over the rather wide range quoted in Bellamy[29] and Sandorfy and Jones,[31] but our spectra cover only the ranges quoted in Table 2.4.

(ii) *α-Branched Aliphatic Primary Amines*

The spectra of α-branched amines also include a characteristic band at 1145–1160 cm^{-1}. As this band appears in the spectra of molecules containing neither an isopropyl group nor a substituted α-methyl group, it is unlikely to be associated with a methyl group and is more likely to be due to coupled skeletal and C—N stretching vibrations.

(iii) *Secondary Aliphatic Amines*

In the spectrum of secondary aliphatic amines, the two main characteristic bands occur at 710–725 cm^{-1} and 1135–1155 cm^{-1}. The former band which is probably due to an NH bending vibration, is asymmetrical with inflections on either side. It appears to get more diffuse and less prominent as the molecular size increases.

The band near 1150 cm^{-1} can be followed through the whole series of simple aliphatic amines from dimethylamine onwards. Unfortunately there has been some disagreement about the assignment of the band in dimethylamine. Stewart[56] assigns it to the antisymmetric CN stretching vibration. Finch, Hyams and Steele[54] and Barcelo and Bellanato[55] assign it to CH_3 deformation vibrations. The theoretical calculations of Finch *et al.* allow a certain amount of CN stretching character to one of the two components which comprise the band. As the band appears consistently in a whole series of symmetrical and (one) asymmetrical secondary amines, it is unlikely to be caused by a simple CH_3 vibration. It is more likely, at least in the case of the large molecules, to be a band caused by the coupling of CH_2 or CH_3 deformation or C—C stretching vibrations with the C—N stretching vibration.

The NH stretching bands are mostly too weak to see at normal vapour pressures, but dimethylamine has a band at 3355 cm^{-1}.[54,55] One of the major differences between α-branched primary amines and the secondary amines is the absence of the NH deformation band around 1620 cm^{-1} in the secondary amines. It has been suggested that it appears weakly on the side of the 1460 cm^{-1} CH deformation band, but it is of no use for identification purposes.

(iv) *Aromatic Amines*

The vapour spectra of the simpler aromatic amines are very similar to their liquid/solid/solution spectra and can be recognised by comparison with those spectra. The NH stretching bands of aromatic amines are stronger than those of aliphatic amines and this is particularly so in the case of the vapour spectra where they are still easily recognisable. The band maxima are at higher frequencies than in the other states, but no figures for the frequency shifts can be given owing to the great variability of the band positions in solutions caused either by solvents of differing polarity, or by differences in the strength of the hydrogen bonds. In the limited number of vapour spectra that we have examined, the NH bending band in primary amines occurs near 1620 cm^{-1} and is not always resolved from the ring vibration near 1600 cm^{-1}, but it is the major component in this absorption. This is probably not true over the whole range of aromatic amines and elsewhere the frequency range is quoted over much wider limits.[29]

The δ(NH) band is very weak in the spectra of aromatic secondary amines. Hadzi and Skrbljak[57] have suggested that this vibration coupled with the C—N stretching vibration gives a band at 1513 and 1514 cm^{-1} in *N*-ethylaniline and diphenylamine respectively. With the broader bands obtained in vapour spectra, it is likely that these bands will generally be obscured by the aromatic band at 1500 cm^{-1}.

Of the typical amine bands observed at lower frequencies, the C—N stretching band between 1250–1350 cm^{-1} is the most consistently prominent. In our vapour spectra it still occurs close to the values found in Bellamy.[29] A medium strong band also occurs between 650–680 cm^{-1} in the vapour spectra of anilines[58] not intramolecularly bonded in the 2 position. The part of the spectrum which does change is the region between 1000–1200 cm^{-1}. Only weak to medium intensity bands (including an NH_2 band[58]) occur in this range and they will only affect the characterisation of individual compounds when the spectra are compared with those obtained from other physical states.

f. Amides

Only the simplest amides have high enough vapour pressures to be examined in the vapour state, even at 200°C, so any observations must be restricted solely to these molecules. The frequencies typical of amides are included in Table 2.5. The vapour and Nujol mull spectra of n-butyramide are compared in Fig. 2.5.

TABLE 2.5. Band maxima in the infrared spectra of amides

Compound	Frequency (cm^{-1})
Bands associated with $\nu(NH)$ vibration	
Primary antisymmetric	3565–3545 (wm)
Primary symmetric	3450–3430 (wm)
Secondary	3495–3485 (m)
Cyclic γ-lactam (pyrrolidone)[59]	3505 (wm)
Cyclic ε-lactam (caprolactam)[59]	3472 (wm)
Amide I[$\nu(C{=}O)$]	
Primary	near 1740 (s)
Secondary	1730–1715 (s)
Tertiary	1710 (s)
Cyclic γ-lactam (pyrrolidone)[59]	1754 (s)
Cyclic ε-lactam (caprolactam)[59]	1712 (s)
Amide II [Partial $\delta(NH)$]	
Primary	1600–1590 (ms)
Secondary	1495–1485 (ms)
Other Bands	
Primary	1385–1375 (ms)
	1320–1265 (ms)

The controversy about the assignments for the bands found in the spectra of amides in the condensed states has been well reviewed in many publications, including Bellamy[29,30] and Jones and Sandorfy[31] from which the pertinent literature references may be obtained.

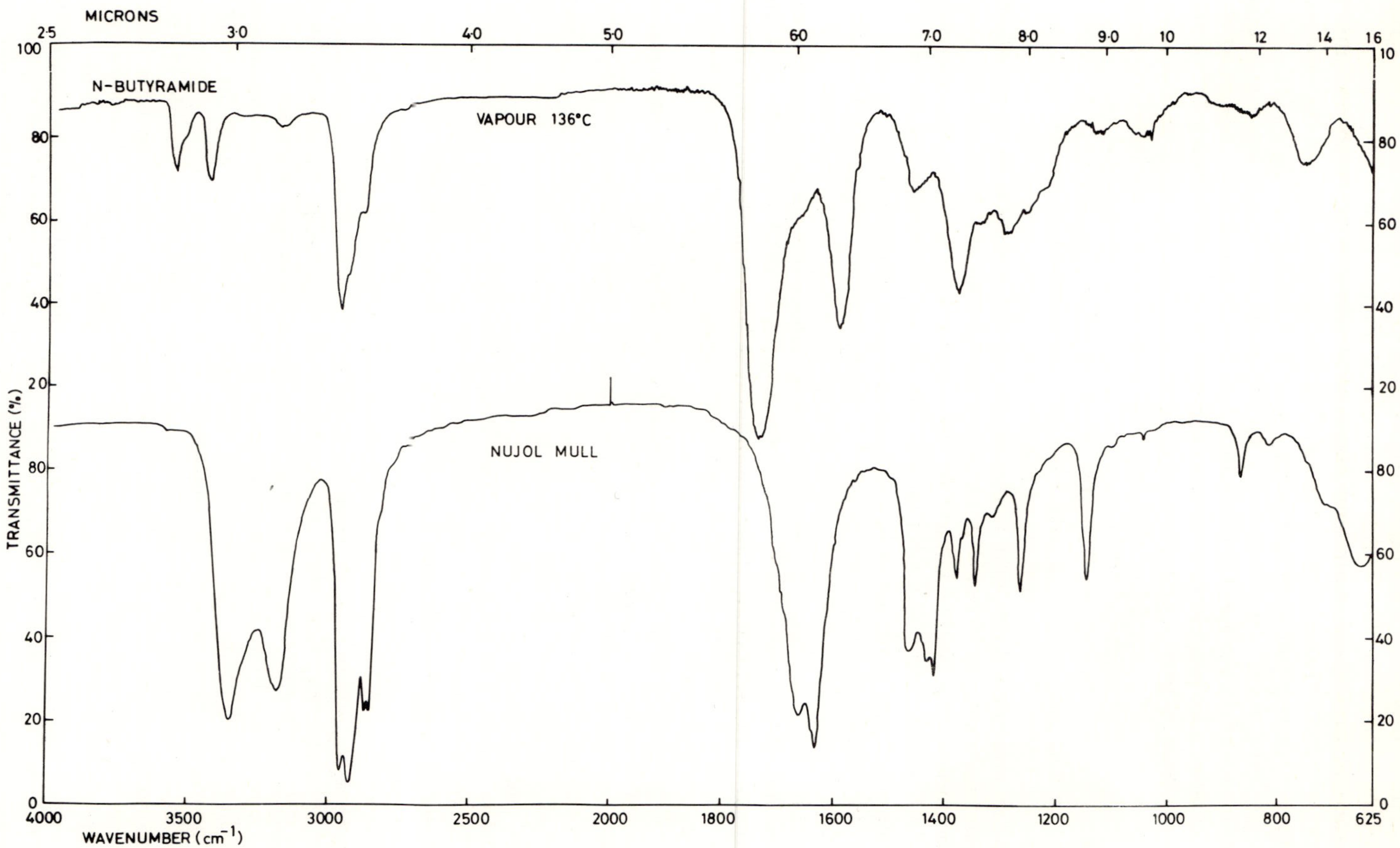

FIG. 2.5. Comparison of the vapour and liquid spectra of n-butyramide

The bands near 3500 cm^{-1} are attributed to the unassociated ν(NH) stretching vibrations. The bands of associated amides may be seen in spectra recorded in the vapour state, but this is probably due to small amounts of material condensed on the cell windows. We have no real proof of the existence of amide dimers in the vapour phase.

Both the primary and secondary ν(NH) bands move to higher frequencies than are found in the spectra of dilute solutions, due to the absence of solvent interactions. The usual antisymmetric and symmetric ν(NH) bands are present in the spectra of primary amides.

Amide I and II Bands

Primary Amides. The Amide I band of primary amides appear near 1740 cm^{-1} in the vapour state compared with 1690 cm^{-1} in dilute solutions and 1650 cm^{-1} in the solid state.[29] The 1690 cm^{-1} band in dilute solution is already attributed to the monomeric amides, and as the change to the vapour state would hardly account for a shift of 50 cm^{-1} it is quite probable that the increased frequency of the monomeric Amide I band in the vapour state is due to an increased ν(C=O) character. The C=O group will not be so heavily coupled to CN vibrations and/or it will have lost much of its ionic character.

The Amide II band of primary amides appears near 1590–1600 cm^{-1} compared with 1590–1620 cm^{-1} in dilute solution and 1620–1650 cm^{-1} in the solid state.[29] It has been suggested[29] that while this vibration may involve both C—N and N—H groups as in secondary amides, in primary amides it is predominantly an NH deformation vibration. It is interesting to note that there is little frequency shift in the band positions in the vapour and dilute solution spectra.

The three primary amides: acetamide, propionamide and butyramide have medium/strong bands near 1375–1385 cm^{-1} and 1265–1320 cm^{-1}, but there is not enough information to correlate them with equivalent bands in the spectra of amides in other physical states. Evans[60] has suggested that the band centred at 1255 cm^{-1} in the vapour spectrum of formamide is due to a CN stretching vibration.

Secondary Amides. The frequency of the Amide I band of secondary amides does not increase so markedly. It appears near 1650 cm^{-1} in the solid state, near 1700 cm^{-1} in dilute solutions and near 1715–1730 cm^{-1} in the vapour state. In the case of the simple molecules that we are dealing with, Miyazawa, Shimanouchi and Mizushima[61] concluded from a normal coordinate analysis of *N*-methyl acetamide that the band at 1653 cm^{-1} in the solid state had 80% carbonyl character. If correct, it is reasonable to expect that the vapour bands near 1715–1730 cm^{-1} have as much, if not more, carbonyl character.

The Amide I band in ε-caprolactam is in line with those of acyclic amides, but that of γ-butyolactam occurs at 1754 cm^{-1} [59] showing that the ring is strained in the same way as for γ-lactones.

There has been much speculation as to the origin of the Amide II band in secondary amides. This is due to its importance in understanding the structure of proteins and polyamides. Here we are dealing only with low molecular weight amides and it can be seen that the Amide II frequency drops from about 1565 cm^{-1} in the solid state and 1535 cm^{-1} in dilute solution[29] to about 1490 cm^{-1} in the vapourstate.[61–66] This value agrees with the postulated allocations of about 1460 cm^{-1} for the δ(NH) band of secondary aliphatic amines and about 1500 cm^{-1} for that of secondary aromatic amines. It is likely, therefore that the Amide II band in the vapour spectra of secondary amides is principally an NH deformation band, with only a small amount of ν(C—N) character.

The Amide I frequency of tertiary amides also increases from about 1650 cm^{-1} to 1710 cm^{-1} in the vapour state. Hydrogen bonding cannot occur in this case, so presumably there is a decrease in C—N coupling and in the ionic character of the OCN group.

g. Esters and lactones

The principle differences between the spectra of esters and lactones in the vapour state and other states is in the ν(C=O) bands. Some typical band maxima are listed in Table 2.6 together with the values of some ν(C—O) bands. The vapour and liquid spectra of ethyl hexanoate are compared in Fig. 2.6.

TABLE 2.6. Band maxima in the infrared spectra of esters and lactones

Compound	No. of spectra observed	Frequency (cm^{-1})
ν(C=O)		
n-Alkyl formates	5	1755–1750 (s)
n-Alkyl acetates	7	1770–1765 (s)
Other non α or α' branched esters	25	1765–1755 (s)
Methyl esters have 6–8 cm^{-1} higher frequencies than other n-alkyl esters		
Isopropyl esters have 5–6 cm^{-1} lower frequencies than other n-alkyl esters		
t-Butyl esters have 7–9 cm^{-1} lower frequencies than other n-alkyl esters		
Isobutyrates have 1–3 cm^{-1} lower frequencies than n-butyrates		
Pivalates have 7–9 cm^{-1} lower frequencies than n-valerates		
$\alpha\beta$ Conj. and aryl esters	6	1760–1740 (s)
Salicylates and anthranilates	4	1720–1700 (s)
Alk-l-enyl or 'phenyl' esters (lowered ~25 cm^{-1} by partial 'quinonoid structure')	6	*ca.* 1780 (s)
Lactones β-saturated[67]	1	1882 centre (s)
γ-saturated	6	1818–1816 (butyrolactone 1824) (s)
δ-saturated	4	near 1780 (s)
ν(C—O)		
n-Alkyl formates[a]	5	1178–1173 (s)
n-Alkyl acetates[a]	7	1240–1237 (s)
Other 'normal' non α or α' branched esters[a]		1190–1170 (s)
Lactones β-saturated[67]	1	1096–1088 (s)
γ-saturated	6	1164–1162 (butyrolactone 1155 (s)
δ-saturated	4	1237–1234 (s)

[a] Me esters do not always agree with series.

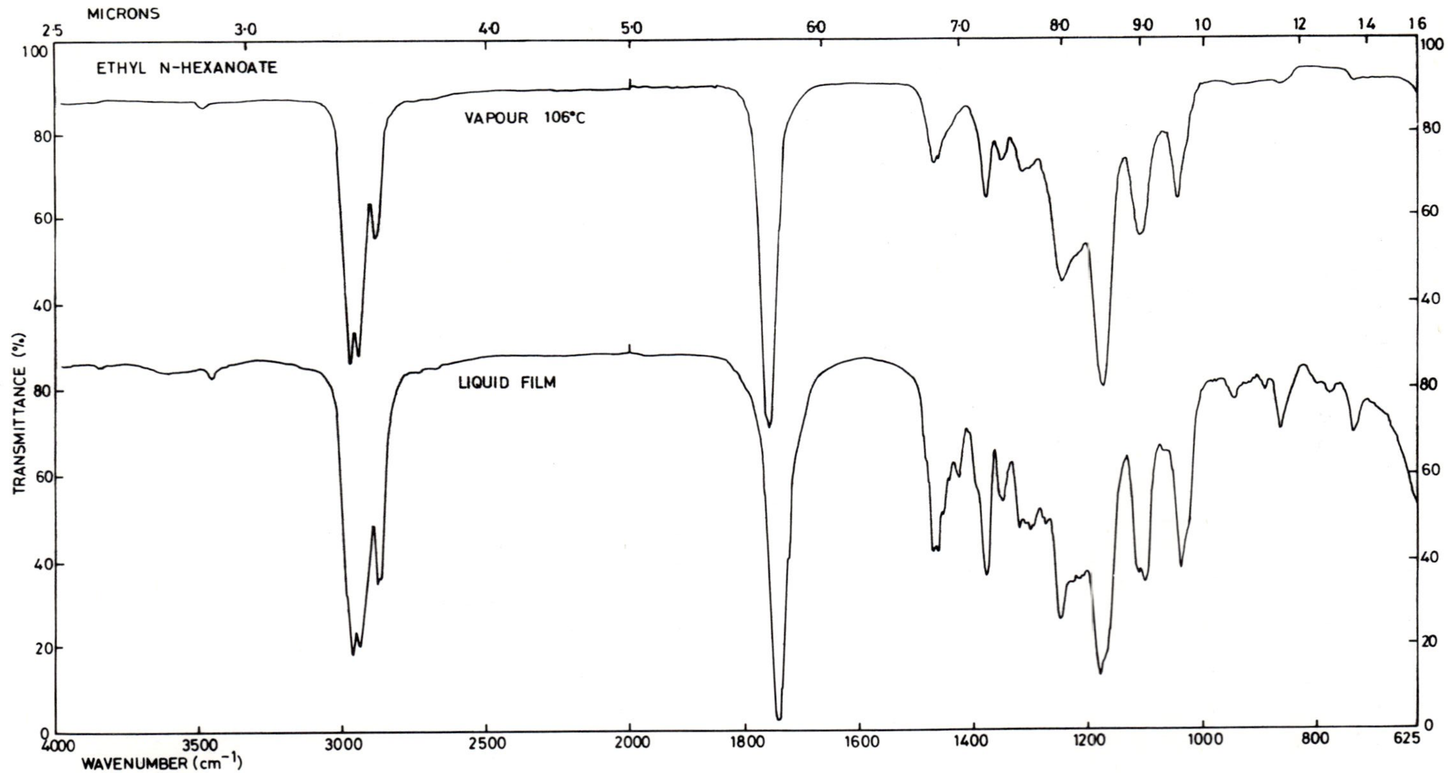

FIG. 2.6. Comparison of the vapour and liquid spectra of ethyl n-hexanoate.

The changes in the ν(C=O) frequencies are only 'solvent' shifts, and the compounds with the biggest changes are those that have the greatest polar interaction with solvents, for instance, the lactones. Table 2.6 gives the details of frequency ranges to be expected from the simpler type of ester. The values will change in a similar manner for esters not mentioned here.

The frequency of the ν(C=O) band of the 2- and 3-hydroxy esters decreases from between 4–10 cm^{-1} when compared with the band given by the unsubstituted esters, but the change is not as significant as that of, for instance, the ν(OH) band of the hydroxy esters. The ν(C=O) band of phenolic esters with an oxygen atom substituted in the *para* position (i.e. the anisyl esters) is up to 20 cm^{-1} lower in frequency than the unsubstituted compounds. We have not examined enough aromatic esters to comment on their C—O stretching bands, but these fall predominantly in the ranges quoted by Bellamy.[29]

The fingerprint regions of each of the simple series of esters (formate to caproate) look very similar within each series. This could be a means of characterising the type of ester, except that the finer differences between the spectra of esters become more blurred in the vapour state. It is probably better to use the evidence in a negative way, i.e. if the 'fingerprint' spectrum is unlike those of other members of a series, then it is unlikely to belong to that series.

h. Aldehydes

The main bands found in normal saturated, 2-ene and 2,4-diene aldehydes are given in Table 2.7. The vapour and liquid spectra of n-pentanal are compared in Fig. 2.7.

TABLE 2.7. Band maxima in the infrared spectra of aldehydes

Compound	No. of spectra observed	Frequency (cm^{-1})
ν(C=O)		
Saturated	9	1750–1745 (s)
$\alpha\beta$-conjugated unsaturated	6	1725–1715 (s)
$\alpha\beta$-$\gamma\delta$-conjugated unsaturated	6	1715–1710 (s)
aromatic	7	1730–1710 (s)
Other bands		
ν(CH) and overtone bands		
Saturated aldehydes	9	2710–2705 (m)
		2810–2805 (wm)
$\alpha\beta$-unsaturated aldehydes	6	2730–2720 (m)
		2810–2800 (wm)
$\alpha\beta$-$\gamma\delta$-unsaturated aldehydes	6	2740–2715 (m)
		2810–2800 (wm)
'Fingerprint' bands $\alpha\beta$-conj.		1145–1135 (ms)
		1100–1070 (v)
$\alpha\beta$-$\gamma\delta$-conj.		1160 (ms)
		1110–1100 (v)

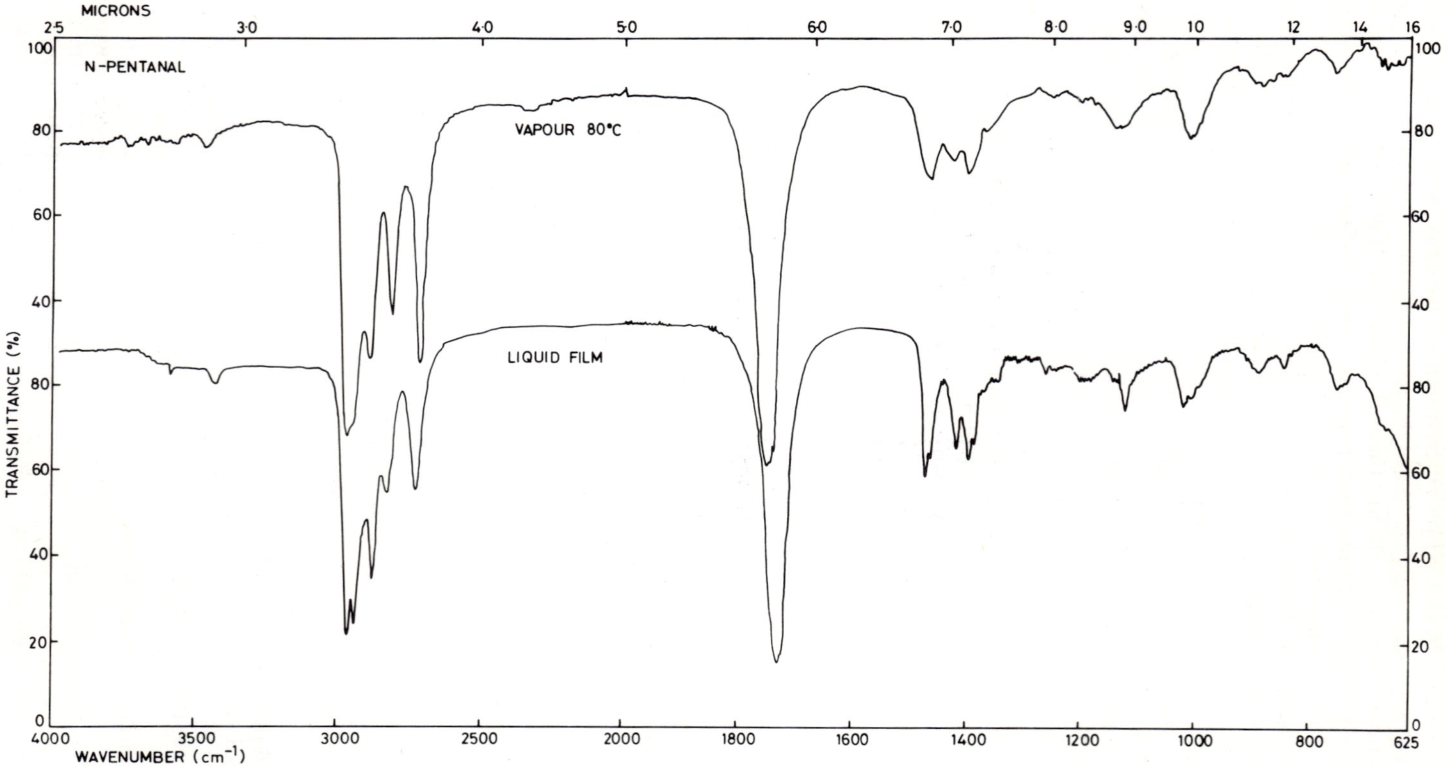

FIG. 2.7. Comparison of the vapour and liquid spectra of n-pentanal.

The simple aldehydes all show a ν(C=O) band shift to higher frequencies. We have not measured enough aromatic aldehyde spectra to make any generalisations other than for the carbonyl bands. In the 2700 cm^{-1} region, benzaldehyde has a doublet centred at 2730 cm^{-1}, cuminic and *p*-tolyl aldehydes have bands near 2720 cm^{-1}, and *p*-anisaldehyde, has a band near 2712 cm^{-1}. The exceptional band positions occur in the cases of salicylaldehyde (2740/2750 cm^{-1}) and 2-methoxybenzaldehyde (2755 cm^{-1}). If these shifts are due to the same cause, it cannot be hydrogen bonding to the carbonyl group, nor can it be due to the substitution of an oxygen atom in the *ortho* position as the *para*-methoxy compound shifts the other way from benzaldehyde. Two possible explanations are either the hydrogen bonding of the aldehyde hydrogen with the 2-position oxygen[68,69] or a steric repulsive effect between the two groups. This is hardly likely in the case of salicylaldehyde as the two substituents are known to hydrogen bond strongly through the carbonyl group (cf. the ν(C=O) frequency).

The spectra of saturated aliphatic aldehydes do not have many typical bands in the fingerprint region. The smaller molecules have a band between 1100–1150 cm^{-1}, but it must not be confused with the band obtained from carboxylic acids, to which the aldehydes are easily oxidised at high temperatures. On the other hand the stronger bands between 1100–1200 cm^{-1} taken in conjunction with the '*trans*' =CH out-of-plane deformation bands near 970 cm^{-1} are typical of $\alpha\beta$ or $\alpha\beta$-$\gamma\delta$-unsaturated aldehydes. Even then some care must be taken over confusion with monomeric acids.

i. Ketones

The main bands found in simple ketones are given in Table 2.8. The vapour and liquid spectra of pentan-3-one are compared in Fig. 2.8.

Table 2.8. Band maxima in the infrared spectra of ketones

Compound	No. of spectra observed	Frequency (cm^{-1})
ν(C=O)		
Saturated aliphatic	13	1740–1730 (s)
[Simple 2-ones	10	~1740–1735 (s)]
Cyclobutanone		1816 (s)
Cyclopentanone		1773 (s)
Cyclohexanone		1739 (s)
$\alpha\beta$-unsaturated	4	1705–1690 (s)
aromatic	4	1710–1705 (s)
Other bands		
2-ones	10	Combination of bands near 1350 and 1160 (m)
Some other aliphatic ketones	3	1150–1100 (m)
Some aromatic ketones	4	1300–1200 (m-s)

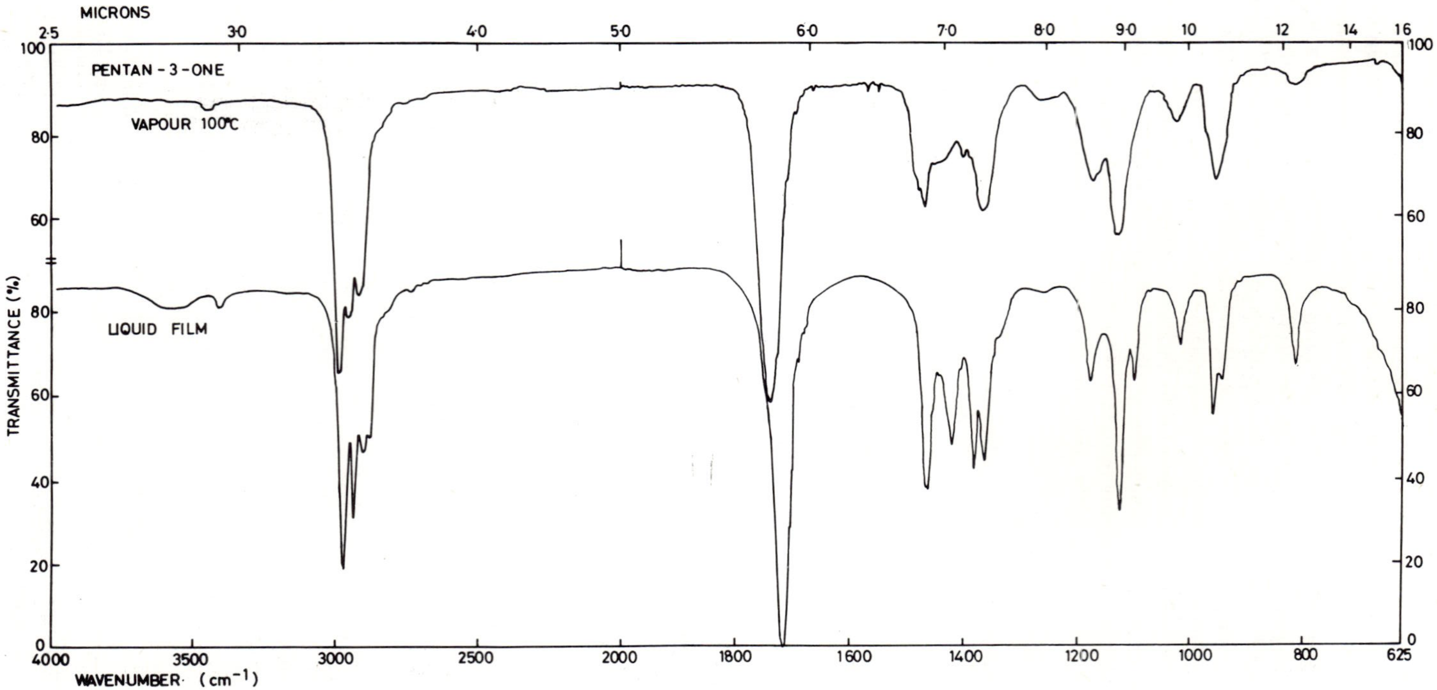

FIG. 2.8. Comparison of the vapour and liquid spectra of pentan-3-one.

As is usual, the carbonyl bands in the vapour spectra have shifted to slightly higher frequencies when compared with the bands obtained from the spectra of other states. The number of $\alpha\beta$-conjugated unsaturated or aromatic ketones that have been examined is small, but the values have been included for the sake of completeness and they are significant.

Although pentane-2,4-dione has a free carbonyl band at 1720/1730 cm^{-1}, it is weak compared to the band at 1632 cm^{-1} which shows that the enolic hydrogen bonding system is predominant, even in the vapour state.

Methyl ketones are distinguished by having distinctive bands near 1360 cm^{-1} and 1160 cm^{-1}. A number of other ketones include a band between 1100–1150 cm^{-1}. The four aromatic compounds that have been examined have a medium-to-strong band between 1200–1300 cm^{-1}.

j. Miscellaneous

ν(≡CH)

This band still appears near 3300 cm^{-1}, but shifts to frequencies that are about 20–40 cm^{-1} higher than those found in the spectra of solutions or liquid states.

ν(=CH) and ν(—CH)

These bands shift to the upper limits of the usual frequency ranges.

ν(S—H)

These bands still appear within the usual range of frequencies, but are shifted about 30 cm^{-1} to higher frequencies, i.e. pentane-3-thiol (2565 → 2597 cm^{-1}) and cyclopentanethiol (2564 → 2591 cm^{-1}) in the DMS spectra[39] from the 'Standard Oil Company, Indiana'.

ν(C≡C)

Only a very small shift to higher frequencies.

ν(C=O) of Acid Halides

Overend and Scherer[70] quote 1869 cm^{-1} for CH_3COF, 1822 cm^{-1} for CH_3COCl and 1821 cm^{-1} for CH_3CO Br.

ν(C=C)

Bands due to unconjugated C=C groups appear near 1640–1650 cm^{-1}, which increase to 1660–1670 cm^{-1} on conjugation to C=O groups in aldehydes and carboxylic acids. Any differences between types of double bonds are not easy to discern.

CH Deformation Bands

Any shifts in either in-plane or out-of-plane deformation bands are negligible in the case of saturated or unsaturated hydrocarbons or alkyl chains.

No attempt is made to discuss the main bulk of halogen, silicon, boron and phosphorous compounds. The index of vapour spectra in the final section includes spectra from a number of commercial collections (see Chapter 6) and some literature references that have been included in the Wyandotte/ASTM[26] collection of punched cards.

3
An introduction to the use of band contours in vapour spectra

One of the most noticeable features of the infrared spectra of vapours is the subdivision of otherwise simple bands into branches. The resultant band contours, which are quite distinctive, can take the form of doublets, triplets with sharp centre branches, quartets with two strong central branches, hybrids of all these forms, or they may appear just as broad badly resolved lumps. In the case of very small molecules the branches are resolved into sequences of well separated rotational lines.

With large molecules and also smaller molecules that have been examined on lower dispersion spectrometers, the rotational lines become less well resolved. Before the stage is reached where the gaseous vibrational band looks like a condensed state band, it is possible to see the rotational branching without seeing the individual rotational lines. The branches are referred to as P, Q and R branches, depending on whether the rotational energy quantum number decreases, stays the same or increases during the vibrational transition.

The appearance of the P, Q and R branches changes for molecules of different symmetry. It is therefore possible in favourable circumstances to deduce the geometry of a molecule, or, conversely, to help allocate bands to specific vibrations when the geometry of a molecule is known, by examining the contours of vibrational bands.

It is usual to discuss the band shapes of molecules according to their symmetry with respect to the three moments of inertia I_A, I_B, I_C about three mutually perpendicular axes. Conventionally, if they are different, $I_A < I_B < I_C$.

The different symmetry types are:—

1. Linear molecules where $I_A = O$, $I_B = I_C$

e.g.

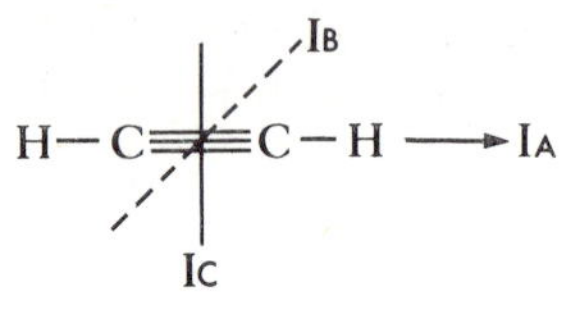

2. Spherical top molecules where $I_A = I_B = I_C$

e.g.

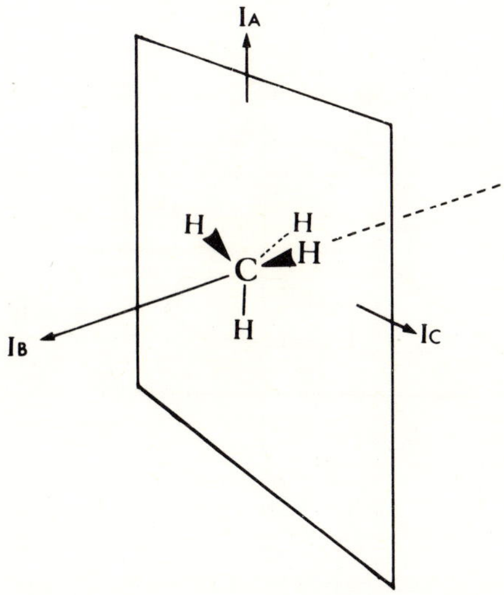

3. Symmetric top molecules which can be divided into two general types,

(*a*) Prolate tops, where $I_A < I_B = I_C$

e.g.

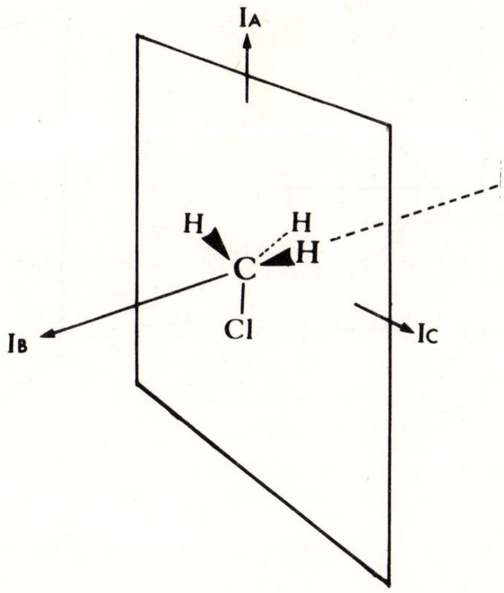

(*b*) Oblate tops, where $I_A = I_B < I_C$

e.g.

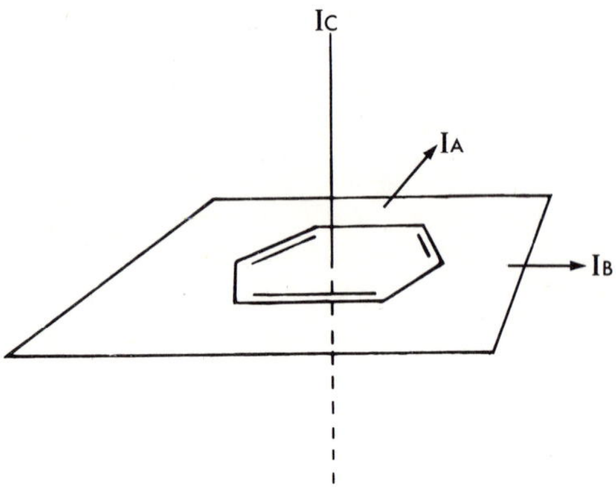

4. Asymmetric top molecules, where $I_A \neq I_B \neq I_C$

e.g.

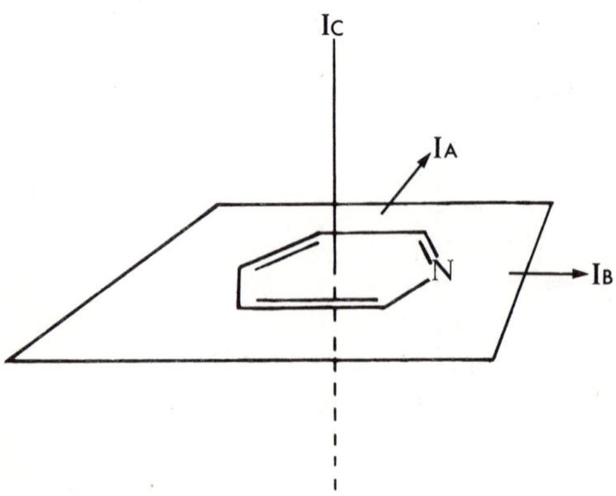

near oblate

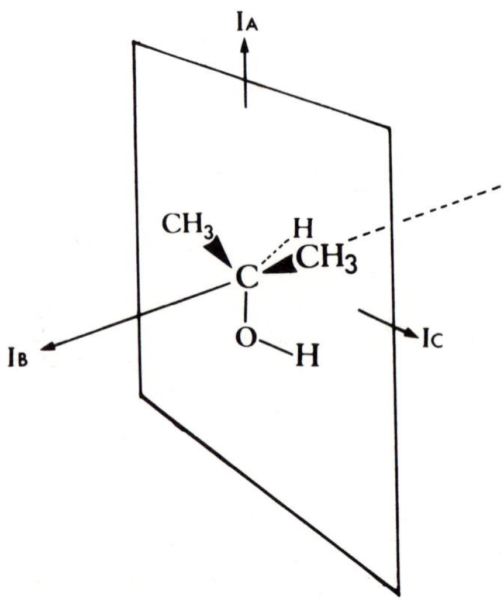

near prolate

a. Linear molecules

When the dipole change is along the symmetry axis, the so-called 'parallel' bands have P and R branches, but no centre Q branch.

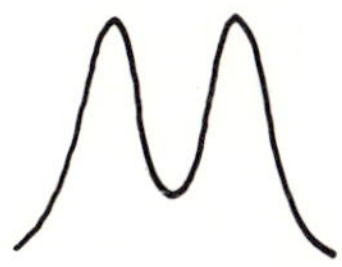

When the dipole change is perpendicular to the symmetry axis, the 'perpendicular' bands have a strong Q branch in addition to the P and R branches.

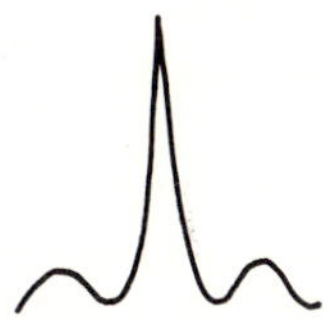

b. Spherical top molecules

These molecules show bands with P, Q and R branches which are almost identical with the perpendicular bands obtained from linear molecules.

c. Symmetric top molecules

'Parallel type' bands are made up of a number of sub-bands. The majority of these sub-bands have Q branches as well as P and R branches. The resultant band with P, Q and R branches is therefore similar to the perpendicular band obtained from linear molecules, but it is more complex due to the superimposition of the many sub-bands. All that can be deduced is that where the contour is of this type and the molecule cannot be of linear or spherical top symmetry, then there is a chance that it has symmetric (or near symmetric) top symmetry. The Q branch increases in intensity relative to the P and R branches as the ratio $I_A/I_B(=\rho)$ increases. As $I_A/I_B \rightarrow O$, i.e. as the molecule becomes more linear, the Q branch of the parallel band disappears completely.

Bands associated with a vibrational transition moment perpendicular to the highest symmetry axis are called perpendicular bands. All the sub-bands have P, Q and R branches, where the Q lines are close-spaced and the P and R lines more separated. The overall effect is to give a series of Q branches with a background of P and R lines between each. As the Q branches are stronger towards the centre of the band, the poorly resolved bands of larger molecules look like a gently curving lump without a sharp centre. With slight asymmetry, they become the Type B bands of asymmetric top molecules (see next sub-section) having the dip in between two Q branches (see Fig. 3.1).

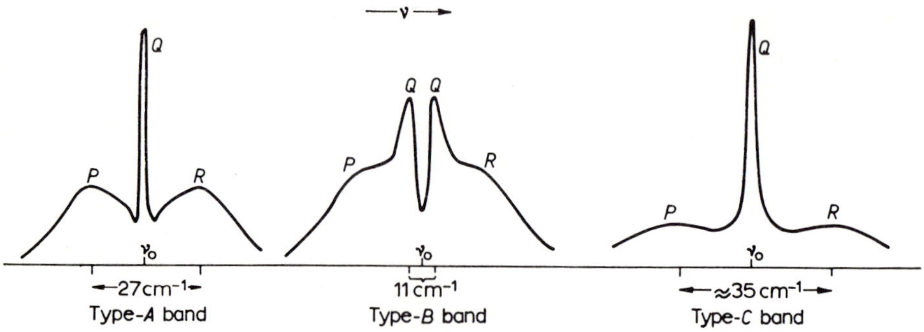

FIG. 3.1. Typical A, B, and C type band contours, drawn to scale for nitromethane. (Permission of W. J. Jones and Elsevier Publishing Co.)

d. Asymmetric top molecules

When discussing asymmetric top molecules, great care must be exercised in using the letters A, B and C.

These molecules, by definition, have the moments of inertia $I_A \neq I_B \neq I_C$ and at the same time $I_A < I_B < I_C$. This means that the rotational constants (A, B, C) are ordered A > B > C. For molecules with some degree of symmetry, the bands are categorised into three types A, B and C, according to whether the dipole moment change occurs along the axis with the moment of inertia I_A, I_B or I_C respectively. The bands of molecules of lower symmetry are hybrids of some or all of these types. Bands of very low symmetry are almost too complex for analysis. The most amenable cases are those where the structure is approaching one of the recognised forms of symmetry.

Jones[71] shows (Fig. 3.1) the typical Type A, B and C bands obtained from nitromethane. Type A and C bands have pronounced Q branches, with those of Type C being the stronger relative to the P and R branches. Type B has the double Q branch. He also gives a table that is reproduced here (Table 3.1),

TABLE 3.1.[a] Symmetric top bands corresponding to the A, B, and C, type bands of planar asymmetric top molecules for $\rho \approx 1$ and $\rho \approx 0$

	Near oblate symmetric top $A \approx B > C, \rho \approx 1$	Near prolate symmetric top $A > B \approx C, \rho \approx 0{\cdot}1$	Near linear molecule $B \approx C, \rho \approx 0{\cdot}0$
Dipole vibration of asymmetric top along the axis (below)	Band type of corresponding oblate symmetric top	Band type of corresponding prolate symmetric top	Band type of corresponding linear molecule
A	$\perp$	$\parallel$	$\parallel$
B	$\perp$	$\perp$	$\perp$
C	$\parallel$	$\perp$	$\perp$

[a] Permission of W. J. Jones and Elsevier Publishing Co.

showing the type of symmetric top band that corresponds to the Type A, B and C bands of planar asymmetric top molecules when $I_A/I_B \sim 1$ and $I_A/I_B \sim 0$.

The Type A band is most typical when I_A/I_B $(=\rho)$ is very small and the band approximates to a symmetric top parallel band, having a strong central maximum. If $\rho = 0$, then the molecule becomes approximately linear and the Type A band becomes similar to a linear parallel band with P and R, but no Q, branches. If $\rho = 1$, then the molecule has symmetric top symmetry and the band is a perpendicular symmetric top band, possessing no characteristic branching features.

The Type B band has two prominent Q branches for larger values of ρ, but it still has a minimum in the centre of the Q branch for lower values of ρ. With insufficient resolution, the two Q branches look like one and the band resembles a Type A band. When $\rho = 1$, the band is identical with the Type A and perpendicular symmetric top bands that are featureless. When $\rho = 0$, the band approximates to a linear perpendicular band with P, Q and R branches.

The Type C band is typical when ρ is large and the band resembles a symmetric top parallel band with a strong Q branch. For a small value of ρ there is no strong central maximum and the shape approximates to that of a Type B or perpendicular symmetric top band, where $I_B \approx I_C$. When $\rho = 0$, the Type C band approximates to a linear perpendicular band with P, Q and R branches.

e. Deductions on molecular geometry from band contours

From the preceding discussion it can be seen that it is possible to tell, from the shape of the vibrational band corollated with it, what relation a bond has to the three main axes in a molecule. Conversely, a band allocation can be confirmed from a knowledge of its contour, if a general idea of the geometry of the molecule has been obtained from other sources. In addition, it is possible to obtain some information on the symmetry of a molecule, by examining the whole spectrum.

It is probably useful therefore to give the possible origins of the main types of band contours, as a corollary to the previous discussion.

(i) *A Band with P and R Branches, but no Q Branch*

It could be a parallel band of a linear molecule, where the dipole change is along the symmetrical axis. But as this type of geometry is rare with polyatomic molecules, it is more likely to be an asymmetric top Type A band associated with a rod like molecule when $\rho = 0$ and the dipole change is along the A axis.

(ii) *A Band with P, Q and R Branches*

It could be a perpendicular band of a linear molecule, where the dipole change is perpendicular to the symmetrical axis, or it could be a parallel band of a symmetric top molecule where the dipole change is along the A, or figure, axis. It is more likely to be an asymmetric top Type A band ($\rho \neq 0$ or 1) where the dipole change is along the A axis, or an asymmetric top Type C band (except for very small values of ρ) when the dipole change is along the C axis. When $\rho = 0$, the Type C band becomes a perpendicular linear band which also has P, Q and R branches. The Q branch of the Type C band is sharper than that of the Type A band and it is also relatively stronger than the associated P and R branches. The asymmetric top Type B band also resembles the perpendicular linear band when $\rho = 0$ and the dipole change is along the B axis.

(iii) *A Band with a Broad Unresolved Maximum*

This shape is associated with symmetrical top perpendicular bands, or Type A ($\rho = 1$), Type B ($\rho = 1$) and Type C (ρ very small, $I_B \approx I_C$) asymmetric top → symmetric top bands.

(iv) *A Band with Two Q Branches*

This shape of band is associated with asymmetric top Type B bands when ρ is large, but not when $\rho = 0$ or 1.

Many bands will not fit any of these clear cut assignments. It is probably better to regard the information gained from the contour of any one band as a lead rather than an unequivocal piece of evidence. Confirmation from other band contours or other techniques is necessary.

While the rotation constants in the upper and lower states are approximately the same for the fundamental bands, this is less true in the case of overtone or combination bands. In the latter case the Q branches can be shaded to longer wavelengths and the band contours will be altered.

No special effort has been made to recognise 'hot bands' in the spectra published here. Hot bands are the result of a transition from a vibrational upper state to yet another higher state, i.e. in terms of the vibrational quantum number, $v = 1 \rightarrow 2$ etc. The probability of a transition from $v = 1$ is the same as for the transition $v = 0 \rightarrow 1$. But as there are many fewer molecules in the $v = 1$ state, the number of molecules making this transition is much smaller and the related bands are much weaker. They appear very close to the $v = 0 \rightarrow 1$ bands, the separation increasing with increasing anharmonicity of the vibration.

The whole subject of band contours is discussed fully by many authors including Herzberg[72] and Jones,[71] from whom further literature references can be found.

4
The combination of gas chromatography and infrared spectroscopy

a. Summary of trapping techniques

Since its introduction in 1954, gas chromatography has been increasingly used as a means of separating and purifying compounds for spectroscopic examinations.

Many methods for collecting gas chromatographic fractions have been described in the literature and they all have a useful place in the context of the work for which they were developed.[73–101] Very few of these methods can be used in all experimental situations. The principal weakness common to the majority of manual trapping techniques is that they are multi-stage processes. First, the fraction is condensed in a tube or trap; adsorbed on to g.l.c. column packing, support or KBr powder; or dissolved in a solvent. Secondly, the fraction then has to be transferred to an infrared cell in one or more steps.

The relative advantages of different techniques become apparent when one considers their trapping efficiencies. This is because the trapping efficiency ultimately controls the smallest sample size that each technique can handle.

For all except the least volatile fractions, the methods which involve condensation on cooled walls or tubes are the least efficient. There can be vapour losses in the trap either when it is at a low temperature or when the sample is warmed for transfer to the infrared cell,[102] see Table 4.1.

The most efficient methods for trapping the sample are those in which the sample is collected in cooled column packing or support,[78,89] or dissolved in a solvent. Very often, however, the latter method requires so much solvent that the solution becomes too dilute. In the case of less volatile materials, it is satisfactory to adsorb or condense the fraction on to cooled KBr powder.[77] Various methods for recovering the sample from packed traps have been described. Spectra can be obtained from samples of 50 μg of oxygenated compounds or 200 μg of hydrocarbons[78] by flushing the trap with argon and condensing both the argon and the sample in a 6 ml cold trap cooled in liquid nitrogen.

The argon is then evaporated and the liquid or solution is transferred to an

TABLE 4.1.[a] Total possible loss of 20 mg fractions of esters condensed in a 50 ml trap at 0°C

Bandwidth		Mass lost as saturated vapour	Mass lost as vapour when trap warms to 20°C	Total possible loss	% recovery of liquid sample
Methyl acetate	0·5 min = 50 ml carrier gas	13·5 mg	34·0 mg	47·5 mg	—
Methyl propionate		5·7 mg	17·1 mg	22·8 mg	—
Methyl butyrate	1·0 min = 100 ml carrier gas	4·9 mg	8·3 mg	13·2 mg	34
Methyl valerate		2·7 mg	4·9 mg	7·6 mg	67
Methyl caproate	2·0 min = 200 ml carrier gas	1·2 mg	0·96 mg	2.2 mg	88
Methyl caprylate		0·4 mg	0·28 mg	0·7 mg	97

[a] Permission of R. P. W. Scott and Varian Aerograph

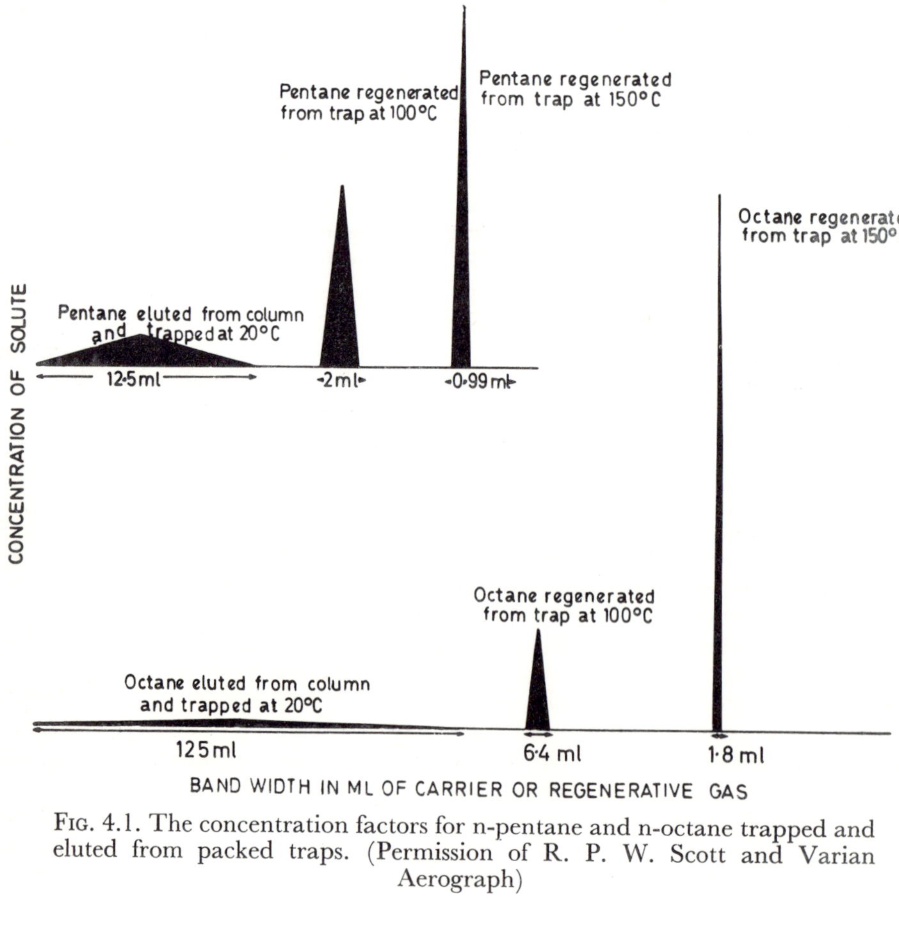

FIG. 4.1. The concentration factors for n-pentane and n-octane trapped and eluted from packed traps. (Permission of R. P. W. Scott and Varian Aerograph)

infrared microcell. A modification of this technique[79] which involves the transfer of the sample directly into a silver chloride microcell enables the same size of sample to be analysed. If a small glass cone is substituted for the silver chloride cell and the solution that is obtained by adding an infrared solvent is transferred to a cavity cell in a beam condenser, then samples of 10 μg of polar compounds can be examined. The most efficient method for recovering samples from packed traps that has been reported so far entails removing the solute under vacuum and adsorbing it on to cooled KBr powder placed in a holder.[80] After pressing, good spectra can be obtained from 10–20 μg of sample using a 1·5 mm diameter disc, or from 1–2 μg using a 0·5 mm diameter disc placed in a beam condenser in both cases.

Curry, Read, Brown and Jenkins[81] have also developed a method for examining high boiling fractions in potassium bromide discs. They condense the sample in a cooled tube, dissolve it in 25 μl of chloroform and transfer it onto some KBr powder that is held on to the tip of the syringe needle used for transferring the sample, by wetting its surface with 0·5 μl of chloroform. After the solvent evaporates, another 0·5 μl of solution is sucked on to the KBr and the process is repeated until all the solution is transferred and the solvent evaporated. The KBr is then mixed with more powder and pressed in a microdie. The authors claim that they can obtain strong spectra of 1 μg samples.

Another way to recover a fraction from a packed trap is to heat the trap and flush the fraction directly into an evacuated infrared gas cell (see Fig. 1.3b). This is the simplest method, which utilises not only the concentrating effect of collecting the sample at low temperatures and eluting at high temperatures,[102] see Fig. 4.1, but also the advantage of the minimum of sample handling. This is similar to Anderson's method[103] of filling a gas cell by heating a cold condensed fraction.

Haslam, Jeffs and Willis[76] and Chang, Ireland and Tai[104] also filled gas cells by variants of these methods. Ballinger, Bartels and Taylor[15] developed a cell (described in the section on sample handling) which after being immersed in a coolant in order to trap a sample, can be heated up to 200° C to regenerate it in the vapour state. These are some of the relatively few examples of the use of vapour spectra in the manual methods of trapping samples, but vapour spectra are necessary in all the direct methods for examining g.l.c. column effluent.

b. Fast scanning spectrometers and some examples of the direct combination of gas chromatography and infrared spectroscopy

The infrared spectroscopist is very envious of the ease with which fast scanning mass spectrometers can examine the effluent from a gas chromatograph column. The main problem in that case is how to increase the solute concentration and at the same time keep the pressure in the mass spectrometer source as low as possible. This has been done by using concentrators which preferentially remove the carrier gas by diffusion processes.[105–108] However, they are not very efficient, as, although the concentration of solute may be increased by a factor of from 10 to 100 times, up to 80% of it can be lost with the removal of the carrier gas. The sensitivity of a mass spectrometer is so much higher than that of an infrared spectrometer that this loss does not always matter. With the concentrators, mass spectrometers can produce spectra from gas chromatographic bands that represent a sample size of 0·1 micrograms or less. As the solute is distributed over the whole band, only a small amount of the sample is used to obtain a spectrum.

Infrared spectroscopy needs both an increased sensitivity and an increase in spectrometer speeds and much interesting work has been done to improve these two factors. The different types of cells that can be used to increase the sensitivity are described in the section on sample handling.

In 1961, Senn[109] and then in 1962, Flett and Hughes[7] described methods for measuring gas chromatographic effluent 'on the fly'. When a fraction was isolated in the gas cell, the following effluent was diverted to the atmosphere through a by-pass valve. The infrared cells could only be used at room temperature. Senn used a minimum volume cell and could trap and identify 0·1–0·2 μl of acetone, using 5$\times$ scale expansion. Flett and Hughes used a multipath gas cell having a pathlength of 240 cm and a volume of 130 ml (P/V ratio = 1·8). A schematic diagram of their apparatus is shown in Fig. 4.2. The sample sizes of the fractions from which they obtained spectra were 700–900 micrograms, but it is clear that they would have obtained spectra good enough for structural analysis work from much less than that in the case of the more polar low molecular weight compounds. The biggest drawback, apart from the use of vapours at room temperature, was the loss of any further fractions that were eluted during the scanning time of the spectrometer, unless they were trapped manually. It was with this in mind that several groups of workers produced methods for increasing the scanning speed of spectrometers.

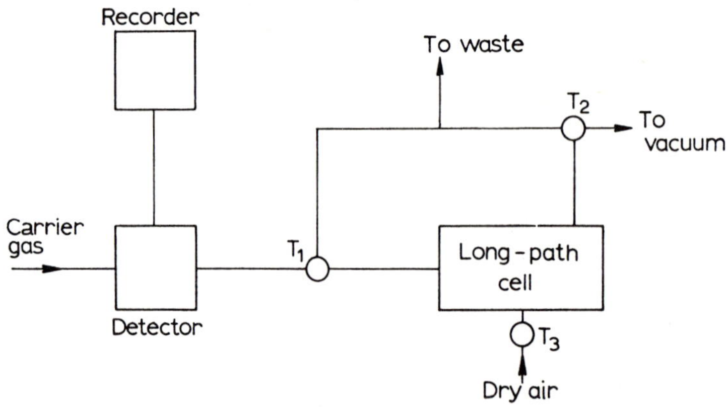

FIG. 4.2. Typical gas handling system for gas chromatographic effluent as used by Flett and Hughes. (Permission of M. St. J. C. Flett and Elsevier Publ. Co.)

The first papers describing faster scanning spectrometers were published simultaneously in 1964 by Bartz and Ruhl[9] and Wilks and Brown.[10]

Bartz and Ruhl[9] used what was effectively two single beam grating monochromators in parallel, see Fig. 4.3a. One spectrometer covered the range from 2·5–7·0 microns, and at the same time the other scanned from 6·5 to 16 microns. Using a high chopping rate of 24 Hz and a fast response recorder, a complete spectrum was obtained in 16 seconds. With a cell-filling time of 12 seconds and a purge time of 4 seconds, the complete cycle of operation took 32 seconds, which allowed the investigation of peaks appearing at approximately half minute intervals. The resolution at the fast speed was not as good as that of conventional small prism instruments. The flow-through gas cell was what has come to be called a 'light-pipe' (see Chapter 1). The inner surface was corrosive resistant and highly reflective and it was claimed to have an effective optical pathlength of 16 inches and a transmission of 40%. It could be heated to a maximum temperature of 250°C. The effluent by-passed the cell while a trapped fraction was being examined. Workable spectra could be obtained from 0·5 mg of moderately polar compounds. An example is shown in Fig. 4.3b.

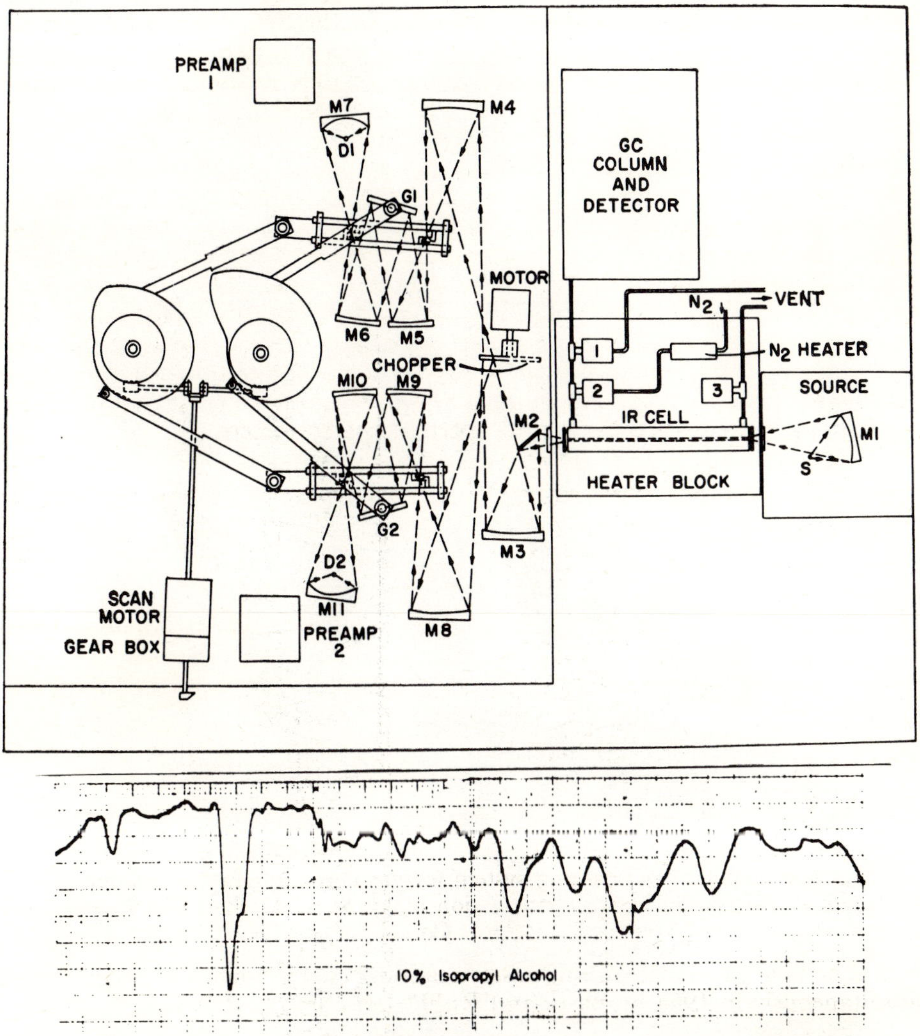

FIG. 4.3. The Bartz and Ruhl spectrometer. (a) Schematic. (b) Spectrum of isopropanol obtained from 10 ml of a 10% solution in toluene. (Reprinted from *Anal. Chem.* **35**, 1892, 1895 (1964). Copyright 1964 by American Chemical Society, reprinted by permission of the copyright owner.)

Wilks and Brown[10] modified a Perkin-Elmer[5] Infracord to scan from 2–15 microns in 45 seconds. The diameter of the servo motor pulley was increased by a factor of two to decrease the spectrometer response time and the slit opening was increased to give more energy, and a resolution of 0·1 microns.

FIG. 4.4. (a) Example of the type of sampling system including light-pipes as used by Wilks and Brown, and Propster. (Permission of Wilks Scientific)

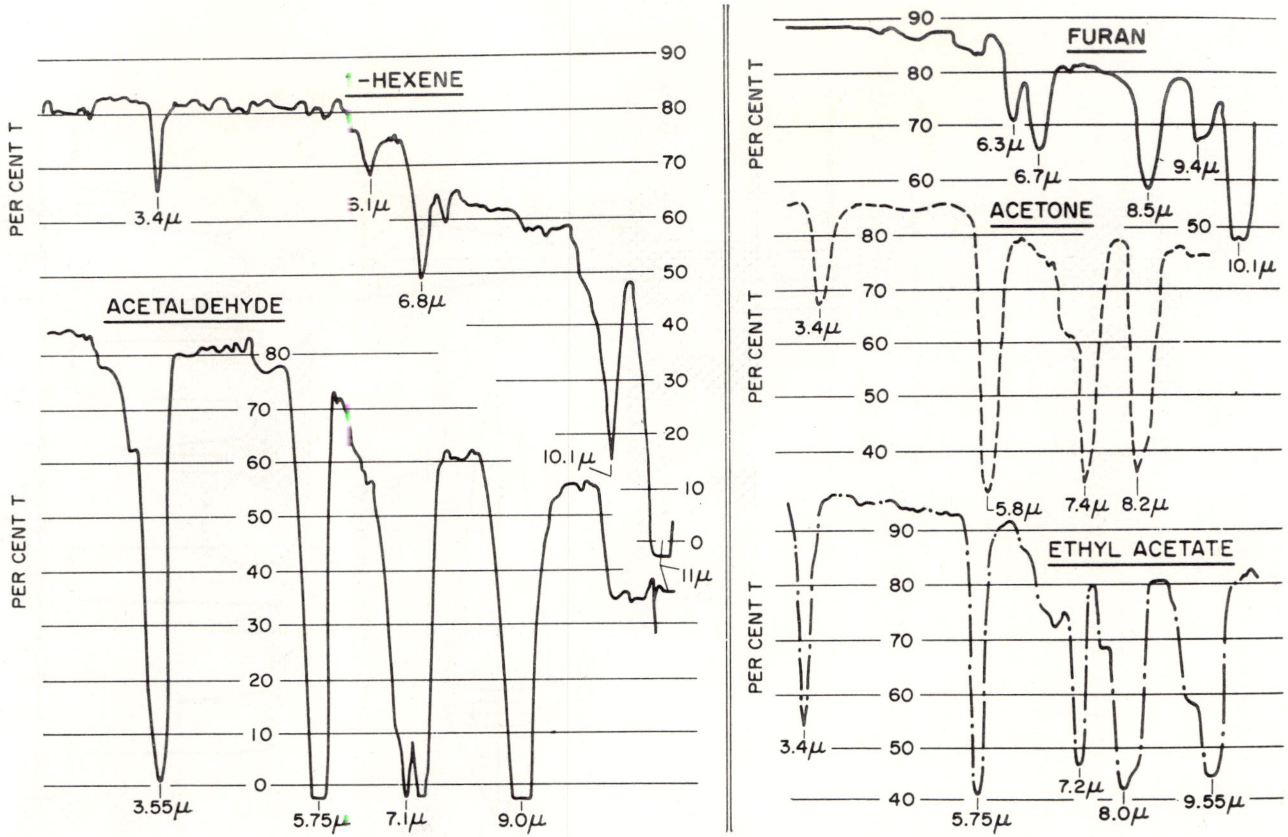

FIG. 4.4. (b) Examples of spectra obtained by Wilks and Brown. (Reprinted from *Anal. Chem.* **15**, 1898 (1964). Copyright 1964 by the American Chemical Society, reprinted by permission of the copyright owner)

The cell space was increased to hold two (sample and reference) 30 cm long light-pipes (see Chapter 1) and their recollimating mirrors by moving the position of the light source and the original collimating mirrors, see Fig. 4.4a. The P/V ratio was not as great as that of the cell used by Bartz and Ruhl and the instrument was not as sensitive. The spectra shown in the paper were obtained from sample sizes of 1·5 μl (examples are shown in Fig. 4.4b). The cells could be heated to 250°C.

Propster[110] described both the Wilks Model 11 Chromatograph-Spectrograph, which is based on the rapid scan spectrometer and also the Model 16 Gas Chromatographic Fraction Analyser attachment for conventional spectrometers, which is basically the extended cell carriage attachment. He also gave a brief comparison of vapour and liquid spectra. The vapour spectra were mostly obtained from 0·5 μl samples on the fast scanning instrument. A spectrum showing the main bands of a 0·18 μl mixture of *meta*- and *para*-xylene is also shown.

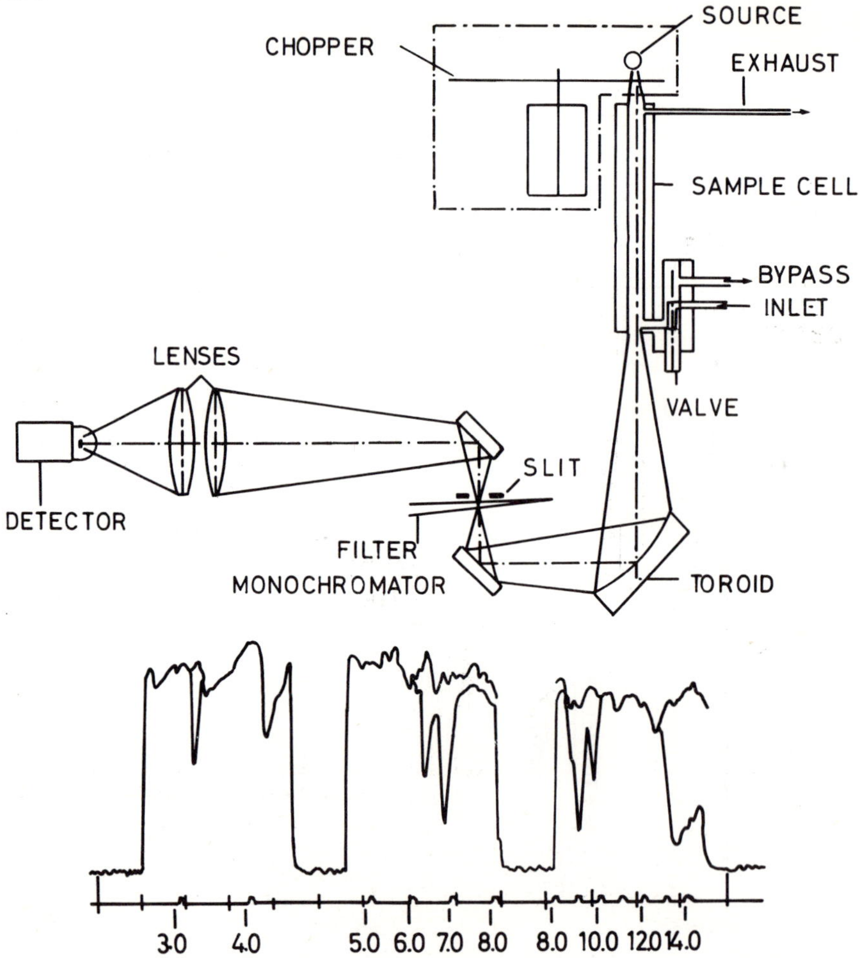

Fig. 4.5. Beckman IR 102. (a) Optical path (b) Spectrum obtained from 0·4 ml chlorobenzene with a 12 second scan. (Permission of Beckman Instruments Inc.)

Beckman[11] produced the IR 102 fast scanning spectrometer especially for gas chromatographic/infrared investigations (see schematic in Fig. 4.5). They were able to obtain scans from 2·5–14·5 microns in 5 or 12·5 seconds by using wedge shaped filters placed as segments on a filter wheel, instead of using a normal prism or grating monochromator. The filters operated over the ranges 2·5–4·5 microns, 4·4–8·0 microns and 7·9–14·5 microns respectively, with a spectral bandwidth of over 40 cm^{-1} at 3 microns and about 18 cm^{-1} at 10 microns. The sample cells were light-pipes of 10, 20 or 30 cm pathlength, which were described in Chapter 1. Sample sizes of 0·4 μl of known polar compounds in a static system gave recognisable spectra. It is doubtful if this sample size is sufficient for unknown gas chromatographic fractions 'on the fly', especially as the overlap between filters was not really sufficient to avoid blank spots and the background spectrum could be misleading at times. Bober and Burner[111] have also published an account of the uses of the IR 102.

The sensitivity of this type of instrument depends to a large degree on the concentration of the fraction in the carrier gas stream and this depends in turn on the chromatographic conditions. Where the mixtures are complex it is hard to keep a high chromatographic resolution when, at the same time, a high temperature programme rate is necessary to keep the concentration up.

It was to get over this difficulty of concentration that Scott, Fowlis, Welti and Wilkins[112] introduced the 'interrupted elution' gas chromatograph combined with a conventional infrared spectrometer. Schematic diagrams are shown in Fig. 4.6. Because the effluent gas stream from the chromatograph can be stopped, there is time for the solute to be trapped and concentrated before it is re-eluted into the infrared gas cell and examined by a spectrometer operating at a normal scanning speed. The concentrator is a short length of g.l.c. column packing similar to that used in the manual trapping of fractions. The solute is trapped in the packing at low temperatures and re-eluted into the gas cell at

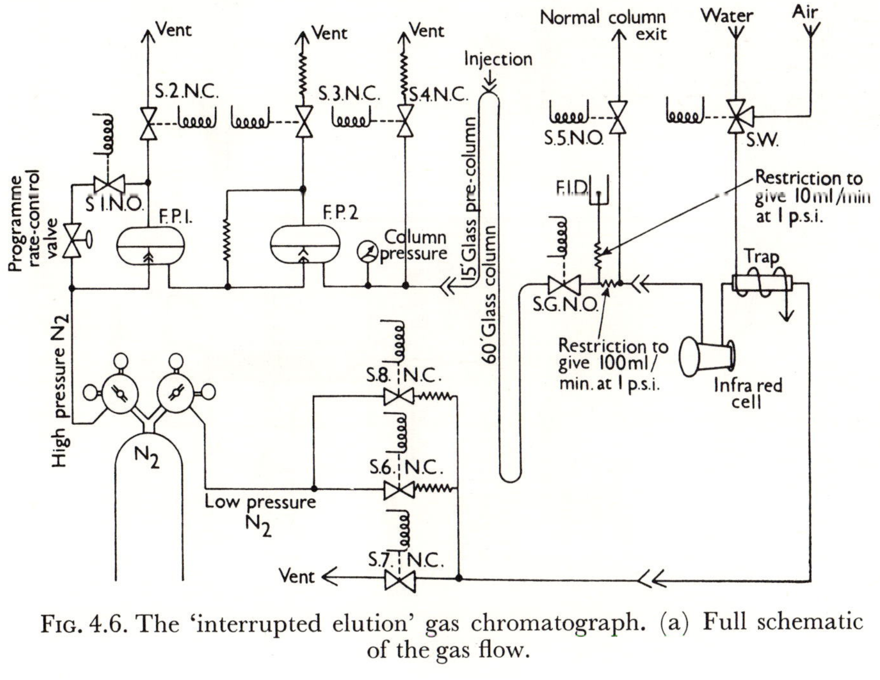

FIG. 4.6. The 'interrupted elution' gas chromatograph. (a) Full schematic of the gas flow.

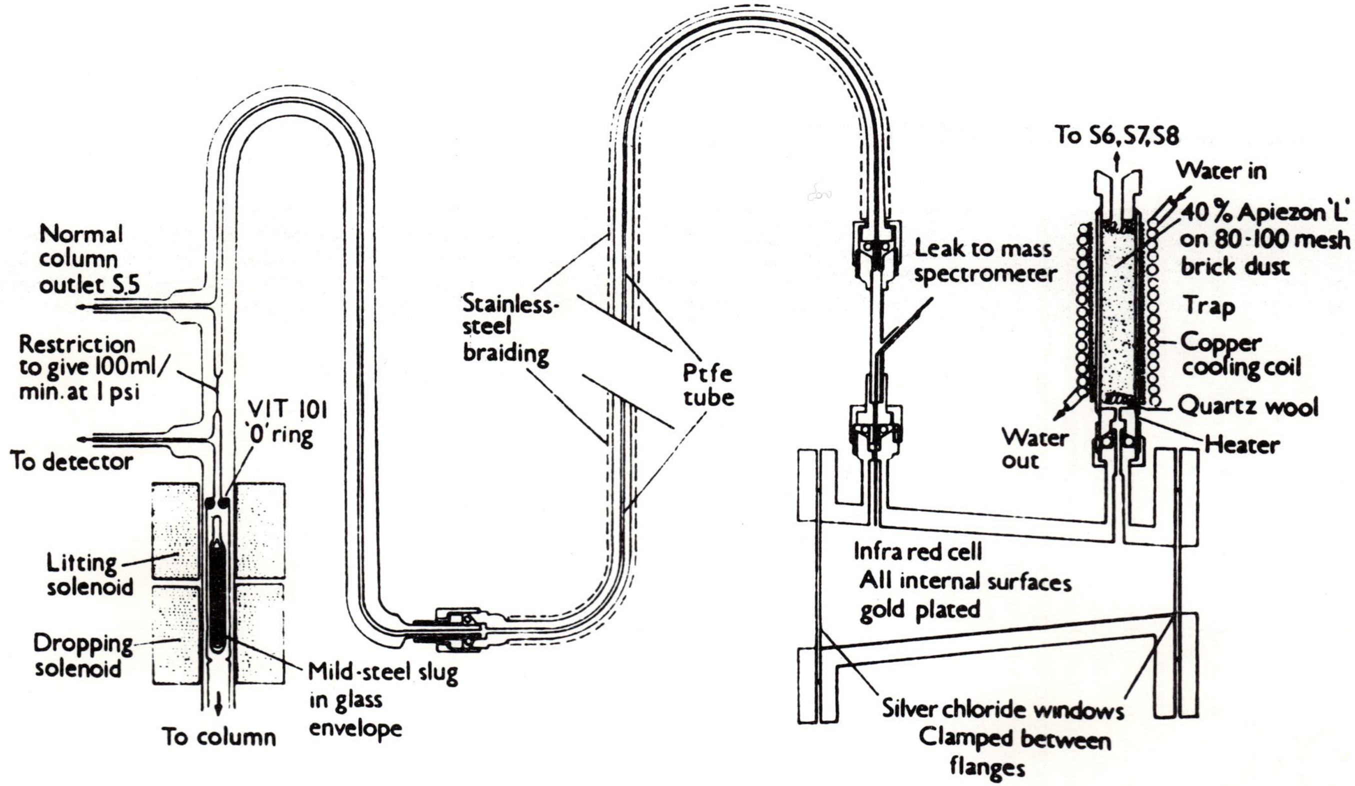

FIG. 4.6. (b) Infrared gas cell and trap/concentrator. (Permission of the Institute of Petroleum)

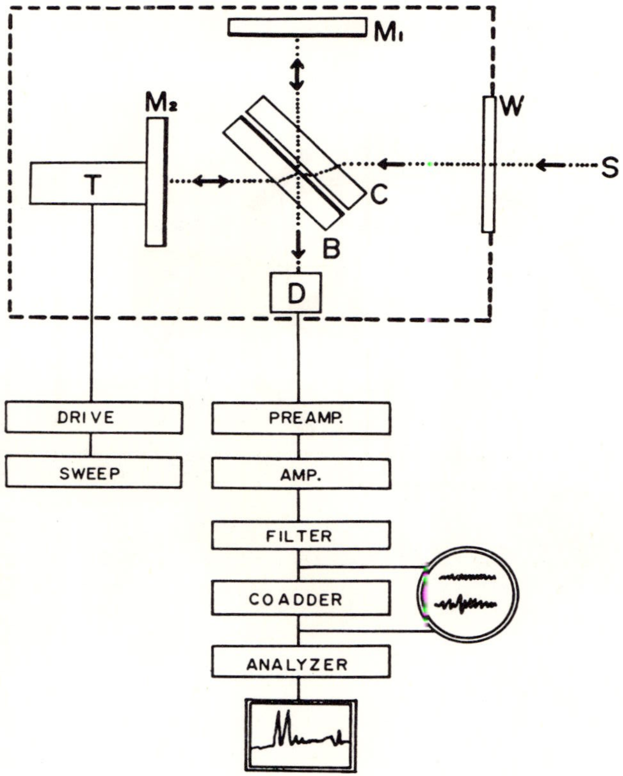

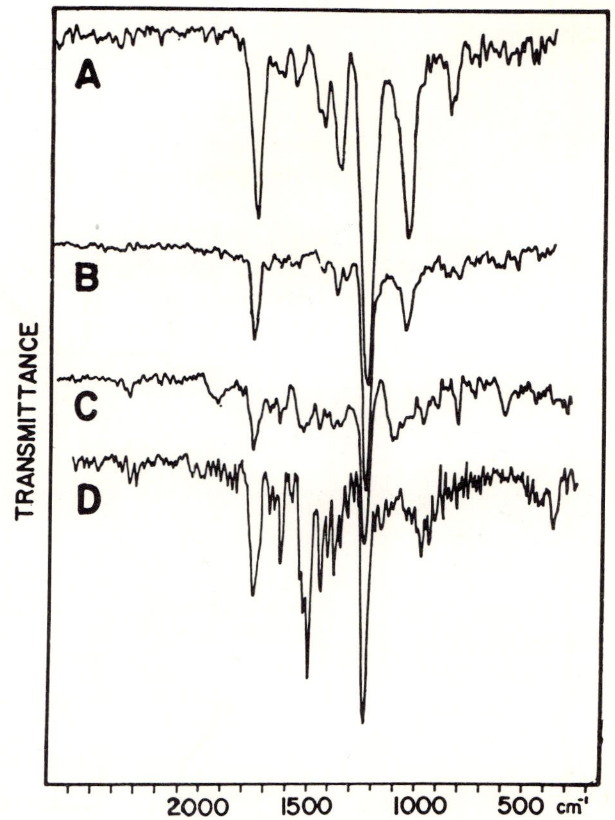

FIG. 4.7. The Block Model 200 interference spectrometer. (a) Block schematic, (b) Spectra of small methyl acetate samples: (A) 30 scans, taken while 1 μl sample was flowing through i.r. cell. (B) 100 scans, 0·1 μl sample trapped in i.r. cell. (C) 300 scans, 0·01 μ l sample trapped in i.r. cell. (D) 300 scans, 0·005 μl sample trapped in i.r. cell. Interferograms stored directly in computer for spectra (A), (B) and (C) but recorded for spectrum (D). (Permission of Pergamon Press)

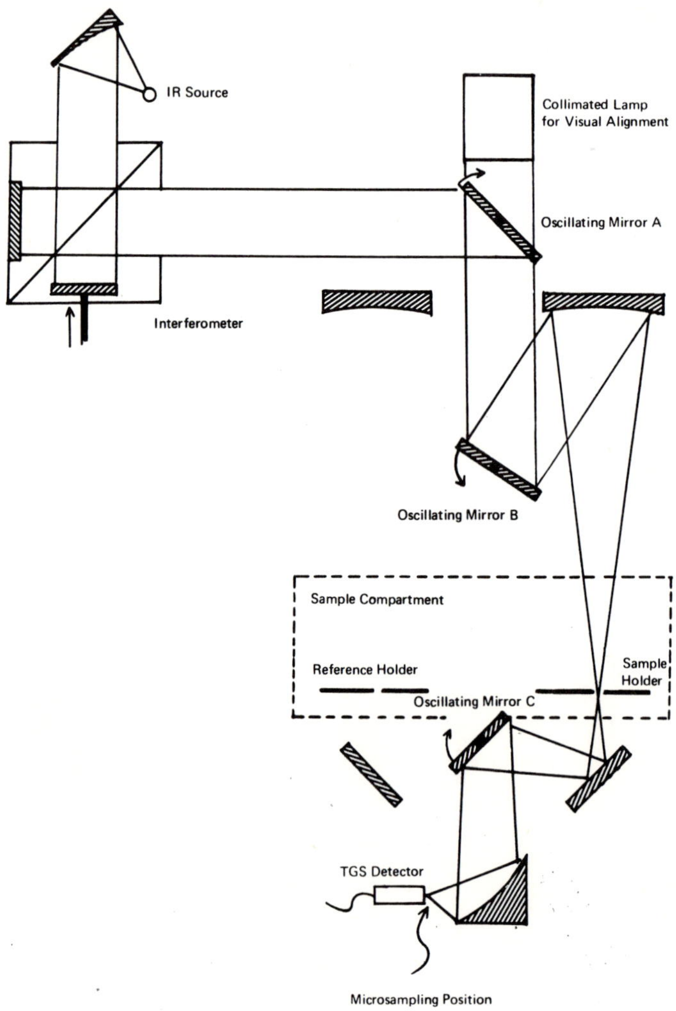

FIG. 4.8. The Block FTS–14 spectrometer system. (a) Diagram of the sampling optics. (Permission of Block Engineering Inc.)

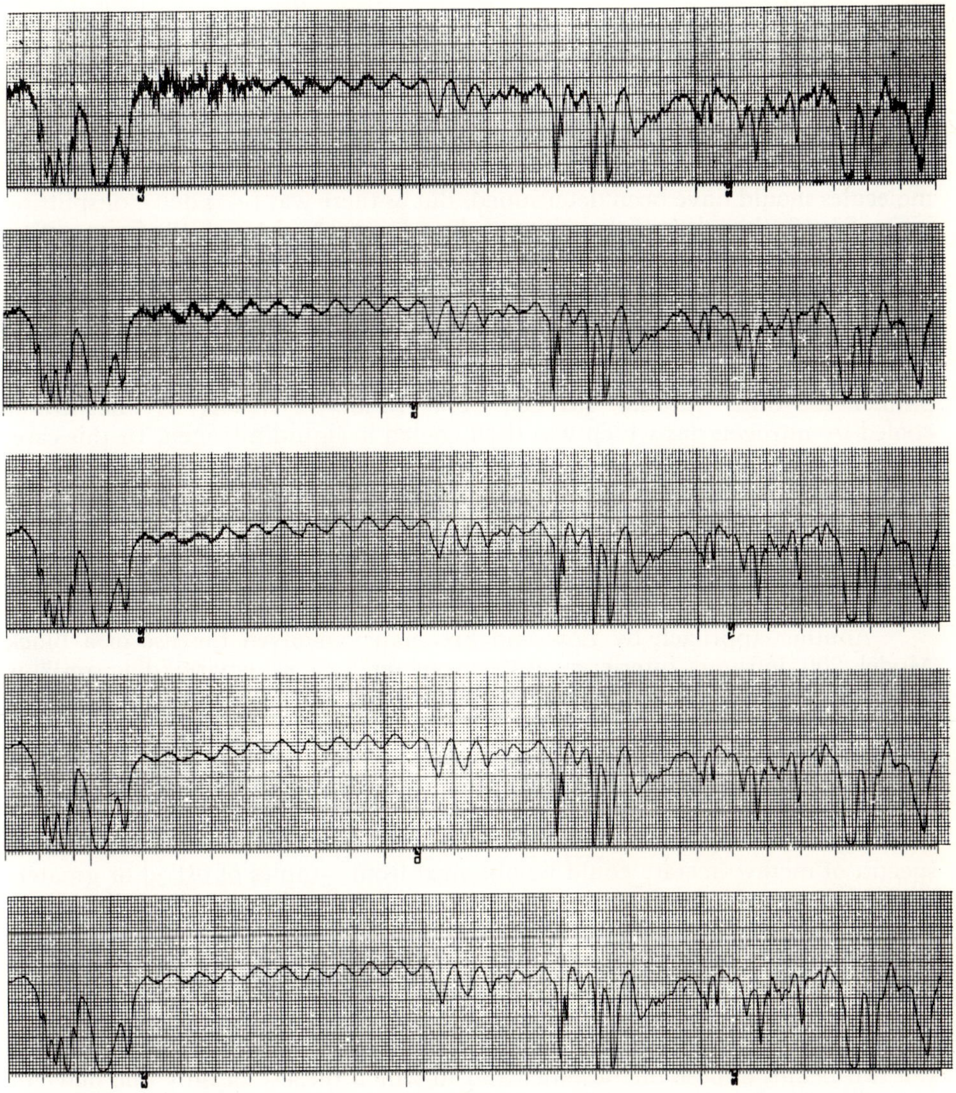

(b) The five spectra above demonstrate the improvement in signal-to-noise gained by averaging increasing numbers of repetitive one-second scans. All are ratio recorded, $2\ cm^{-2}$ resolution spectra of polystyrene film, representing in sequence one, four, sixteen, 64 and 256 scans, each of both the samples and reference beams. The one-scan measurement was completed in under four seconds, the 256 scan measurement (Class II quality) took about 20 minutes.

Note carefully that neither the resolution nor the wavelength accuracy is degraded in the one-second scan. This spectrum is most significant because it demonstrates the capability of the FTS–14 to measure qualitative spectra of short-lived species of gas chromatograph effluent peaks 'on the fly'.

higher temperatures, with the gas stream flowing in the opposite direction. The concentration factor is roughly the ratio of the partition coefficients of the solute at the two temperatures. Fig. 4.1 shows a representation of this effect for two hydrocarbons as originally described by Scott.[102] For the automated system, the fractions were trapped at a temperature of 12°C and re-eluted at a temperature of 180°C under a nitrogen flow rate of 10 ml/min. The original cell was a normal minimum volume cell of 9 cm pathlength and 25 ml volume, giving a very low *P*/*V* ratio of 0·36. With the automatic built-in scale expansion, sample sizes (in the cell) of 25 μg for small polar molecules or 100 μg of non-polar molecules should have been determined. But as there was (a) a detector splitting ratio of 1:10, (b) a loss of the wings of the g.l.c. peaks through the finite setting of the sensing system, and (c) the concentration procedure could not be ideal for every component, the realistic minimum values were more like 50 μg and 200 μg. The sensitivity was increased by using larger pathlength, lower volume vapour cells, but then the concentration factor and the ratio of partition coefficients had to be increased. A greater efficiency was obtained with a 10 cm pathlength, 5 ml volume gas cell (*P*/*V* ratio = 2) using a concentrator cooled by nitrogen gas which was itself cooled in liquid nitrogen. In this case, however, the concentration was more likely to be due to condensation and evaporation than to any partition phenomenon. In practice, this technique is more useful for the automatic separation and elucidation of crude synthetic organic mixtures where a relatively large amount of easily resolvable components is available, than it is for the examination of a multiple small component mixture of, for instance, an essential oil.

Another approach by Low and Freeman[12] involved the use of a Block Model 200 infrared interference spectrometer[17] (a schematic diagram[16] is given in Fig. 4.7a). The spectrometer scans from 2500–250 cm^{-1} in 1 second with a spectral resolution of 18 cm^{-1}. The signals of consecutive scans can be recorded on magnetic tape or stored in the core memory of a computer that is an integral part of the instrument. The g.l.c. effluent can be examined as it passes through a 5 cm long, 4 mm diameter, 0·6 ml volume (*P*/*V* ratio = 8·3) gas cell, which is heated to 150°C; or it can be trapped within the cell. Usable spectra of methyl acetate could be obtained from samples of 0·1 μl or greater. When the sample size was reduced, the weaker bands tended to disappear, as all strong emission signals became equal after their attenuation to bring them within the dynamic range of the electronics (see Fig. 4.7b). Parts of the interferogram were thus diminished below the threshold voltage required for distinguishing absorption bands for digitisation and storage in the core memory.

Low[113] has also described a dual beam interference spectrometer in which the two beams oppose one another and in which the use of only one detector introduces negligible errors. Because he used two light sources instead of one common light source, Low compensated for the residual off-balance signal by electronically accumulating a number of blank dual beam spectra which he then subtracted from the equivalent number of accumulated dual beam sample spectra. He was able to show a spectrum obtained from an isolated 10 μg sample of methyl stearate following an analogue wave analysis of the compensated accumulation of 50 one-second interferograms. He did not give the spectral bandwidth of the spectrometer.

Block[17] have recently introduced the FTS–14 spectrometer which combines a Model 296 high resolution Fourier transform optical system with a small computer. The spectrometer works over a total range of approx. 15000—10 cm^{-1}

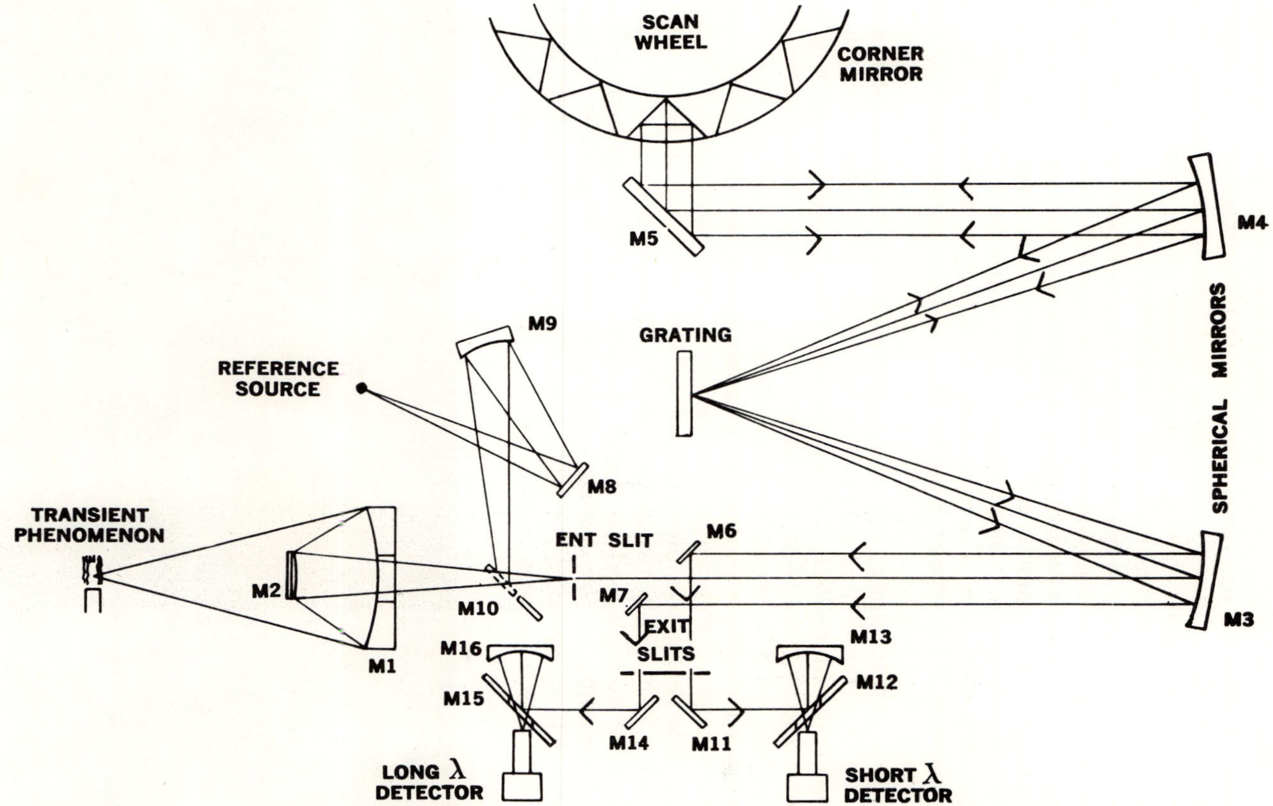

FIG. 4.9. The Warner and Swasey Model 501 spectrometer. (a) Optical path.

but with KBr optics this is restricted to the very convenient range of 3800–400 cm^{-1}. The diagram of the sampling optics (Fig. 4.8a) shows that the sample is remote from the source and that only one fast response triglycine sulphate detector is used for either a single or double beam system. There is provision for a collimated lamp for the visual alignment of the spectrometer. A single scan takes one second, but the S/N ratio can be increased by accumulating successive spectra. Fig. 4.8b shows that in a series of accumulated spectra of polystyrene the one-second scan loses no resolution or wavelength accuracy. A spectrum with 0·5 cm^{-1} resolution can be processed in the small computer in 120 seconds, and a spectrum with 2 cm^{-1} resolution can be processed in 30 seconds. The latter spectrum is of Coblentz Society Class II Research Grade quality. Block quote that for gas chromatographic 'on the fly' operations, a 4 cm^{-1} resolution spectrum can be obtained in 5 seconds with a print-out time of 20 seconds, using a high speed digital plotter. This instrument would appear to be the most suitable so far developed for the direct combination of gas chromatography and infrared spectroscopy.

The fastest scanning infrared spectrometer that has been developed for use with g.l.c. is the Warner and Swasey Model 503.[114] A schematic of the basic instrument, the Model 501, is shown in Fig. 4.9a. Its application to g.l.c. has been described by Dolin, Kruegle and Krakow.[115] The scanning system depends on the revolving of a turntable of twenty-four corner mirrors placed in the intermediate focal plane of a double pass monochromator.[116] The grating and all the other optical components are stationary. The positioning of the mirrors after the first pass allows the dispersed light to be shifted in both directions, giving what is in effect, two parallel scanning regions. Dolin *et al.* quote an overall range of 2·7–9 μ which is split roughly into the sub-ranges 3–5 μ and 5–9 μ. Depending on the grating and filters, other ranges can be covered. Overall repetitive scanning speeds of 1·6, 4 and 8 scans per second are available, of which 80% of the time is actually spent on wavelength scanning. The normal

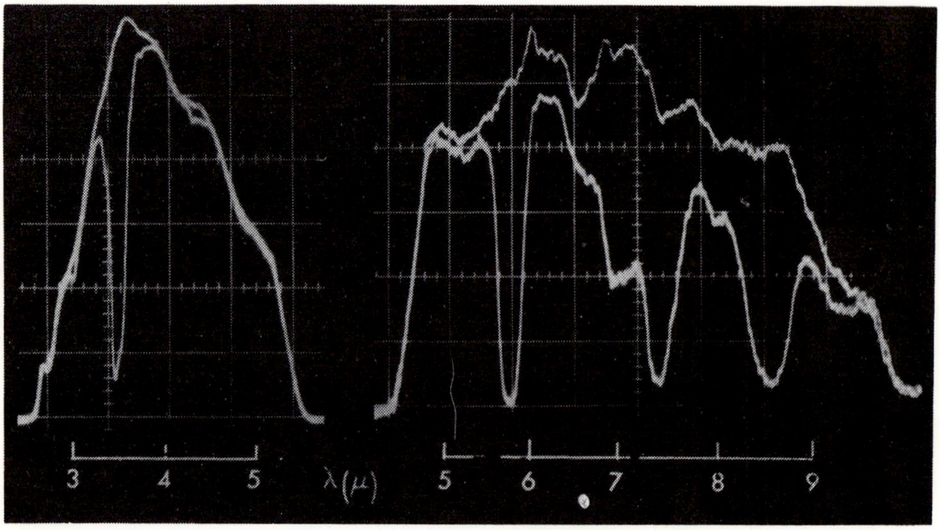

FIG. 4.9. (b) Spectrum of butan-2-one scanned in 0·5 seconds. (Permission of the Warner and Swasey Co.)

cell is a light-pipe 10 cm long and 1 ml volume (P/V ratio = 10) which can be heated to 250°C but cells up to 30 cm long can be used. The spectra are superimposed on spectral backgrounds that are U or V shaped and the instrument has a spectral bandwidth that varies within a factor of two of 20 cm^{-1}. The published spectra were obtained from sample sizes of over 100 μg (see Fig. 4.9b), although the main bands from 25 μg acetone were easily recognisable. The spectral bandwidth can be reduced and the resolution increased if the scanning range is severely restricted.

A series of papers on fast scanning spectrometers have been published in *Applied Optics*. Three were concerned with infrared spectroscopy,[117–119] although no particular applications to gas chromatography were described.

c. Criteria for the direct combination of gas chromatography and infrared spectroscopy

The progress made on the direct combination of these techniques up to the time of writing can be summarised as follows:

1. Spectrometers capable of scanning fast enough have been developed, but with the exception of the Block FTS–14 spectrometer, their spectral bandwidths, base lines and spectral presentation are not really good enough yet for the structural analysis of more difficult unknown compounds. It is possible that with the fast response detector, some of the low cost, low resolution spectrometers might be capable of a reduced scanning time of less than 20 seconds.
2. Higher sensitivity gas cells with increased pathlength to volume ratios have been developed, but they are not yet used to their best advantage since their volumes are generally too small to accommodate the necessary carrier gas volumes. This is particularly true when long, high efficiency conventional packed columns are being used.

Ultimately a successful combination technique will need the development of a spectrometer with the performance of a small grating spectrometer, but with a scanning speed sufficiently fast to complete a scan and allow the cell to be purged before the elution of another peak from a chromatograph. *Fast scanning must not be an end in itself.* It must be combined with low spectral bandwidths and acceptable resolution. If the choice has to be made, in most cases a less efficient monitoring system using conventional spectrometers, which loses a proportion of the fractions, would be more acceptable for structural identification purposes. Bearing in mind that the chromatographic conditions can be controlled, a scanning time of 5 seconds would be ideal and even a scanning time of 15 seconds would be sufficient in the majority of cases.

If a high efficiency conventional column is used, it would be necessary to collect fractions eluted in 100 ml of carrier gas. To accommodate this volume in a light-pipe with a P/V ratio of 10, the cell path length would have to be 10 metres, which is impracticable. It would be preferable to use a cell of, for instance, 40 cm pathlength and 4 ml volume after having concentrated the sample twenty-five times. We are then back to cold condensation or trapping as the means of concentration and the cell would have to have its own solute detector.

A better approach is to adapt the gas chromatograph column to suit the requirements of the infrared spectrometers. Whatever the scanning speed of the spectrometer, it will always require the highest partial vapour pressure of the solute in the most efficient cell, which will most probably be of low volume. What is needed therefore is a chromatograph that has a high efficiency, giving a high solute concentration in the effluent gas stream, and which at the same time, has a high column capacity of the solute with no loss of sample due to splitting during the injection on to the column. Long high efficiency packed columns give too low a solute concentration in the effluent gas stream and too great a volume of carrier gas associated with an eluted fraction. Open tube capillary columns have too low a sample capacity and too much sample is lost in the splitter at the front of the column. The columns that are most likely to be of use are either support coated open tube (SCOT) columns, or very narrow (0·5–2·0 mm i.d.) conventional columns. Typical solute capacities vary with the resolution of the g.l.c. column, but for a single component they are of the order of 0·05 μl for SCOT columns and 0·1 μl for the narrow columns. If, for example, an 18 ft, 1 mm i.d. column packed with 85–100 mesh brickdust carrying 10% stationary phase is temperature programmed at a comparatively slow rate of 1°C/minute, with a linear carrier gas velocity of 2·5–3·0 cm/sec, then the carrier gas bandwidth for an eluted fraction is about 2·0/3·0 ml.* A cell of 2 ml volume would contain most of the fraction, provided that there was very little vapour diffusion between the column and the cell. With a P/V ratio of 10, a pathlength of 20 cm and a cross-section of 0·1 cm^2, the cell should be capable of examining most 10–100 microgram samples. It would then be necessary to adapt the spectrometer to hold a pair of 20 cm pathlength sample and reference cells together with the optical systems for condensing and recollimating the light beams.

If it should still be necessary to have a concentration device, the most simple one would consist of a low volume cold trap fitted with a quick-acting heater to regenerate the vapour.

If a conventional spectrometer is being used, some of the peaks from a complex chromatogram would still be lost, unless the fractions were switched between two or more spectrometers. This would be more economic than buying a fast scanning spectrometer that produced poor spectra.

A 'tailor-made' chromatographic system will be just as necessary when the ideal fast scanning spectrometer has been developed, for it, too, will need the highest solute vapour pressure that is obtainable in a high P/V ratio gas cell.

* R. P. W. Scott. Private communication.

5
Vapour spectra

a. Experimental conditions

All the spectra that are included in this section were obtained on a Perkin-Elmer 257 grating spectrophotometer. The majority were run at medium speed, with the knowledge that the calibration was within specification. Where exact values have been quoted in the text (and in a previous paper by Welti and Stephany[20]), the apparent values have been corrected. The frequency calibrations were carried out with indene, ammonia and water vapour and routinely with polystyrene using the values supplied by the IUPAC commission on spectroscopy.[120] The abscissa repeatability is quoted as 3 cm^{-1} at 3000 cm^{-1}, 1·5 cm^{-1} at 1750 cm^{-1} and 1 cm^{-1} at 625 cm^{-1}. For the spectra not having quoted band values, the instrumental specification for abscissa accuracy is ± 5 cm^{-1} at 3000 cm^{-1}, $\pm 2{\cdot}5$ cm^{-1} at 1750 cm^{-1} and ± 2 cm^{-1} at 625 cm^{-1}. However, it must be remembered that this instrument is open to the atmosphere and that where band maxima coincide with strong water vapour lines the calibrations show that the maxima may appear up to 4 cm^{-1} low because of the temporary loss of energy.

The samples, which came from both research and commercial sources, gave the correct liquid, solid or solution spectra. The carboxylic acids were checked by gas chromatography.

The cell conditions and sample size are given on each spectrum. The cell had a volume of approximately 25 ml and a pathlength of 9·2 cm. The silver chloride windows were self-sealing between the body of the cell and the flanges, eliminating the use of O-rings. The cell could be heated to 200°C, but in fact it was only taken to a temperature that was suitable for the stability, vapour pressure and sample size for any compound. As mentioned previously, primary amines should be examined at the lowest possible temperature because of their tendency to liberate ammonia. Acids such as lactic or pyruvic will decompose at too high a temperature. But of course, with acids one must work at a temperature compatible with the existence of the monomer or dimer, whichever you wish. The carboxylic acids shown here are all in the monomeric form, except for formic and acetic acids which are presented in both forms.

The samples were injected into the cell in the specified amounts with a 10 μl Hamilton syringe. This works on a 'difference' principle, leaving a residue in the needle itself. As the needle is inserted into the hot cell, varying amounts of this residue will be volatilised according to the volatility of the compound. The specified amounts are therefore all approximate and to varying degrees below the real injected sample size. A total delivery syringe would have been preferred.

The cell was purged with nitrogen and the samples were examined in the presence of nitrogen. The samples were also diluted with nitrogen where it was necessary to reduce the intensity of very strong bands.

Some of the spectra include base lines where there was reason to believe it was not absolutely straight. These are confined to the carboxylic acids which were obtained at an earlier date than the rest of the spectra and which also show an increased amount of noise at the low frequency end of the spectrum. In some cases also, a slow scan through the monomeric OH stretching band is included when more exact calibrated frequency values were needed for the alcohols and phenols.

Some of the primary amine spectra have a break in the spectra near 1200 cm^{-1}. This was caused by starting the scans near 1200 cm^{-1}, so that they would pass through the strong ammonia band centred at 950 cm^{-1} before the ammonia had a chance to develop.

The Spectra Indices

The spectra are presented in groups of compound types. These are sub-divided according to the position and degree of unsaturation in each molecule. Each member of a sub-group is placed in an order of increasing chain-length or complexity.

The numerical spectrum index is therefore divided into molecular types.

The molecular formula index has been ordered in conventional form as explained more fully for the commercial reference spectrum index.

b. Numerical index – in chemical classes

Spectrum Number	Name
	ALCOHOLS
1	Methanol
2	Ethanol
3	Propan-1-ol
4	Butan-1-ol
5	Pentan-1-ol
6	Hexan-1-ol
7	3-Methylbutan-1-ol
8	Hex-5-en-1-ol
9	Pent-4-en-1-ol
10	*cis*-Hex-3-en-1-ol
11	*trans*-Hex-3-en-1-ol
12	2-Phenylethanol
13	Prop-2-en-1-ol
14	*cis*-Hex-2-en-1-ol
15	*trans*-Hex-2-en-1-ol
16	Benzyl alcohol
17	Prop-2-yn-1-ol
18	Furfuryl alcohol
19	Tetrahydrofurfuryl alcohol
20	Propan-2-ol
21	Butan-2-ol
22	Pentan-2-ol
23	Hexan-2-ol
24	Octan-2-ol
25	Pentan-3-ol
26	Octan-3-ol
27	Octan-4-ol
28	Decan-5-ol
29	Cyclopentanol
30	Cyclohexanol
31	But-1-en-3-ol
32	Hex-1-en-3-ol
33	Oct-1-en-3-ol
34	1-Phenylethanol
35	Pent-4-en-2-ol
36	Hept-1-en-4-ol
37	*t*-Butanol
38	2-Methylpentan-2-ol
39	1,1-Dimethyl-2-phenylethanol
40	2-Methylbut-3-en-2-ol
41	Citronellol
42	Geraniol
43	Linalool
44	α-Terpineol
45	2-Methoxyethanol
46	2-(2-ethoxy)ethoxyethanol
47	Ethane-1,2-diol
48	Propane-1,2-diol

Spectrum Number	Name
49	Propane-1,3-diol
50	*N,N*-Dimethyl-2-aminoethanol
51	*N,N*-Diethyl-2-aminoethanol
	PHENOLS
52	Phenol
53	*o*-Cresol
54	*m*-Cresol
55	*p*-Cresol
56	2,3-Xylenol
57	2,4-Xylenol
58	2,5-Xylenol
59	2,6-Xylenol
60	3,4-Xylenol
61	3,5-Xylenol
62	2-Methoxyphenol (Guaicol)
63	2,6-Dimethoxyphenol
64	2-Chlorophenol
65	2-Bromophenol
66	4-Chlorophenol
67	4-Bromophenol
68	2-Isopropyl-5-methylphenol (Thymol)
69	4-Allyl-2-methoxyphenol (Eugenol)
70	1,2-Dihydroxybenzene (Catechol)
71	1,2-Dihydroxy-3-methylbenzene(3-Methylcatechol)
72	1,2-Dihydroxy-4-methylbenzene (4-Methylcatechol)
	CARBOXYLIC ACIDS
73	Formic acid (40°C)
74	Formic acid (130°C)
75	Acetic acid (35°C)
76	Acetic acid (130°C)
77	Propanoic acid
78	n-Butanoic acid
79	n-Pentanoic acid
80	n-Hexanoic acid
81	n-Heptanoic acid
82	n-Octanoic acid
83	n-Nonanoic acid
84	2-Methylpropanoic acid
85	3-Methylbutanoic acid
86	2-Methylbutanoic acid
87	4-Methylpentanoic acid
88	3-Methylpentanoic acid
89	2-Ethylbutanoic acid
90	2-Ethylhexanoic acid
91	But-2-enoic acid
92	Hex-2-enoic acid
93	Hept-2-enoic acid

Spectrum Number	Name
94	Oct-2-enoic acid
95	Non-2-enoic acid
96	Dec-2-enoic acid
97	Hexa-2,4-dienoic acid
98	2-Methylprop-2-enoic acid
99	3-Methylbut-2-enoic acid
100	2-Methylbut-2-enoic acid
101	But-3-enoic acid
102	Hex-3-enoic acid
103	Oct-3-enoic acid
104	Undec-10-enoic acid
105	Benzoic acid
106	2-Oxo-propanoic acid (Pyruvic)
107	2-Hydroxypropanoic acid (Lactic)
	AMINES
108	1-Aminoethane
109	1-Aminopropane
110	1-Aminobutane
111	1-Aminopentane
112	1-Aminohexane
113	1-Aminoheptane
114	1-Aminooctane
115	1,1,3,3-Tetramethyl-1-aminobutane
116	1-Aminononane
117	Piperidine
118	1-Aminoprop-2-ene
119	Benzylamine
120	2-Aminopropane
121	2-Aminobutane
122	2-Aminopentane
123	2-Aminooctane
124	3-Aminopentane
125	Diethylamine
126	Dipropylamine
127	Dibutylamine
128	Dipentylamine
129	Dihexylamine
130	*N*-Methylbutylamine
131	Tetrahydropyrrole (Pyrrolidine)
132	Aniline
133	*N*-Methylaniline
134	2-Methylaniline
135	3-Methylaniline
136	2,4-Dimethylaniline
137	2-Methoxyaniline
138	4-Ethoxyaniline
139	2-Chloroaniline
140	3-Chloroaniline
141	Pyrrole
142	Acetamide

Spectrum Number	Name
143	Propionamide
144	Butyramide
145	*N,N*-Diethylaniline
146	*N,N*-Dimethylformamide
147	Pyridine
148	2-Methylpyridine
149	3-Methylpyridine
150	4-Methylpyridine
151	Quinoline
	ALDEHYDES
152	Acetaldehyde
153	Propanal
154	n-Butanal
155	n-Pentanal
156	n-Hexanal
157	n-Heptanal
158	n-Octanal
159	n-Nonanal
160	n-Decanal
161	But-2-enal
162	Hex-2-enal
163	*trans*-Oct-2-enal
164	*trans*-Non-2-enal
165	*trans*-Dec-2-enal
166	*trans*-Undec-2-enal
167	Hexa-2,4-dienal
168	*trans, trans*-Hepta-2,4-dienal
169	*trans, cis*-Nona-2,4-dienal
170	*trans, trans*-Deca-2,4-dienal
171	*trans, trans*-Undec-2,4-dienal
172	*trans, trans*-Dodeca-2,4-dienal
173	Benzaldehyde
174	4-Isopropylbenzaldehyde (Cuminic)
175	Furfural
176	5-Methylfurfural
177	2-Methoxybenzaldehyde
178	4-Methoxybenzaldehyde
179	Salicylaldehyde
	KETONES
180	Propan-2-one
181	Butan-2-one
182	Pentan-2-one
183	Hexan-2-one
184	3-Methylpentan-2-one
185	Heptan-2-one
186	Octan-2-one
187	Nonan-2-one
188	Decan-2-one
189	Undecan-2-one
190	Pentan-3-one
191	Octan-3-one
192	Undecan-5-one

Spectrum Number	Name
193	Butan-2,3-dione
194	3-Hydroxybutan-2-one
195	Pentan-2,4-dione
196	Cyclopentanone
197	Cyclohexanone
198	*dl*-Piperitone
199	*d*-Carvone
200	Pulegone
201	β-Ionone
202	Acetophenone
203	4-Methylacetophenone
204	Propiophenone
205	2-Acetylfuran
	ESTERS
206	Methyl formate
207	Ethyl formate
208	Propyl formate
209	Butyl formate
210	'Amyl' formate
211	Methyl acetate
212	Ethyl acetate
213	Propyl acetate
214	Butyl acetate
215	'Amyl' acetate
216	Hexyl acetate
217	Octyl acetate
218	Ethyl propanoate
219	Propyl propanoate
220	Butyl propanoate
221	'Amyl' propanoate
222	Allyl propanoate
223	Methyl butanoate
224	Ethyl butanoate
225	Propyl butanoate
226	Butyl butanoate
227	'Amyl' butanoate
228	Allyl butanoate
229	Methyl pentanoate
230	Propyl pentanoate
231	Butyl pentanoate
232	'Amyl' pentanoate
233	Ethyl 3-methylbutanoate
234	Methyl 2-methylbutanoate
235	Ethyl 2-methylbutanoate
236	Methyl hexanoate
237	Ethyl hexanoate
238	'Amyl' hexanoate
239	Allyl hexanoate
240	Dimethyl malonate

Spectrum Number	Name
241	Diethyl malonate
242	Dimethyl succinate
243	Ethyl 3-oxobutanoate
244	Methyl 2-hydroxypropanoate
245	Ethyl 2-hydroxypropanoate
246	Butyl 2-hydroxypropanoate
247	Methyl 2-hydroxy-2-methylpropanoate
248	Ethyl 2-hydroxy-2-methylpropanoate
249	Ethyl 3-hydroxybutanoate
250	Ethyl 3-aminobutanoate
251	Ethyl but-2-enoate
252	Methyl oct-2-ynoate
253	Methyl benzoate
254	Ethyl benzoate
255	Methyl furoate
256	Ethyl furoate
257	Vinyl acetate
258	Isopropenyl acetate
259	Linalyl acetate
260	Phenyl acetate
261	*p*-Cresyl 2-methylpropanoate
262	4-Methoxyphenyl formate
263	4-Methoxyphenyl acetate
264	Methyl salicylate
265	Methyl anthranilate
266	Ethyl anthranilate
267	Methyl *N*-methylanthranilate
	LACTONES
268	γ-Butyrolactone
269	γ-Pentalactone
270	γ-Hexalactone
271	γ-Heptalactone
272	γ-Octalactone
273	γ Nonalactone
274	δ-Heptalactone
275	δ-Octalactone
276	δ-Isooctalactone
277	δ-Nonalactone
	ETHERS
278	Diethyl ether
279	Dibutyl ether
280	Ethyl vinyl ether
281	1,1-Diethoxyethane
282	Dioxane
283	2,3-Dihydropyran
284	1,8-Cineole
285	Methoxybenzene (Anisole)
286	1,2-Dimethoxybenzene (Veratrole)
287	2-Methylfuran

Spectrum Number	Name
	HYDROCARBONS
288	n-Heptane
289	n-Nonane
290	n-Decane
291	n-Undecane
292	n-Dodecane
293	Dec-1-ene
294	Dodec-1-ene
295	*cis*-Ocimene
296	*d*-Limonene
297	*β*-Pinene
298	*dl*-Camphene
299	Caryophyllene
300	Pent-1-yne
301	Phenylacetylene
302	Benzene
303	Ethylbenzene
304	n-Octylbenzene
305	1-Isopropyl-4-methylbenzene (*p*-Cymene)
306	1,3,5-Trimethylbenzene (Mesitylene)

c. Molecular formula index

C	H	BR	CL	F	I	B	N	O	P	S	SI	M	NAME	SPECTRUM NO.	LOCA- TION
1	2							2					FORMIC ACID (130 DEG C)	74	BA
1	2							2					FORMIC ACID (40 DEG C)	73	BA
1	4							1					METHANOL	1	BA
2	4							1					ACETALDEHYDE	152	BA
2	4							2					ACETIC ACID (130 DEG C)	76	BA
2	4							2					ACETIC ACID (35 DEG C)	75	BA
2	4							2					METHYL FORMATE	206	BA
2	5						1	1					ACETAMIDE	142	BA
2	6							1					ETHANOL	2	BA
2	6							2					ETHANE-1,2-DIOL	47	BA
2	7						1						1-AMINO ETHANE	108	BA
3	4							1					PROP-2-YN-1-OL	17	BA
3	4							3					2-OXO PROPANOIC ACID (PYRUVIC)	106	BA
3	6							1					PROP-2-EN-1-OL	13	BA
3	6							1					PROPANAL	153	BA
3	6							1					PROPAN-2-ONE	180	BA
3	6							2					ETHYL FORMATE	207	BA
3	6							2					METHYL ACETATE	211	BA
3	6							2					PROPANOIC ACID	77	BA
3	6							3					2-HYDROXY PROPANOIC ACID (LACTIC)	107	BA
3	7						1						1-AMINO PROP-2-ENE	118	BA
3	7						1	1					N,N-DIMETHYL FORMAMIDE	146	BA
3	7						1	1					PROPIONAMIDE	143	BA
3	8							1					PROPAN-2-OL	20	BA
3	8							1					PROPAN-1-OL	3	BA
3	8							2					2-METHOXY ETHANOL	45	BA
3	8							2					PROPANE-1,3-DIOL	49	BA
3	8							2					PROPANE-1,2-DIOL	48	BA
3	9						1						2-AMINO PROPANE	120	BA
3	9						1						1-AMINO PROPANE	109	BA
4	5						1						PYRROLE	141	BA
4	6							1					BUT-2-ENAL	161	BA
4	6							2					BUT-2-ENOIC ACID	91	BA
4	6							2					BUT-3-ENOIC ACID	101	BA
4	6							2					BUTAN-2,3-DIONE	193	BA
4	6							2					GAMMA-BUTYROLACTONE	268	BA
4	6							2					2-METHYL PROP-2-ENOIC ACID	98	BA
4	6							2					VINYL ACETATE	257	BA
4	8							1					BUT-1-EN-3-OL	31	BA
4	8							1					BUTAN-2-ONE	181	BA
4	8							1					N-BUTANAL	154	BA
4	8							1					ETHYL VINYL ETHER	280	BA
4	8							2					N-BUTANOIC ACID	78	BA
4	8							2					DIOXANE	282	BA
4	8							2					ETHYL ACETATE	212	BA
4	8							2					3-HYDROXY BUTAN-2-ONE	194	BA
4	8							2					2-METHYL PROPANOIC ACID	84	BA
4	8							2					PROPYL FORMATE	208	BA
4	8							3					METHYL 2-HYDROXY PROPANOATE	244	BA
4	9						1						TETRAHYDROPYRROLE (PYRROLIDINE)	131	BA
4	9						1	1					BUTYRAMIDE	144	BA
4	10							1					TERT-BUTANOL	37	BA
4	10							1					BUTAN-2-OL	21	BA
4	10							1					BUTAN-1-OL	4	BA
4	10							1					DIETHYL ETHER	278	BA
4	11						1						1-AMINO BUTANE	110	BA
4	11						1						2-AMINO BUTANE	121	BA
4	11						1						DIETHYLAMINE	125	BA
4	11						1	1					2-(N,N-DIMETHYL)AMINO ETHANOL	50	BA
5	4							2					FURFURAL	175	BA
5	5						1						PYRIDINE	147	BA
5	6							1					2-METHYL FURAN	287	BA
5	6							2					FURFURYL ALCOHOL	18	BA
5	8												PENT-1-YNE	300	BA
5	8							1					CYCLOPENTANONE	196	BA
5	8							1					2,3-DIHYDROPYRAN	283	BA
5	8							2					ISOPROPENYL ACETATE	258	BA
5	8							2					2-METHYL BUT-2-ENOIC ACID	100	BA
5	8							2					3-METHYL BUT-2-ENOIC ACID	99	BA
5	8							2					GAMMA-PENTALACTONE	269	BA
5	8							2					PENTA-2,4-DIONE	195	BA
5	8							4					DIMETHYL MALONATE	240	BA
5	10							1					CYCLOPENTANOL	29	BA
5	10							1					2-METHYL BUT-3-EN-2-OL	40	BA
5	10							1					PENTAN-2-ONE	182	BA
5	10							1					PENTAN-3-ONE	190	BA
5	10							1					PENT-1-EN-4-OL	35	BA
5	10							1					PENT-4-EN-1-OL	9	BA
5	10							1					N-PENTANAL	155	BA
5	10							2					BUTYL FORMATE	209	BA
5	10							2					ETHYL PROPANOATE	218	BA
5	10							2					METHYL BUTANOATE	223	BA
5	10							2					2-METHYL BUTANOIC ACID	86	BA
5	10							2					3-METHYL BUTANOIC ACID	85	BA
5	10							2					N-PENTANOIC ACID	79	BA
5	10							2					PROPYL ACETATE	213	BA
5	10							2					TETRAHYDROFURFURYL ALCOHOL	19	BA
5	10							3					ETHYL 2-HYDROXY PROPANOATE	245	BA
5	10							3					METHYL 2-HYDROXY 2-METHYL PROPANOATE	247	BA
5	11						1						PIPERIDINE	117	BA
5	12							1					3-METHYL BUTAN-1-OL	7	BA
5	12							1					PENTAN-1-OL	5	BA
5	12							1					PENTAN-2-OL	22	BA
5	12							1					PENTAN-3-OL	25	BA
5	13						1						3-AMINO PENTANE	124	BA
5	13						1						2-AMINO PENTANE	122	BA
5	13						1						1-AMINO PENTANE	111	BA
5	13						1						N-METHYL BUTYLAMINE	130	BA
6	5	1						1					4-BROMOPHENOL	67	BA
6	5	1						1					2-BROMOPHENOL	65	BA

C H BR CL F I B N O P S SI M

C	H	BR	CL	F	I	B	N	O	P	S	SI	M	NAME	SPECTRUM NO.	LOCATION
6	5		1					1					4-CHLOROPHENOL	66	BA
6	5		1					1					2-CHLOROPHENOL	64	BA
6	6												BENZENE	302	BA
6	6		1				1						3-CHLORO ANILINE	140	BA
6	6		1				1						2-CHLORO ANILINE	139	BA
6	6							1					PHENOL	52	BA
6	6							2					2-ACETYL FURAN	205	BA
6	6							2					1,2-DIHYDROXY BENZENE (CATECHOL)	70	BA
6	6							2					5-METHYL FURFURAL	176	BA
6	6							3					METHYL FUROATE	255	BA
6	7						1						ANILINE	132	BA
6	7						1						3-METHYL PYRIDINE	149	BA
6	7						1						2-METHYL PYRIDINE	148	BA
6	7						1						4-METHYL PYRIDINE	150	BA
6	8							1					HEXA-2,4-DIENAL	167	BA
6	8							2					HEXA-2,4-DIENOIC ACID	97	BA
6	10							1					CYCLOHEXANONE	197	BA
6	10							1					HEX-2-ENAL	162	BA
6	10							2					ALLYL PROPANOATE	222	BA
6	10							2					ETHYL BUT-2-ENOATE (CROTONATE)	251	BA
6	10							2					HEX-2-ENOIC ACID	92	BA
6	10							2					HEX-3-ENOIC ACID	102	BA
6	10							2					GAMMA-HEXALACTONE	270	BA
6	10							3					ETHYL 3-OXOBUTANOATE	243	BA
6	10							4					DIMETHYL SUCCINATE	242	BA
6	12							1					CYCLOHEXANOL	30	BA
6	12							1					CIS-HEX-3-EN-1-OL	10	BA
6	12							1					HEX-5-EN-1-OL	8	BA
6	12							1					HEX-1-EN-3-OL	32	BA
6	12							1					TRANS-HEX-2-EN-1-OL	15	BA
6	12							1					CIS-HEX-2-EN-1-OL	14	BA
6	12							1					TRANS-HEX-3-EN-1-OL	11	BA
6	12							1					HEXAN-2-ONE	183	BA
6	12							1					N-HEXANAL	156	BA
6	12							1					3-METHYL PENTAN-2-ONE	184	BA
6	12							2					AMYL FORMATE	210	BA
6	12							2					BUTYL ACETATE	214	BA
6	12							2					ETHYL BUTANOATE	224	BA
6	12							2					2-ETHYL BUTANOIC ACID	89	BA
6	12							2					N-HEXANOIC ACID	80	BA
6	12							2					3-METHYL PENTANOIC ACID	88	BA
6	12							2					4-METHYL PENTANOIC ACID	87	BA
6	12							2					METHYL PENTANOATE	229	BA
6	12							2					METHYL 2-METHYL BUTANOATE	234	BA
6	12							2					PROPYL PROPANOATE	219	BA
6	12							3					ETHYL 3-HYDROXY BUTANOATE	249	BA
6	12							3					ETHYL 2-HYDROXY 2-METHYL PROPANOATE	248	BA
6	13						1	2					ETHYL 3-AMINO BUTANOATE	250	BA
6	14							1					HEXAN-2-OL	23	BA
6	14							1					HEXAN-1-OL	6	BA
6	14							1					2-METHYL PENTAN-2-OL	38	BA
6	14							2					1,1-DIETHOXY ETHANE	281	BA
6	14							3					2-(2-ETHOXY)ETHOXY ETHANOL	46	BA
6	15						1						1-AMINO HEXANE	112	BA
6	15						1						DIPROPYLAMINE	126	BA
6	15						1	1					2-(N,N-DIETHYL)AMINO ETHANOL	51	BA
7	6							1					BENZALDEHYDE	173	BA
7	6							2					BENZOIC ACID	105	BA
7	6							2					SALICYLALDEHYDE	179	BA
7	8							1					BENZYL ALCOHOL	16	BA
7	8							1					M-CRESOL	54	BA
7	8							1					O-CRESOL	53	BA
7	8							1					P-CRESOL	55	BA
7	8							1					METHOXY BENZENE (ANISOLE)	285	BA
7	8							2					1,2-DIHYDROXY 3-METHYL BENZENE (3-METHYL CATECHOL)	71	BA
7	8							2					1,2-DIHYDROXY 4-METHYL BENZENE (4-METHYL CATECHOL)	72	BA
7	8							2					2-METHOXY PHENOL(GUAICOL)	62	BA
7	8							3					ETHYL FUROATE	256	BA
7	9						1						BENZYLAMINE	119	BA
7	9						1						3-METHYL ANILINE	135	BA
7	9						1						2-METHYL ANILINE	134	BA
7	9						1						N-METHYL ANILINE	133	BA
7	9						1	1					2-METHOXY ANILINE	137	BA
7	10							1					TR,TR-HEPTA-2,4-DIENAL	168	BA
7	12							2					ALLYL BUTANOATE	228	BA
7	12							2					DELTA-HEPTALACTONE	274	BA
7	12							2					GAMMA-HEPTALACTONE	271	BA
7	12							2					HEPT-2-ENOIC ACID	93	BA
7	12							4					DIETHYL MALONATE	241	BA
7	14							1					HEPTAN-2-ONE	185	BA
7	14							1					HEPT-1-EN-4-OL	36	BA
7	14							1					N-HEPTANAL	157	BA
7	14							2					AMYL ACETATE	215	BA
7	14							2					BUTYL PROPANOATE	220	BA
7	14							2					ETHYL 3-METHYL BUTANOATE (ISOVALERATE)	233	BA
7	14							2					ETHYL 2-METHYL BUTANOATE	235	BA
7	14							2					N-HEPTANOIC ACID	81	BA
7	14							2					METHYL HEXANOATE	236	BA
7	14							2					PROPYL BUTANOATE	225	BA
7	14							3					BUTYL 2-HYDROXY PROPANOATE	246	BA
7	16												N-HEPTANE	288	BA
7	17						1						1-AMINO HEPTANE	113	BA
8	6												PHENYL ACETYLENE	301	BA
8	8							1					ACETOPHENONE	202	BA
8	8							2					METHYL BENZOATE	253	BA
8	8							2					4-METHOXY BENZALDEHYDE	178	BA
8	8							2					2-METHOXY BENZALDEHYDE	177	BA
8	8							2					PHENYL ACETATE	260	BA

C H BR CL F I B N O P S SI M

C	H	BR	CL	F	I	B	N	O	P	S	SI	M	NAME	SPECTRUM NO.	LOCA-TION
8	8							3					METHYL SALICYLATE	264	BA
8	8							3					P-METHOXY PHENYL FORMATE	262	BA
8	9						1	2					METHYL ANTHRANILATE	265	BA
8	10												ETHYL BENZENE	303	BA
8	10							1					1-PHENYL ETHANOL	34	BA
8	10							1					2-PHENYL ETHANOL	12	BA
8	10							1					3,5-XYLENOL	61	BA
8	10							1					3,4-XYLENOL	60	BA
8	10							1					2,6-XYLENOL	59	BA
8	10							1					2,5-XYLENOL	58	BA
8	10							1					2,4-XYLENOL	57	BA
8	10							1					2,3-XYLENOL	56	BA
8	10							2					1,2-DIMETHOXY BENZENE (VERATROLE)	286	BA
8	10							3					2,6-DIMETHOXY PHENOL	63	BA
8	11						1						2,4-DIMETHYL ANILINE	136	BA
8	11						1	1					4-ETHOXY ANILINE	138	BA
8	14							1					TRANS-OCT-2-ENAL	163	BA
8	14							2					DELTA-ISOOCTALACTONE	276	BA
8	14							2					OCT-2-ENOIC ACID	94	BA
8	14							2					OCT-3-ENOIC ACID	103	BA
8	14							2					DELTA-OCTALACTONE	275	BA
8	14							2					GAMMA-OCTALACTONE	272	BA
8	16							1					OCT-1-EN-3-OL	33	BA
8	16							1					N-OCTANAL	158	BA
8	16							1					OCTAN-2-ONE	186	BA
8	16							1					OCTAN-3-ONE	191	BA
8	16							2					AMYL PROPANOATE	221	BA
8	16							2					BUTYL BUTANOATE	226	BA
8	16							2					ETHYL HEXANOATE	237	BA
8	16							2					2-ETHYL HEXANOIC ACID	90	BA
8	16							2					HEXYL ACETATE	216	BA
8	16							2					N-OCTANOIC ACID	82	BA
8	16							2					PROPYL PENTANOATE	230	BA
8	18							1					DIBUTYL ETHER	279	BA
8	18							1					OCTAN-3-OL	26	BA
8	18							1					OCTAN-2-OL	24	BA
8	18							1					OCTAN-4-OL	27	BA
8	19						1						1-AMINO OCTANE	114	BA
8	19						1						2-AMINO OCTANE	123	BA
8	19						1						DIBUTYLAMINE	127	BA
8	19						1						1,1,3,3-TETRAMETHYL 1-AMINO BUTANE	115	BA
9	7						1						QUINOLINE	151	BA
9	10							1					4-METHYLACETOPHENONE	203	BA
9	10							1					PROPIOPHENONE	204	BA
9	10							2					ETHYL BENZOATE	254	BA
9	10							3					P-METHOXY PHENYL ACETATE	263	BA
9	11						1	2					ETHYL ANTHRANILATE	266	BA
9	11						1	2					METHYL N-METHYL ANTHRANILATE	267	BA
9	12												1,3,5-TRIMETHYL BENZENE (MESITYLENE)	306	BA
9	14							1					TR,CIS-NONA-2,4-DIENAL	169	BA
9	14							2					METHYL OCT-2-YN OATE	252	BA
9	16							1					TRANS-NON-2-ENAL	164	BA
9	16							2					ALLYL HEXANOATE	239	BA
9	16							2					NON-2-ENOIC ACID	95	BA
9	16							2					GAMMA-NONALACTONE	273	BA
9	16							2					DELTA-NONALACTONE	277	BA
9	18							1					NONAN-2-ONE	187	BA
9	18							1					N-NONANAL	159	BA
9	18							2					AMYL BUTANOATE	227	BA
9	18							2					BUTYL PENTANOATE	231	BA
9	18							2					N-NONANOIC ACID	83	BA
9	20												N-NONANE	289	BA
9	21						1						1-AMINO NONANE	116	BA
10	12							1					4-ISOPROPYL BENZALDEHYDE(CUMINIC)	174	BA
10	12							2					4-ALLYL 2-METHOXY PHENOL (EUGENOL)	69	BA
10	14												1-ISOPROPYL 4-METHYL BENZENE(P-CYMENE)	305	BA
10	14							1					D-CARVONE	199	BA
10	14							1					1,1-DIMETHYL 2-PHENYL ETHANOL	39	BA
10	14							1					2-ISOPROPYL 5-METHYL PHENOL (THYMOL)	68	BA
10	15						1						N,N-DIETHYL ANILINE	145	BA
10	16												DL-CAMPHENE	298	BA
10	16												D-LIMONENE	296	BA
10	16												CIS-OCIMENE	295	BA
10	16												BETA-PINENE	297	BA
10	16							1					TR,TR-DECA-2,4-DIENAL	170	BA
10	16							1					DL-PIPERITONE	198	BA
10	16							1					PULEGONE	200	BA
10	18							1					1,8-CINEOLE	284	BA
10	18							1					TRANS-DEC-2-ENAL	165	BA
10	18							1					GERANIOL	42	BA
10	18							1					LINALOOL	43	BA
10	18							1					A-TERPINEOL	44	BA
10	18							2					DEC-2-ENOIC ACID	96	BA
10	20												DEC-1-ENE	293	BA
10	20							1					CITRONELLOL	41	BA
10	20							1					N-DECANAL	160	BA
10	20							1					DECAN-2-ONE	188	BA
10	20							2					AMYL PENTANOATE	232	BA
10	20							2					OCTYL ACETATE	217	BA
10	22												N-DECANE	290	BA
10	22							1					DECAN-5-OL	28	BA
10	23						1						DIPENTYLAMINE	128	BA
11	14							2					P-CRESYL 2-METHYL PROPANOATE	261	BA
11	18							1					TR,TR-UNDECA-2,4-DIENAL	171	BA
11	20							1					TRANS-UNDEC-2-ENAL	166	BA
11	20							2					UNDEC-10-ENOIC ACID	104	BA
11	22							1					UNDECAN-5-ONE	192	BA
11	22							1					UNDECAN-2-ONE	189	BA
11	22							2					AMYL HEXANOATE	238	BA
11	24												N-UNDECANE	291	BA

C H BR CL F I B N O P S SI M

C	H	BR	CL	F	I	B	N	O	P	S	SI	M	NAME	SPECTRUM NO.	LOCATION
12	20							1					TR,TR-DODECA-2,4-DIENAL	172	BA
12	20							2					LINALYL ACETATE	259	BA
12	24												DODEC-1-ENE	294	BA
12	26												N-DODECANE	292	BA
12	27						1						DIHEXYLAMINE	129	BA
13	20							1					BETA-IONONE	201	BA
14	22												N-OCTYL BENZENE	304	BA
15	24												CARYOPHYLLENE	299	BA

d. Spectra

"The following 102 pages contain 306 Vapour Spectra in numerical serial order."

COMPOUND	MOL. FORMULA	B. PT.	INDEX No.
Methanol	CH_4O	65°C	**1**

CELL TEMP.	APPROX AMOUNT	CELL PATH	REMARKS
~130°C	2μl	Length 9·2cm Volume 25ml	

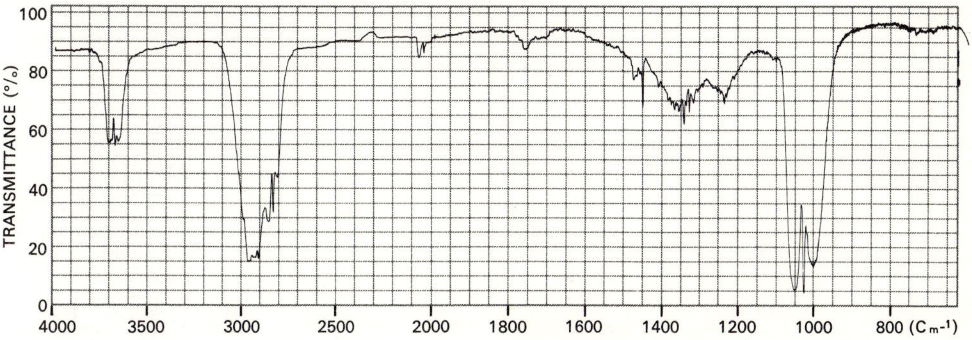

COMPOUND	MOL. FORMULA	B. PT.	INDEX No.
Ethanol	C_2H_6O	78·5°C	**2**

CELL TEMP.	APPROX AMOUNT	CELL PATH	REMARKS
~150°C	4μl	Length 9·2cm Volume 25ml	

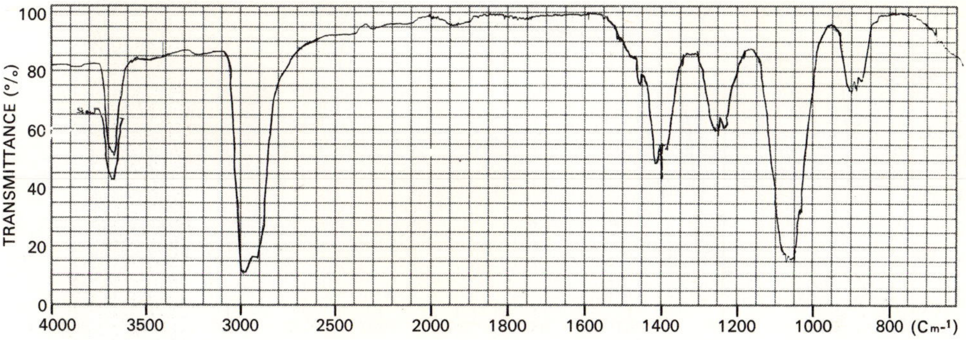

COMPOUND	MOL. FORMULA	B. PT.	INDEX No.
Propan-1-ol	C_3H_8O	97°C	**3**

CELL TEMP.	APPROX AMOUNT	CELL PATH	REMARKS
~ 155°C	4μl	Length 9·2cm Volume 25ml	

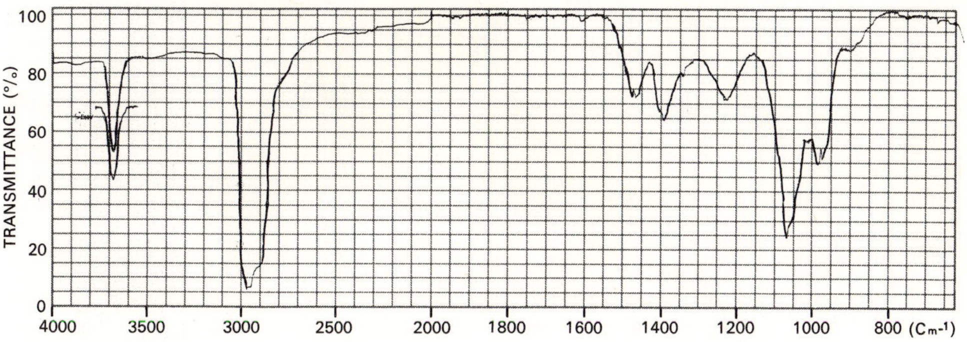

COMPOUND	MOL. FORMULA	B. PT.	INDEX No.
Butan-1-ol	$C_4H_{10}O$	118°C	**4**

CELL TEMP.	APPROX AMOUNT	CELL PATH	REMARKS
~155°C	4μl	Length 9·2cm Volume 25ml	

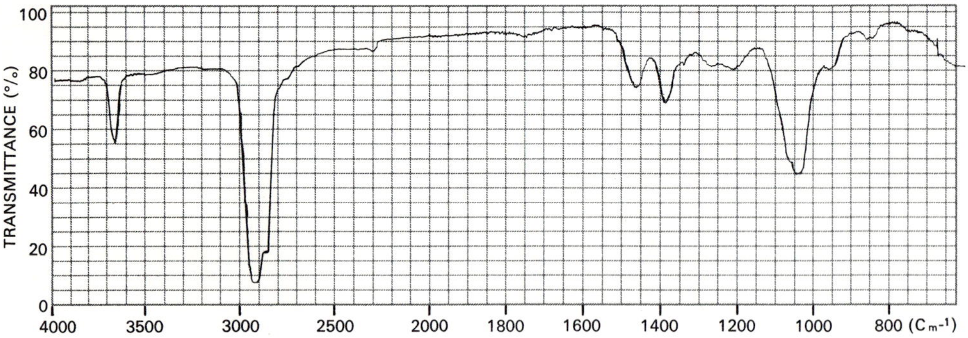

COMPOUND	MOL. FORMULA	B. PT.	INDEX No.
Pentan-1-ol	$C_5H_{12}O$	137°C	**5**

CELL TEMP.	APPROX AMOUNT	CELL PATH	REMARKS
155°C	6μl	Length 9·2cm Volume 25ml	

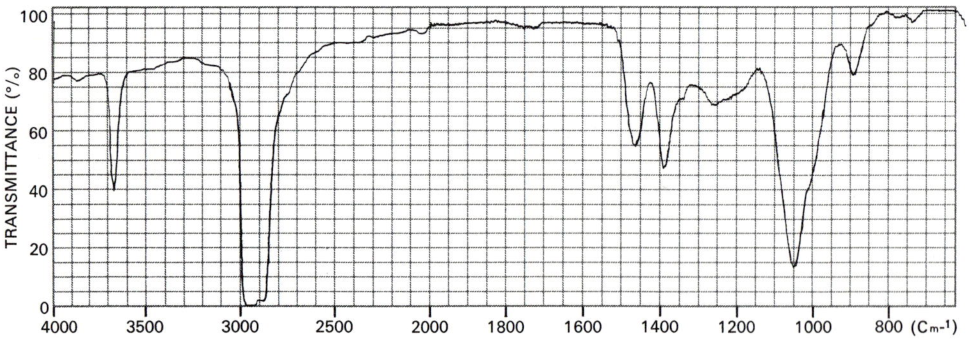

COMPOUND	MOL. FORMULA	B. PT.	INDEX No.
Hexan-1-ol	$C_6H_{14}O$	158°C	**6**

CELL TEMP.	APPROX AMOUNT	CELL PATH	REMARKS
~155°C	4μl	Length 9·2cm Volume 25ml	

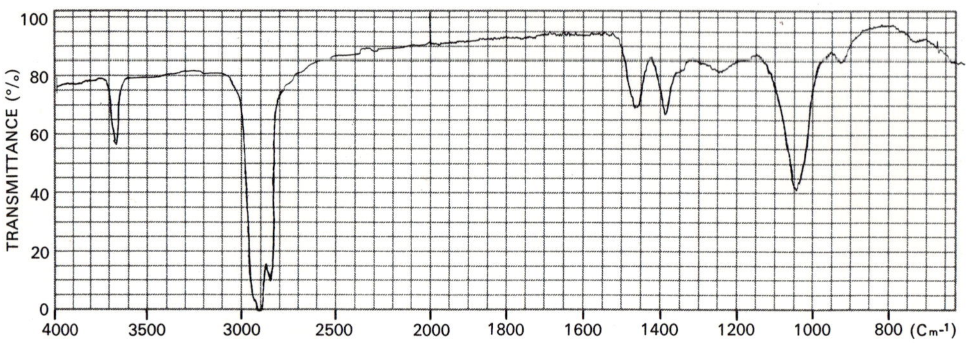

COMPOUND	MOL. FORMULA	B. PT.	INDEX No.
3-Methylbutan-1-ol	$C_5H_{12}O$	132°C	**7**

CELL TEMP.	APPROX AMOUNT	CELL PATH	REMARKS
~135°C	4μl	Length 9·2cm Volume 25ml	

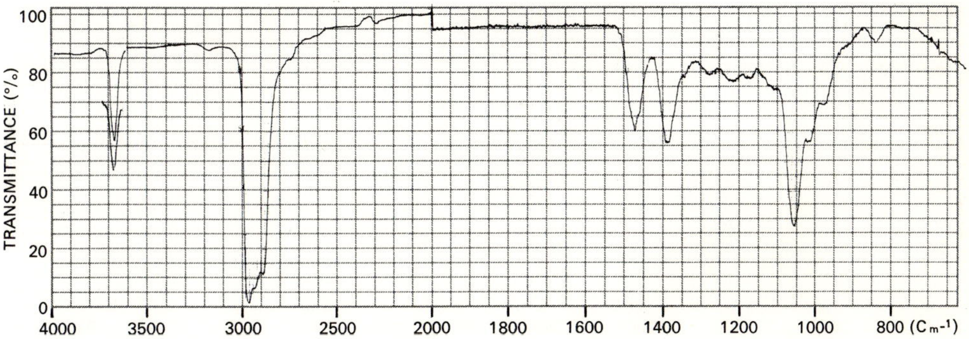

COMPOUND	MOL. FORMULA	B. PT.	INDEX No.
Hex-5-en-1-ol	$C_6H_{12}O$	57-58°C/15mm	**8**

CELL TEMP.	APPROX AMOUNT	CELL PATH	REMARKS
~125°C	2μl	Length 9·2cm Volume 25ml	

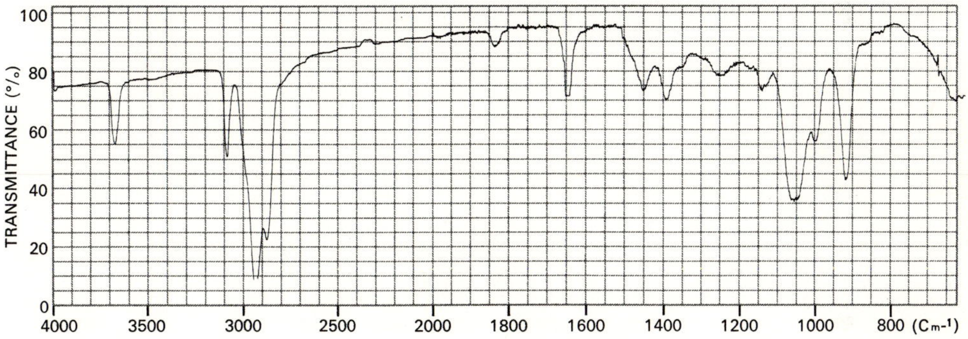

COMPOUND	MOL. FORMULA	B. PT.	INDEX No.
Pent-4-en-1-ol	$C_5H_{10}O$	142°C	**9**

CELL TEMP.	APPROX AMOUNT	CELL PATH	REMARKS
~130°C	2μl	Length 9·2cm Volume 25ml	

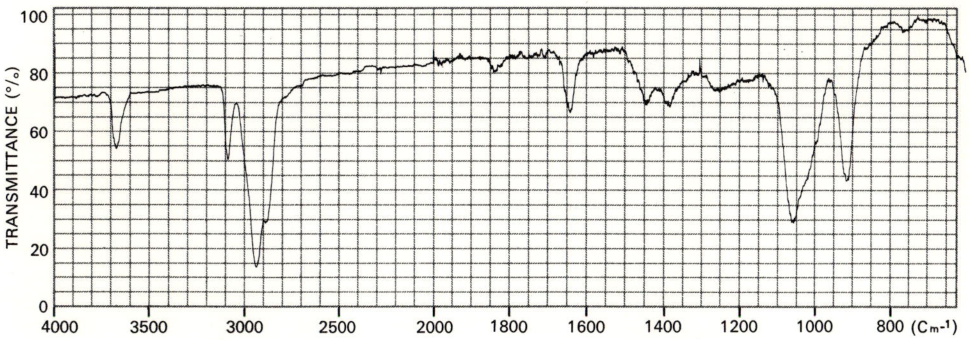

COMPOUND		MOL. FORMULA	B. PT.	INDEX No.
cis-Hex-3-en-1-ol		$C_6H_{12}O$	156°C	**10**

CELL TEMP.	APPROX AMOUNT	CELL PATH	REMARKS
~135°C	2·5μl	Length 9·2cm Volume 25ml	

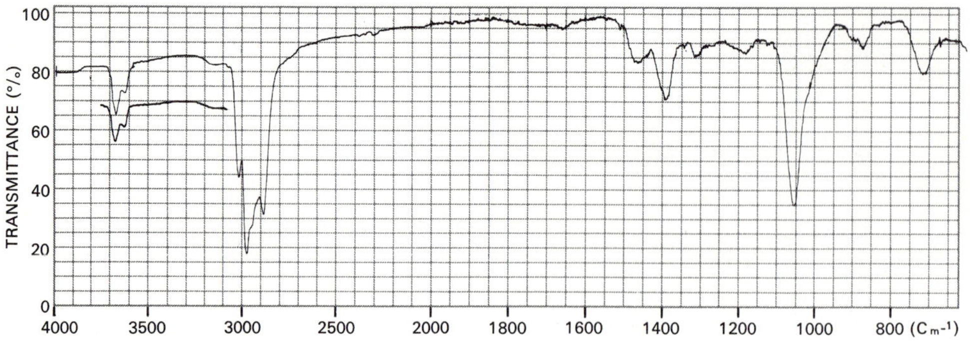

COMPOUND		MOL. FORMULA	B. PT.	INDEX No.
trans-Hex-3-en-1-ol		$C_6H_{12}O$	-	**11**

CELL TEMP.	APPROX AMOUNT	CELL PATH	REMARKS
~140°C	4μl	Length 9·2cm Volume 25ml	

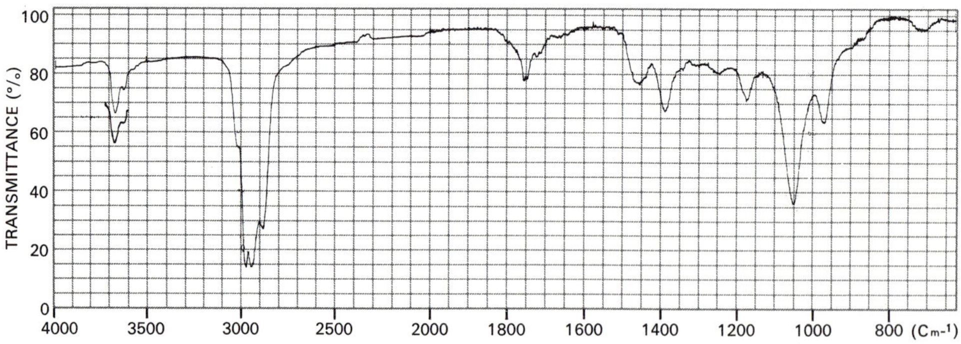

COMPOUND		MOL. FORMULA	B. PT.	INDEX No.
2-Phenylethanol		$C_8H_{10}O$	219°C	**12**

CELL TEMP.	APPROX AMOUNT	CELL PATH	REMARKS
~145°C	2μl	Length 9·2cm Volume 25ml	

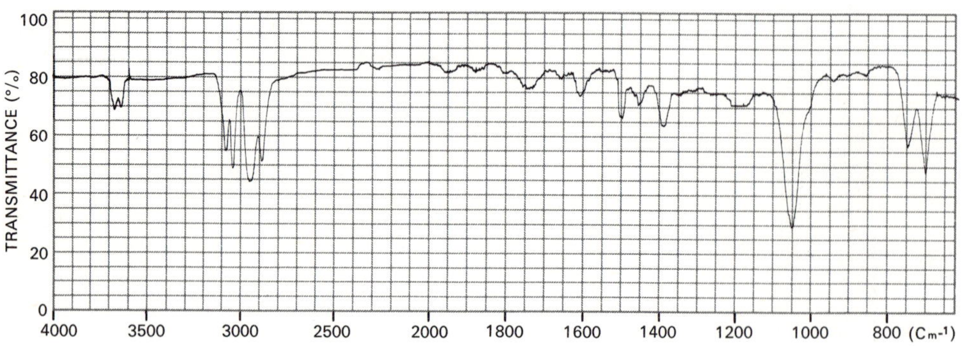

COMPOUND	MOL. FORMULA	B. PT.	INDEX No.
Prop-2-en-1-ol	C_3H_6O	97° C	**13**

CELL TEMP.	APPROX AMOUNT	CELL PATH	REMARKS
~130° C	2μl	Length 9·2cm Volume 25ml	

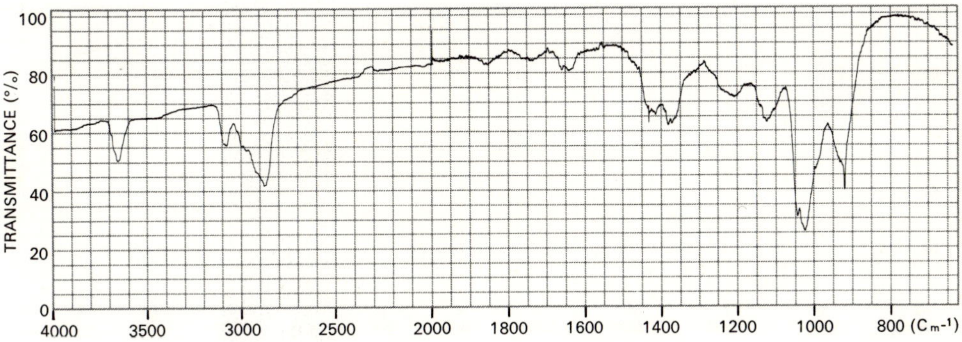

COMPOUND	MOL. FORMULA	B. PT.	INDEX No.
cis-Hex-2-en-1-ol	$C_6H_{12}O$	-	**14**

CELL TEMP.	APPROX AMOUNT	CELL PATH	REMARKS
~125°C	2μl	Length 9·2cm Volume 25ml	

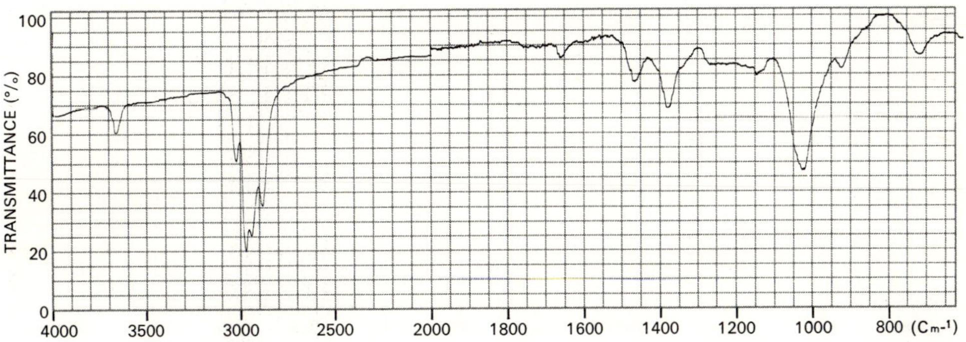

COMPOUND	MOL. FORMULA	B. PT.	INDEX No.
trans-Hex-2-en-1-ol	$C_6H_{12}O$	-	**15**

CELL TEMP.	APPROX AMOUNT	CELL PATH	REMARKS
~130°C	3μl	Length 9·2cm Volume 25ml	

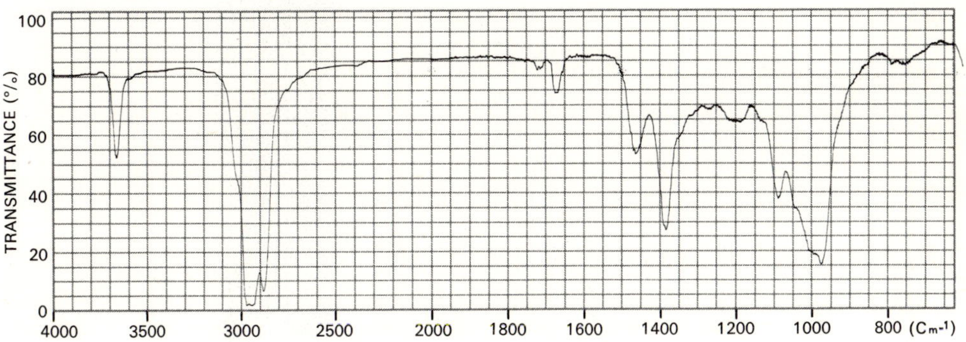

COMPOUND	MOL. FORMULA	B. PT.	INDEX No.
Benzyl alcohol	C_7H_8O	205° C	**16**

CELL TEMP.	APPROX AMOUNT	CELL PATH	REMARKS
~130° C	3μl	Length 9·2cm Volume 25ml	

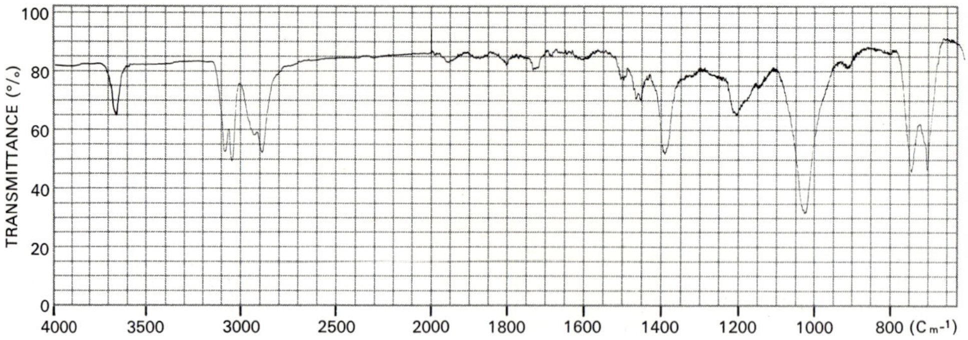

COMPOUND	MOL. FORMULA	B. PT.	INDEX No.
Prop-2-yn-1-ol	C_3H_4O	115°C	**17**

CELL TEMP.	APPROX AMOUNT	CELL PATH	REMARKS
~145° C	2μl	Length 9·2cm Volume 25ml	

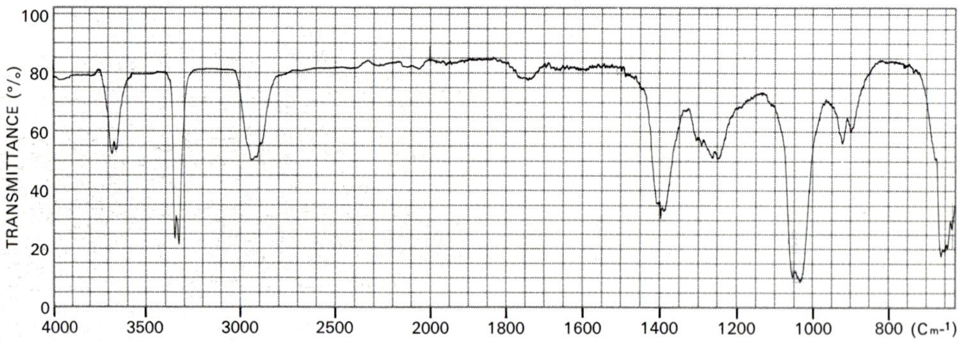

COMPOUND	MOL. FORMULA	B. PT.	INDEX No.
Furfuryl alcohol	$C_5H_6O_2$	171°C	**18**

CELL TEMP.	APPROX AMOUNT	CELL PATH	REMARKS
~140°C	4μl	Length 9·2cm Volume 25ml	

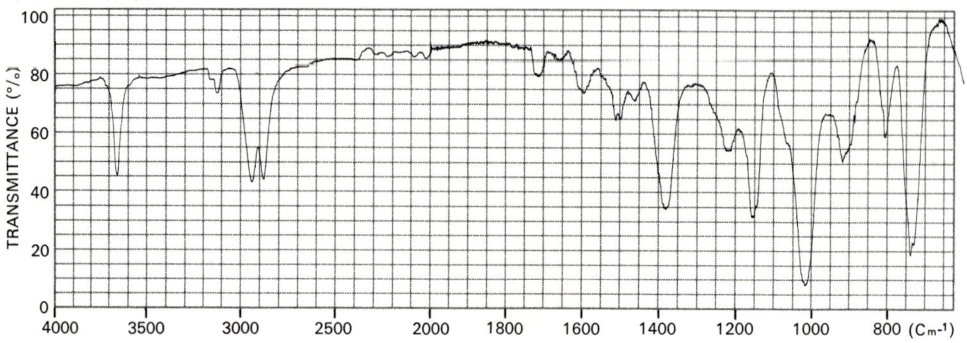

COMPOUND	MOL. FORMULA	B. PT.	INDEX No.
Tetrahydrofurfuryl alcohol	$C_5H_{10}O_2$	178° C	**19**

CELL TEMP.	APPROX AMOUNT	CELL PATH	REMARKS
~130°C	2·5μl	Length 9·2cm Volume 25ml	

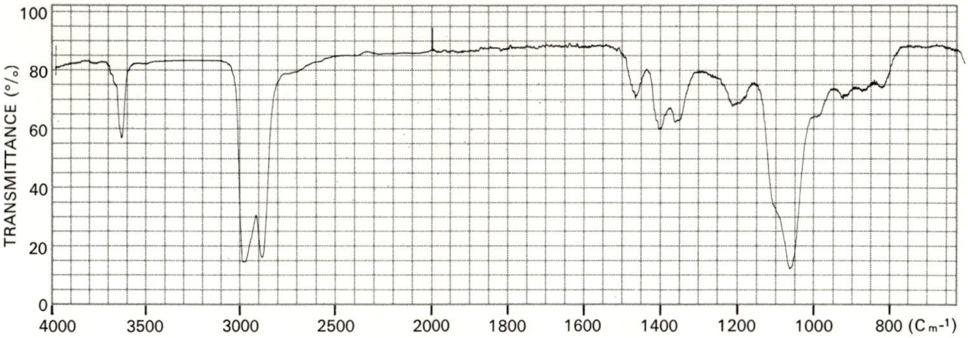

COMPOUND	MOL. FORMULA	B. PT.	INDEX No.
Propan-2-ol	C_3H_8O	82° C	**20**

CELL TEMP.	APPROX AMOUNT	CELL PATH	REMARKS
~125° C	2μl	Length 9·2cm Volume 25ml	

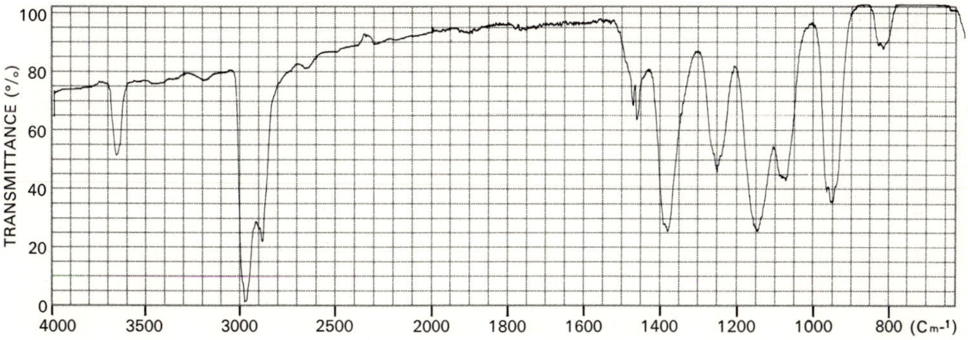

COMPOUND	MOL. FORMULA	B. PT.	INDEX No.
Butan-2-ol	$C_4H_{10}O$	100° C	**21**

CELL TEMP.	APPROX AMOUNT	CELL PATH	REMARKS
~130°C	2μl	Length 9·2cm Volume 25ml	

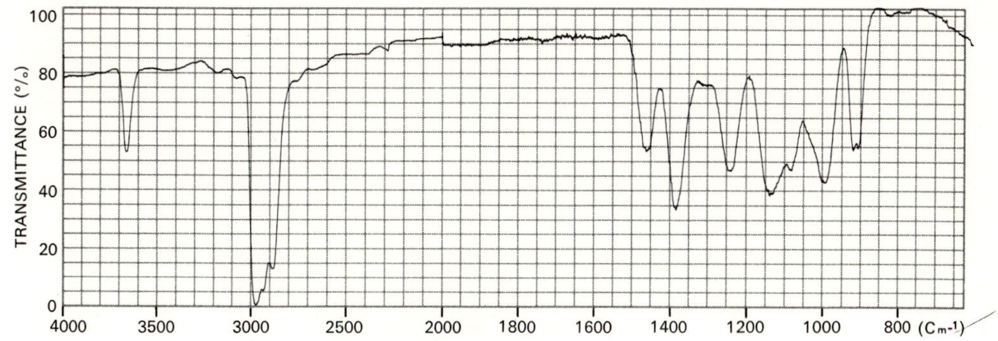

COMPOUND	MOL. FORMULA	B. PT.	INDEX No.
Pentan-2-ol	$C_5H_{12}O$	119°C	**22**

CELL TEMP.	APPROX AMOUNT	CELL PATH	REMARKS
~125°C	2µl	Length 9·2cm Volume 25ml	

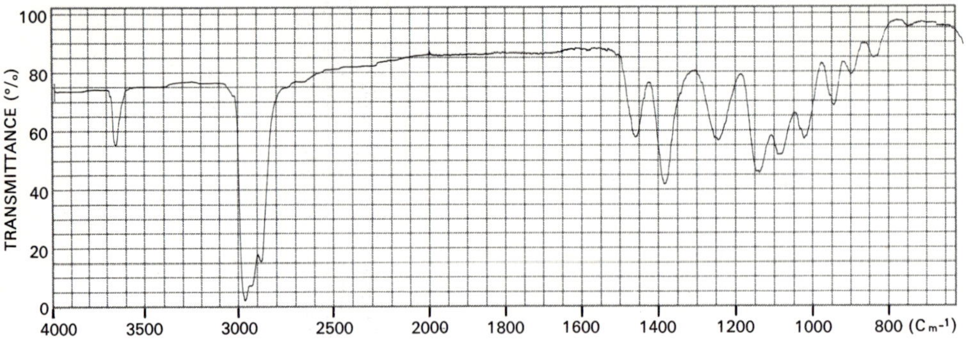

COMPOUND	MOL. FORMULA	B. PT.	INDEX No.
Hexan-2-ol	$C_6H_{14}O$	140°C	**23**

CELL TEMP.	APPROX AMOUNT	CELL PATH	REMARKS
~135°C	2µl	Length 9·2cm Volume 25ml	

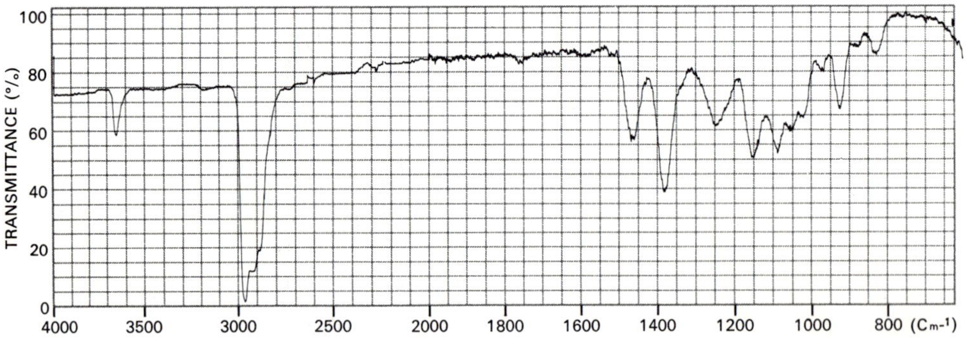

COMPOUND	MOL. FORMULA	B. PT.	INDEX No.
Octan-2-ol	$C_8H_{18}O$	180°C	**24**

CELL TEMP.	APPROX AMOUNT	CELL PATH	REMARKS
~130°C	2·5µl	Length 9·2cm Volume 25ml	

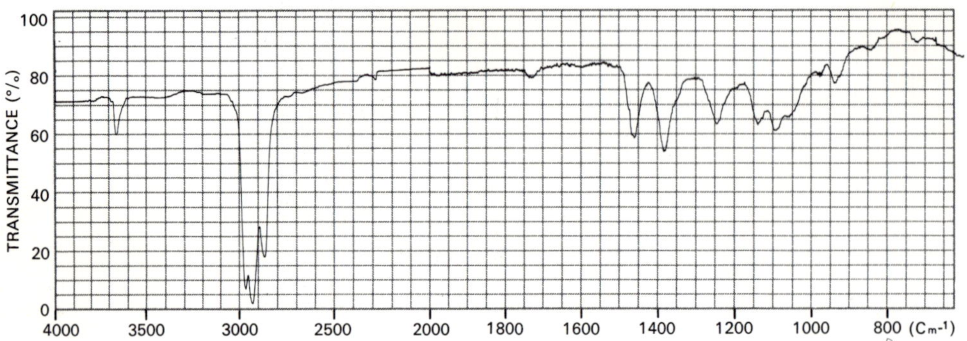

COMPOUND	MOL. FORMULA	B. PT.	INDEX No.
Pentan-3-ol	$C_5H_{12}O$	116°C	**25**

CELL TEMP.	APPROX AMOUNT	CELL PATH	REMARKS
~125°C	2μl	Length 9·2cm Volume 25ml	

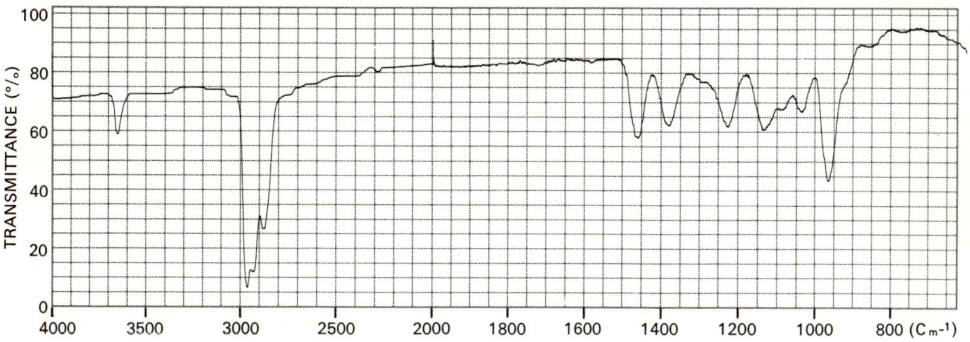

COMPOUND	MOL. FORMULA	B. PT.	INDEX No.
Octan-3-ol	$C_8H_{18}O$	173°C	**26**

CELL TEMP.	APPROX AMOUNT	CELL PATH	REMARKS
~130°C	5μl	Length 9·2cm Volume 25ml	

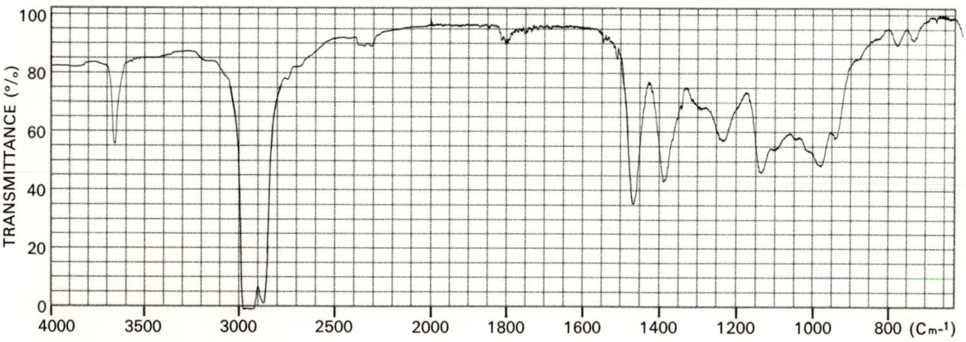

COMPOUND	MOL. FORMULA	B. PT.	INDEX No.
Octan-4-ol	$C_8H_{18}O$	176°C	**27**

CELL TEMP.	APPROX AMOUNT	CELL PATH	REMARKS
~130°C	3μl	Length 9·2cm Volume 25ml	

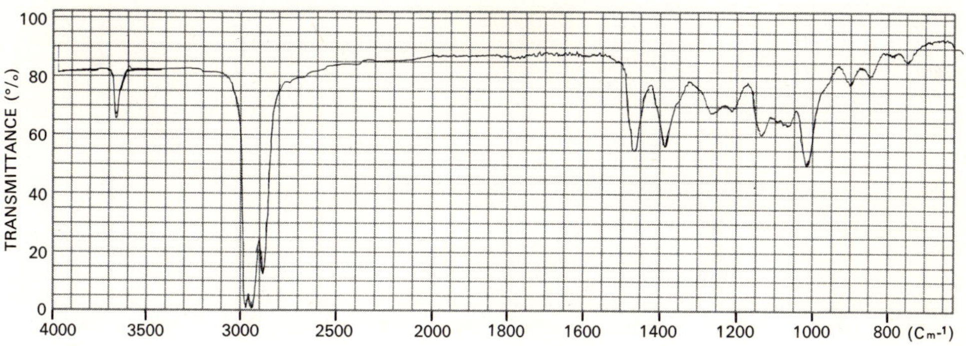

COMPOUND	MOL. FORMULA	B. PT.	INDEX No.
Decan-5-ol	$C_{10}H_{22}O$	-	**28**

CELL TEMP.	APPROX AMOUNT	CELL PATH	REMARKS
~125°C	2μl	Length 9·2cm Volume 25ml	

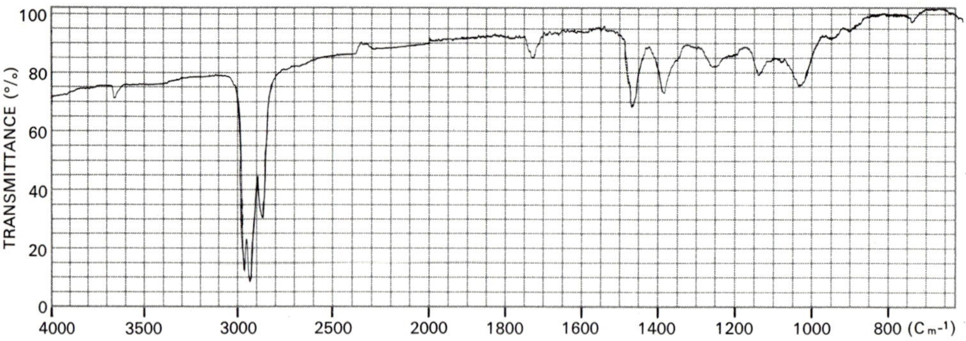

COMPOUND	MOL. FORMULA	B. PT.	INDEX No.
Cyclopentanol	$C_5H_{10}O$	141°C	**29**

CELL TEMP.	APPROX AMOUNT	CELL PATH	REMARKS
~130°C	2·5μl	Length 9·2cm Volume 25ml	

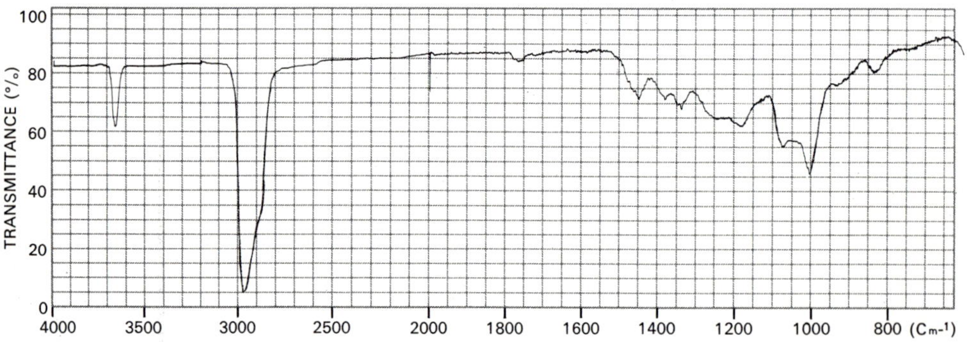

COMPOUND	MOL. FORMULA	B. PT.	INDEX No.
Cyclohexanol	$C_6H_{12}O$	161°C	**30**

CELL TEMP.	APPROX AMOUNT	CELL PATH	REMARKS
~130°C	3μl	Length 9·2cm Volume 25ml	

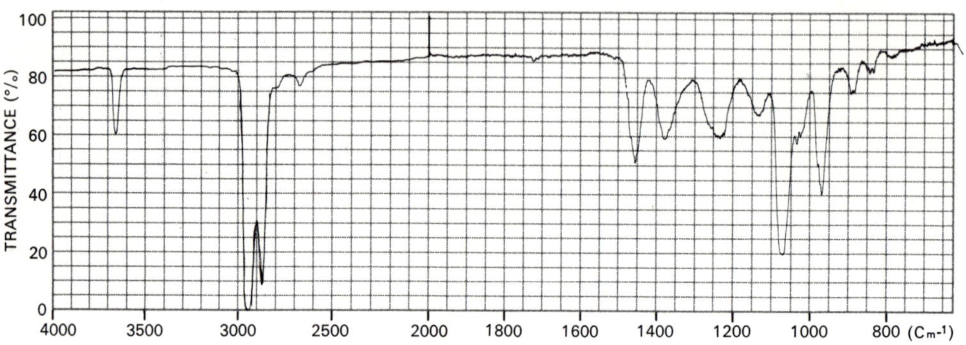

COMPOUND	MOL. FORMULA	B. PT.	INDEX No.
But-1-en-3-ol	C_4H_8O	97°C	**31**

CELL TEMP.	APPROX AMOUNT	CELL PATH	REMARKS
~120°C	2µl	Length 9·2cm Volume 25ml	

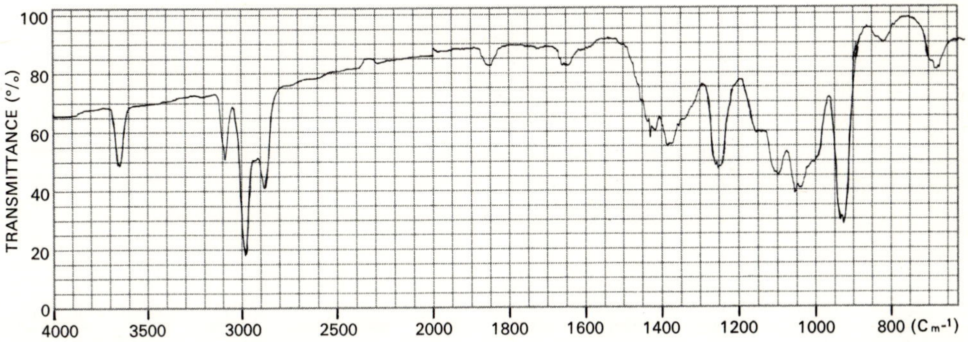

COMPOUND	MOL. FORMULA	B. PT.	INDEX No.
Hex-1-en-3-ol	$C_6H_{12}O$	134°C	**32**

CELL TEMP.	APPROX AMOUNT	CELL PATH	REMARKS
~140°C	3µl	Length 9·2cm Volume 25ml	

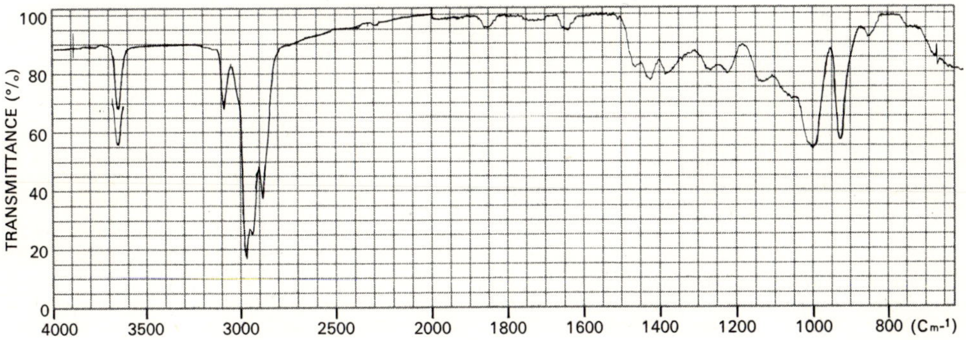

COMPOUND	MOL. FORMULA	B. PT.	INDEX No.
Oct-1-en-3-ol	$C_8H_{16}O$	-	**33**

CELL TEMP.	APPROX AMOUNT	CELL PATH	REMARKS
~145°C	3•5µl	Length 9·2cm Volume 25ml	

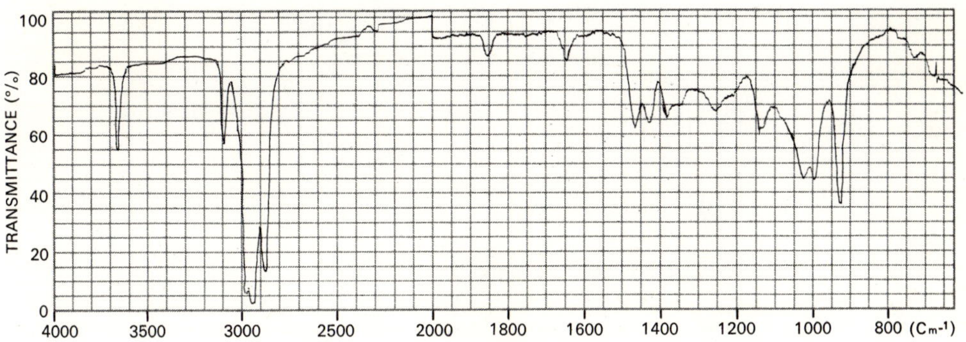

COMPOUND	MOL. FORMULA	B. PT.	INDEX No.
1-Phenylethanol	$C_8H_{10}O$	204°C	**34**

CELL TEMP.	APPROX AMOUNT	CELL PATH	REMARKS
~130°C	3μl	Length 9·2cm Volume 25ml	

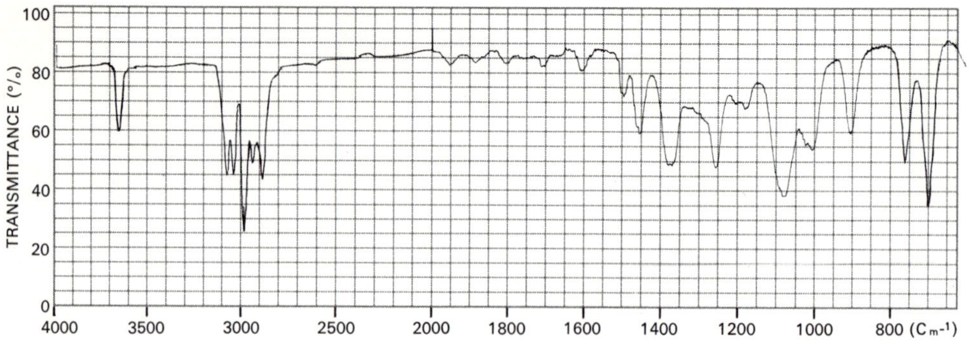

COMPOUND	MOL. FORMULA	B. PT.	INDEX No.
Pent-1-en-4-ol	$C_5H_{10}O$	115°C	**35**

CELL TEMP.	APPROX AMOUNT	CELL PATH	REMARKS
~125°C	2μl	Length 9·2cm Volume 25ml	

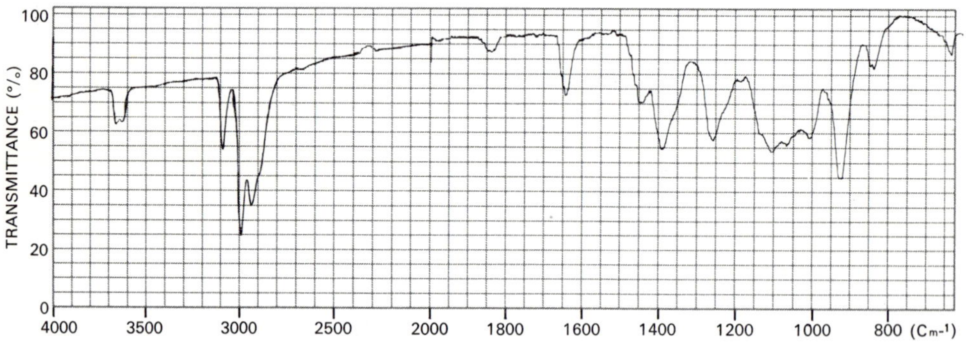

COMPOUND	MOL. FORMULA	B. PT.	INDEX No.
Hept-1-en-4-ol	$C_7H_{14}O$	152°C	**36**

CELL TEMP.	APPROX AMOUNT	CELL PATH	REMARKS
~120°C	2μl	Length 9·2cm Volume 25ml	

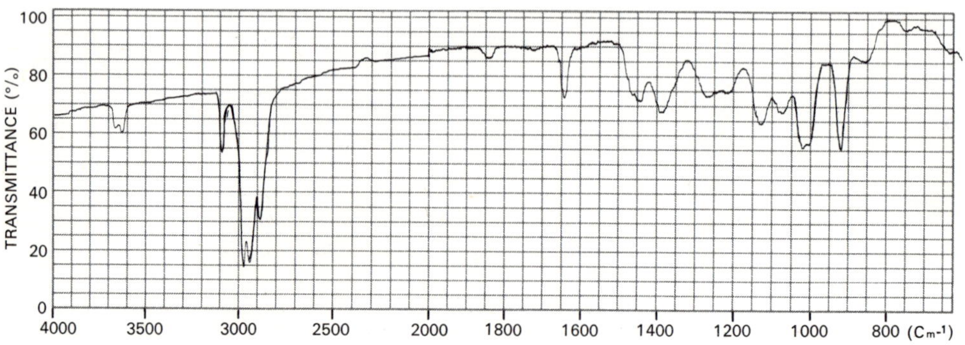

COMPOUND	MOL. FORMULA	B. PT.	INDEX No.
t-Butanol	$C_4H_{10}O$	83°C	**37**

CELL TEMP.	APPROX AMOUNT	CELL PATH	REMARKS
~130°C	2μl	Length 9·2cm Volume 25ml	

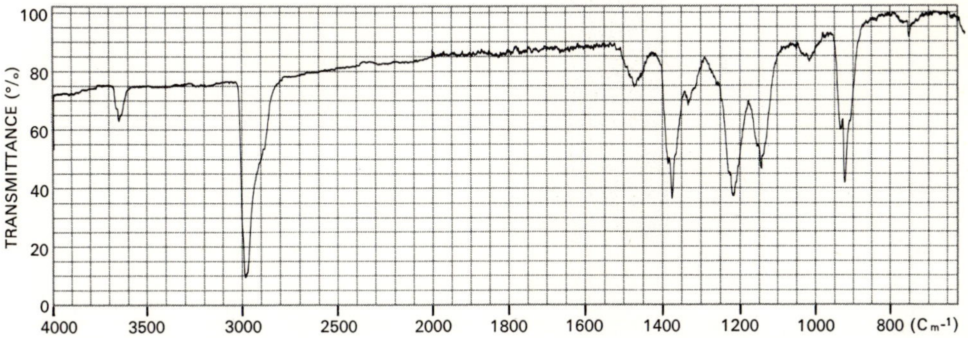

COMPOUND	MOL. FORMULA	B. PT.	INDEX No.
2-Methylpentan-2-ol	$C_6H_{14}O$	108°C	**38**

CELL TEMP.	APPROX AMOUNT	CELL PATH	REMARKS
~130°C	2μl	Length 9·2cm Volume 25ml	

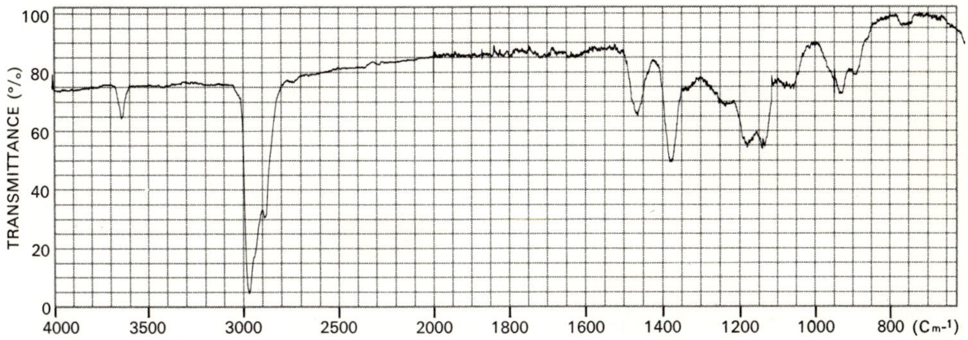

COMPOUND	MOL. FORMULA	B. PT.	INDEX No.
1,1-Dimethyl-2-phenylethanol	$C_{10}H_{14}O$	215°C	**39**

CELL TEMP.	APPROX AMOUNT	CELL PATH	REMARKS
~140°C	2μl	Length 9·2cm Volume 25ml	

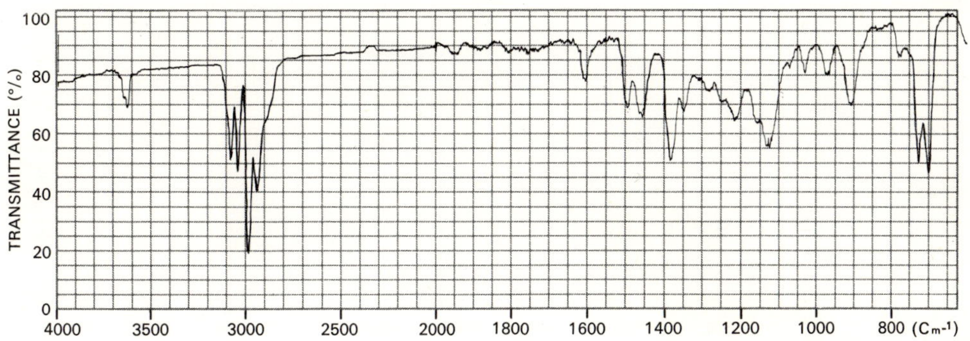

COMPOUND	MOL. FORMULA	B. PT.	INDEX No.
2-Methylbut-3-en-2-ol	$C_5H_{10}O$	95-99°C	**40**

CELL TEMP.	APPROX AMOUNT	CELL PATH	REMARKS
~130°C	3μl	Length 9·2cm Volume 25ml	

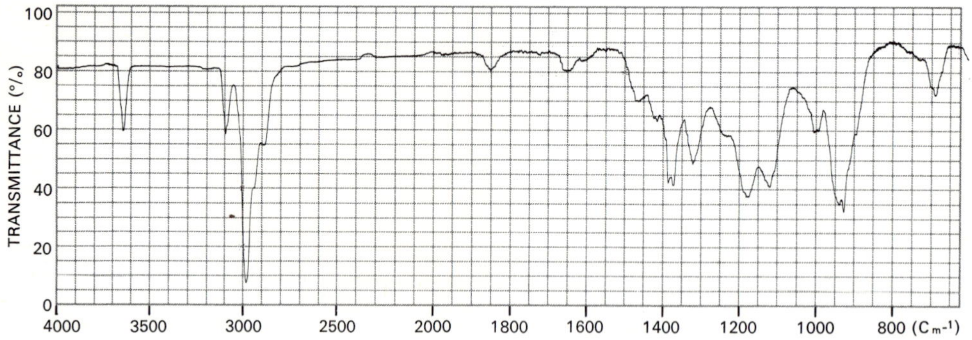

COMPOUND	MOL. FORMULA	B. PT.	INDEX No.
Citronellol	$C_{10}H_{20}O$	245°C	**41**

CELL TEMP.	APPROX AMOUNT	CELL PATH	REMARKS
~135°C	2μl	Length 9·2cm Volume 25ml	

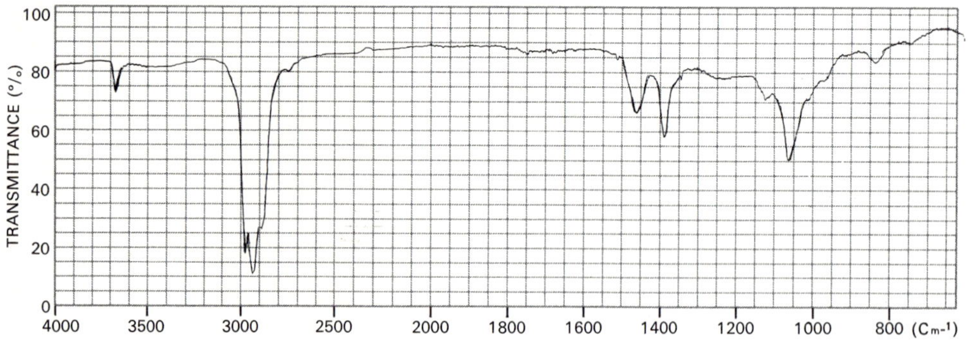

COMPOUND	MOL. FORMULA	B. PT.	INDEX No.
Geraniol	$C_{10}H_{18}O$	-	**42**

CELL TEMP.	APPROX AMOUNT	CELL PATH	REMARKS
~145°C	2μl	Length 9·2cm Volume 25ml	

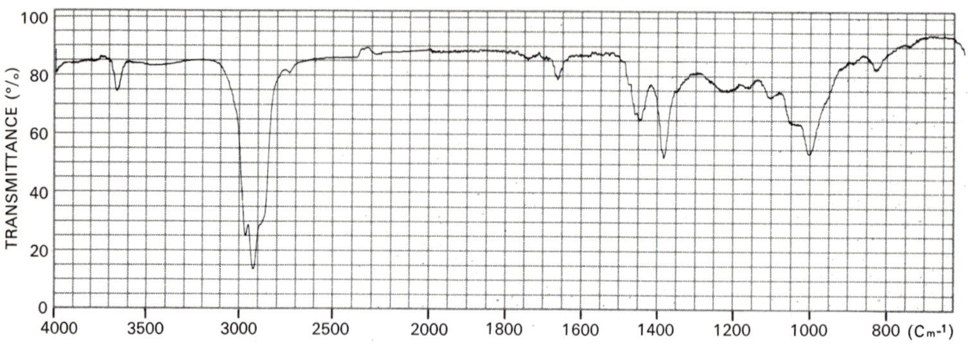

COMPOUND	MOL. FORMULA	B. PT.	INDEX No.
Linalool	$C_{10}H_{18}O$	200°C	**43**

CELL TEMP.	APPROX AMOUNT	CELL PATH	REMARKS
~145°C	2μl	Length 9·2cm Volume 25ml	

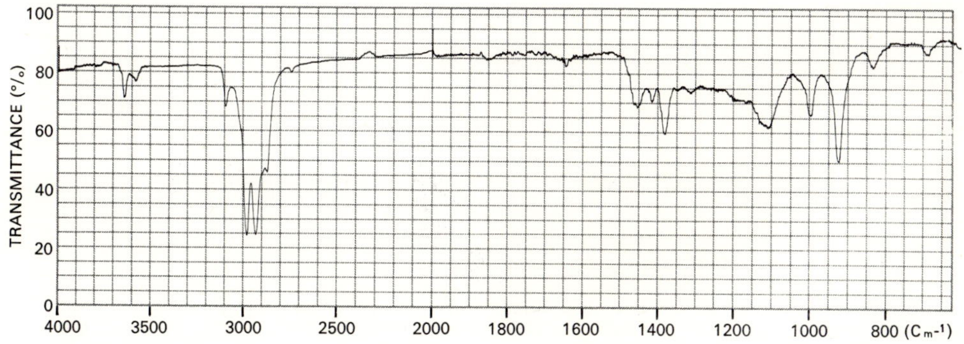

COMPOUND	MOL. FORMULA	B. PT.	INDEX No.
α-Terpineol	$C_{10}H_{18}O$	220°C	**44**

CELL TEMP.	APPROX AMOUNT	CELL PATH	REMARKS
~145°c	A few mg	Length 9·2cm Volume 25ml	

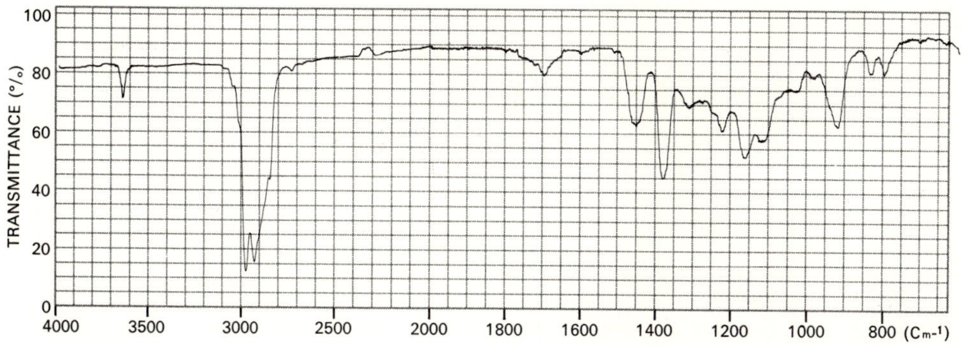

COMPOUND	MOL. FORMULA	B. PT.	INDEX No.
2-Methoxyethanol	$C_3H_8O_2$	125°C	**45**

CELL TEMP.	APPROX AMOUNT	CELL PATH	REMARKS
~150°C	2μl	Length 9·2cm Volume 25ml	

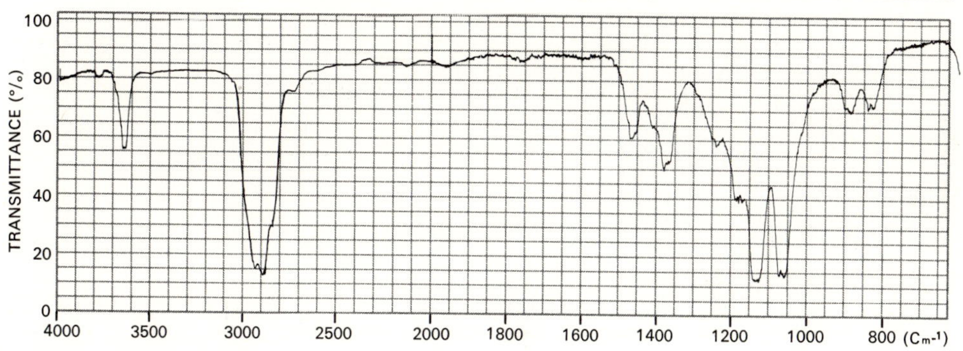

COMPOUND	MOL. FORMULA	B. PT.	INDEX No.
2-(2-Ethoxy)ethoxyethanol	$C_6H_{14}O_3$	201°C	**46**

CELL TEMP.	APPROX AMOUNT	CELL PATH	REMARKS
~135°C	2μl	Length 9·2cm Volume 25ml	

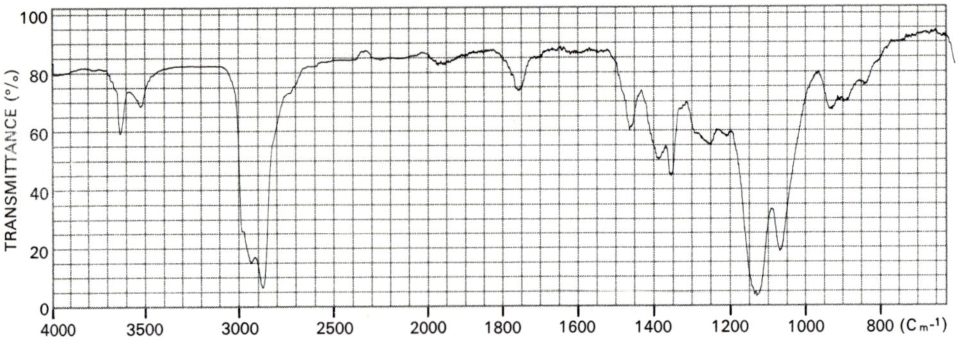

COMPOUND	MOL. FORMULA	B. PT.	INDEX No.
Ethane-1,2-diol	$C_2H_6O_2$	198°C	**47**

CELL TEMP.	APPROX AMOUNT	CELL PATH	REMARKS
~140°C	1μl	Length 9·2cm Volume 25ml	

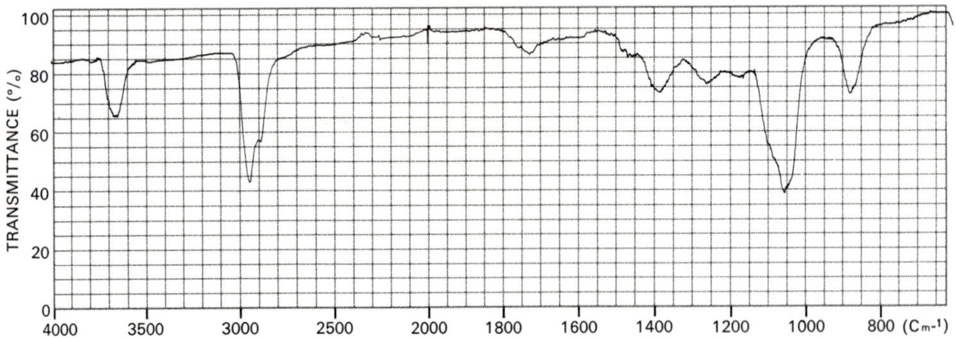

COMPOUND	MOL. FORMULA	B. PT.	INDEX No.
Propane-1,2-diol	$C_3H_8O_2$	189°C	**48**

CELL TEMP.	APPROX AMOUNT	CELL PATH	REMARKS
~145°C	2μl	Length 9·2cm Volume 25ml	

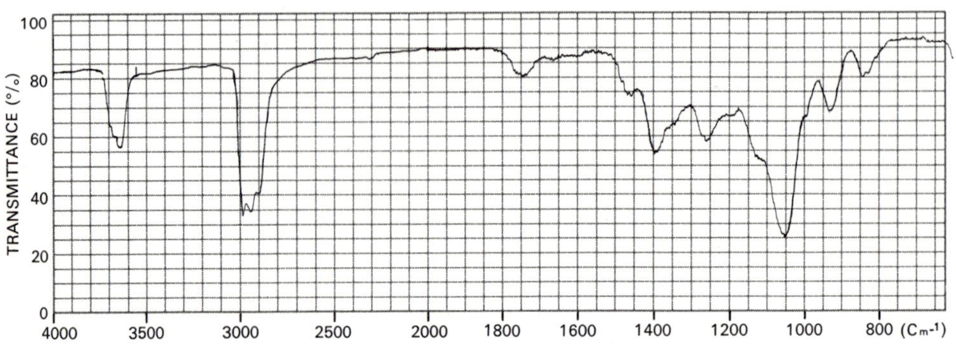

COMPOUND	MOL. FORMULA	B. PT.	INDEX No.
Propane-1,3-diol	$C_3H_8O_2$	214°C	**49**

CELL TEMP.	APPROX AMOUNT	CELL PATH	REMARKS
~ 145°C	1μl	Length 9·2cm Volume 25ml	

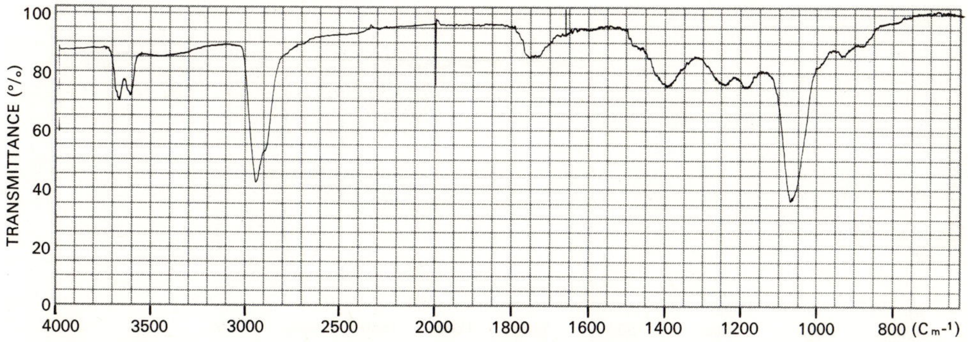

COMPOUND	MOL. FORMULA	B. PT.	INDEX No.
2-(N,N-Dimethyl)aminoethanol	$C_4H_{11}NO$	135°C	**50**

CELL TEMP.	APPROX AMOUNT	CELL PATH	REMARKS
~ 150°C	2μl	Length 9·2cm Volume 25ml	

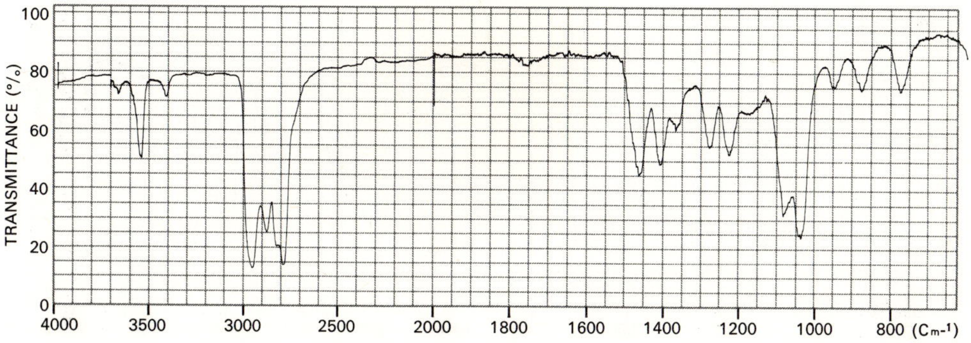

COMPOUND	MOL. FORMULA	B. PT.	INDEX No.
2-(N,N-Diethyl)aminoethanol	$C_6H_{15}NO$	163°C	**51**

CELL TEMP.	APPROX AMOUNT	CELL PATH	REMARKS
~ 130°C	2μl	Length 9·2cm Volume 25ml	

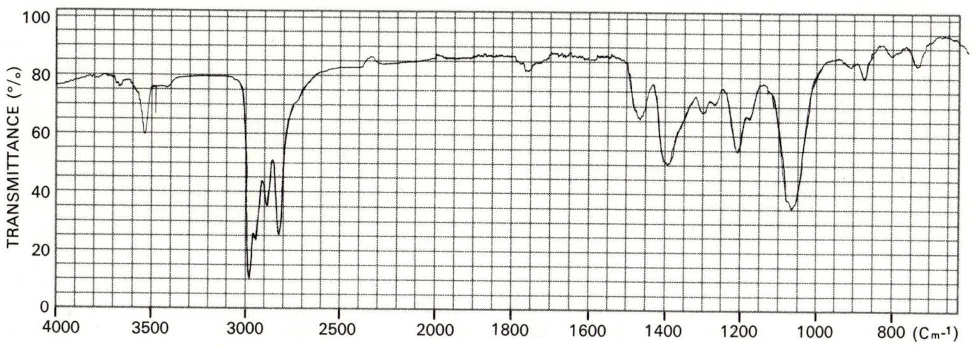

COMPOUND	MOL. FORMULA	B. PT.	INDEX No.
Phenol	C_6H_6O	182°C	**52**

CELL TEMP.	APPROX AMOUNT	CELL PATH	REMARKS
~130°C	A few mg	Length 9·2cm Volume 25ml	

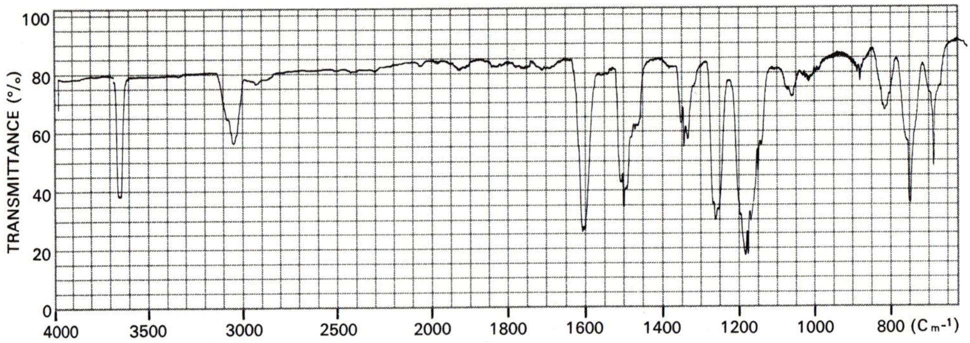

COMPOUND	MOL. FORMULA	B. PT.	INDEX No.
o-Cresol	C_7H_8O	191°C	**53**

CELL TEMP.	APPROX AMOUNT	CELL PATH	REMARKS
~130°C	2·5μl	Length 9·2cm Volume 25ml	

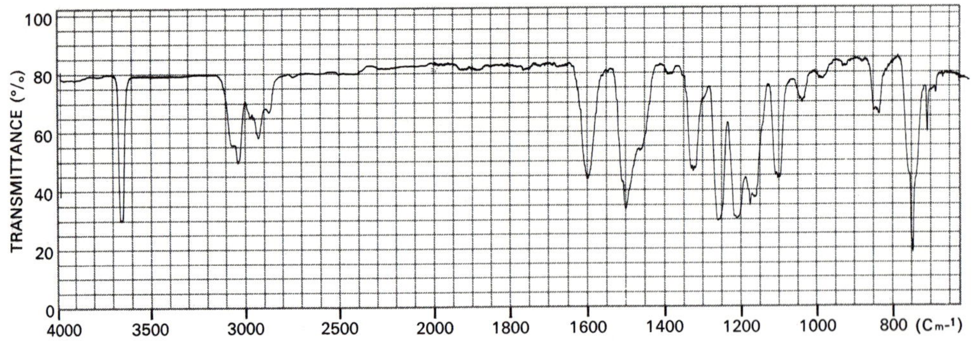

COMPOUND	MOL. FORMULA	B. PT.	INDEX No.
m-Cresol	C_7H_8O	203°C	**54**

CELL TEMP.	APPROX AMOUNT	CELL PATH	REMARKS
~130°C	2·5μl	Length 9·2cm Volume 25ml	

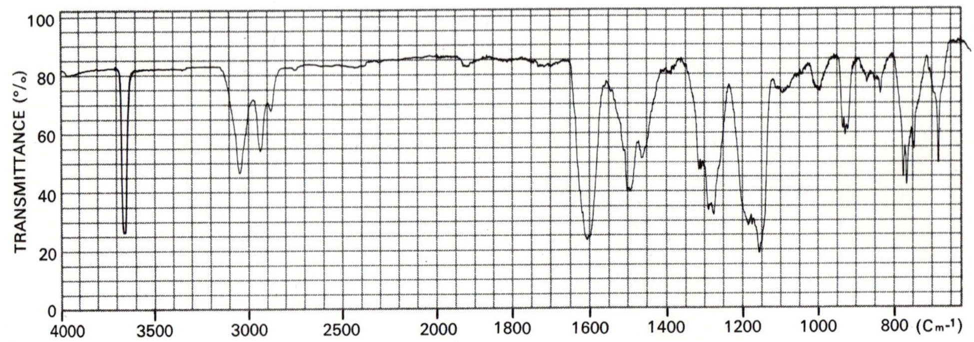

COMPOUND	MOL. FORMULA	B. PT.	INDEX No.
p-Cresol	C_7H_8O	210°C	**55**

CELL TEMP.	APPROX AMOUNT	CELL PATH	REMARKS
~145°C	3μl	Length 9·2cm Volume 25ml	

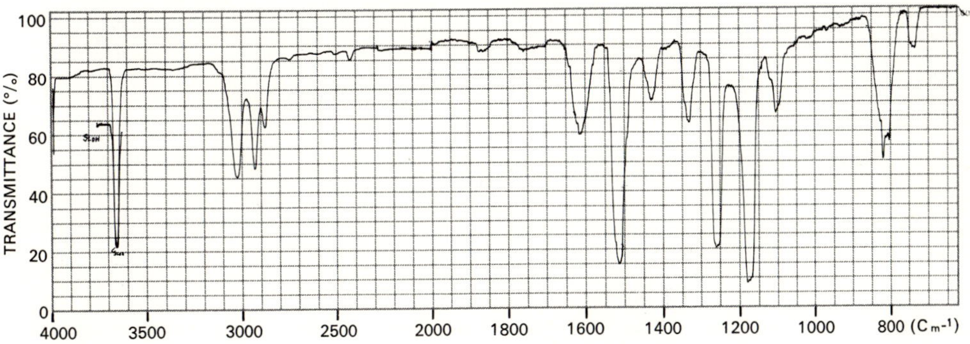

COMPOUND	MOL. FORMULA	B. PT.	INDEX No.
2,3-Xylenol	$C_8H_{10}O$	218°C	**56**

CELL TEMP.	APPROX AMOUNT	CELL PATH	REMARKS
~130°C	A few mg	Length 9·2cm Volume 25ml	

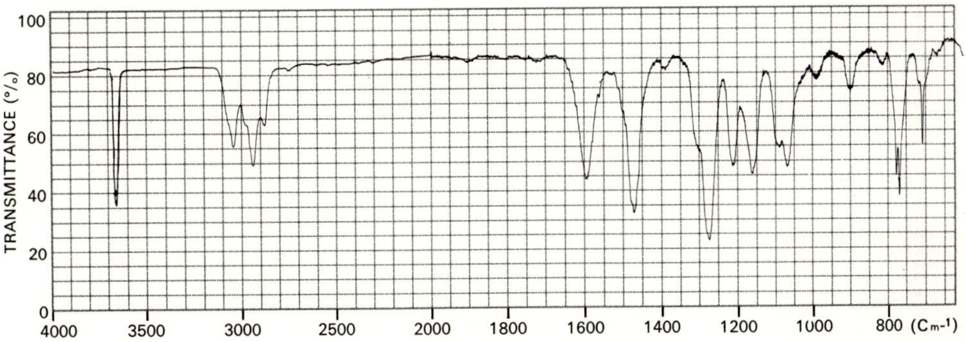

COMPOUND	MOL. FORMULA	B. PT.	INDEX No.
2,4-Xylenol	$C_8H_{10}O$	212°C	**57**

CELL TEMP.	APPROX AMOUNT	CELL PATH	REMARKS
~135°C	3μl	Length 9·2cm Volume 25ml	

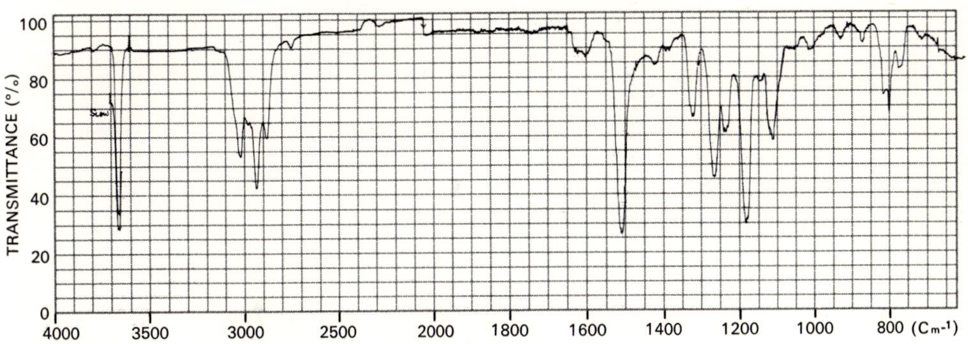

COMPOUND	MOL. FORMULA	B. PT.	INDEX No.
2,5-Xylenol	$C_8H_{10}O$	242°C	**58**

CELL TEMP.	APPROX AMOUNT	CELL PATH	REMARKS
~150°C	A few mg	Length 9·2cm Volume 25ml	

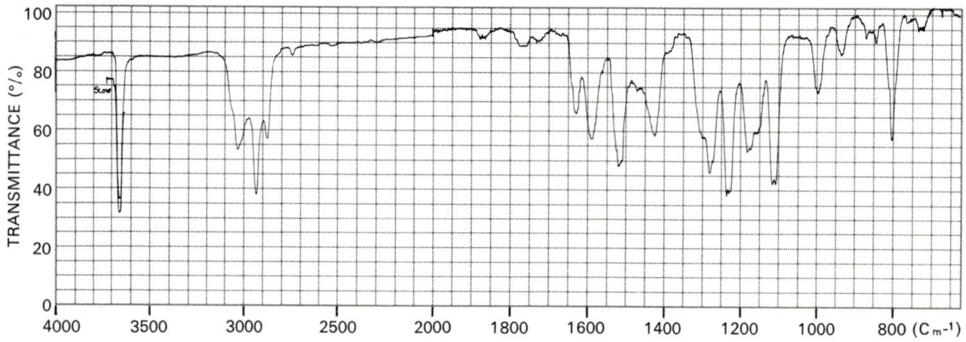

COMPOUND	MOL. FORMULA	B. PT.	INDEX No.
2,6-Xylenol	$C_8H_{10}O$	212°C	**59**

CELL TEMP.	APPROX AMOUNT	CELL PATH	REMARKS
~130°C	A few mg	Length 9·2cm Volume 25ml	

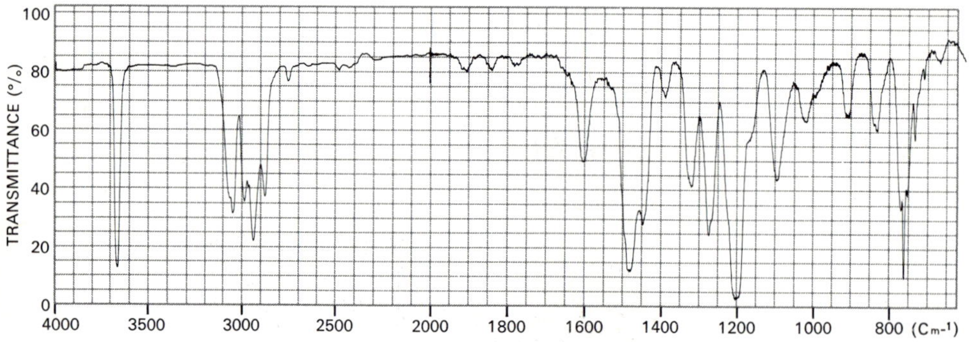

COMPOUND	MOL. FORMULA	B. PT.	INDEX No.
3,4-Xylenol	$C_8H_{10}O$	225°C	**60**

CELL TEMP.	APPROX AMOUNT	CELL PATH	REMARKS
~130°C	A few mg	Length 9·2cm Volume 25ml	

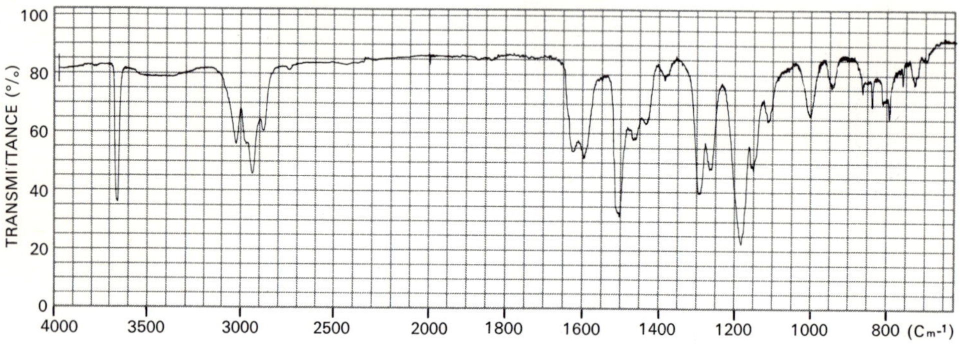

COMPOUND	MOL. FORMULA	B. PT.	INDEX No.
3, 5-Xylenol	$C_8H_{10}O$	219°C	**61**

CELL TEMP.	APPROX AMOUNT	CELL PATH	REMARKS
~ 130°C	A few mg	Length 9·2cm Volume 25ml	

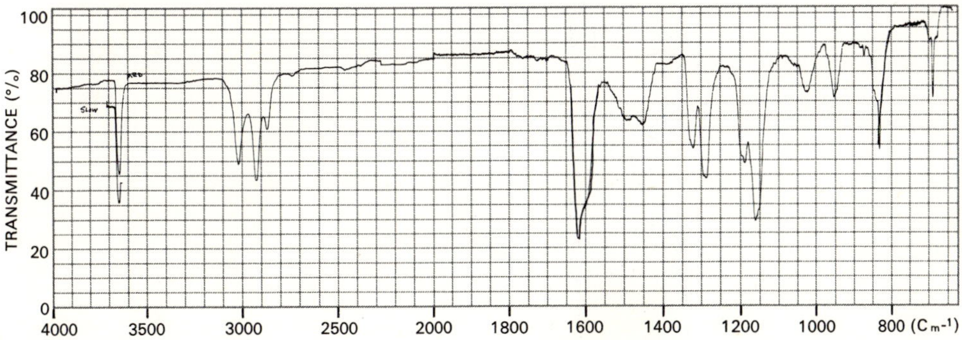

COMPOUND	MOL. FORMULA	B. PT.	INDEX No.
2-Methoxyphenol	$C_7H_8O_2$	205°C	**62**

CELL TEMP.	APPROX AMOUNT	CELL PATH	REMARKS
~130°C	A few mg	Length 9·2cm Volume 25ml	

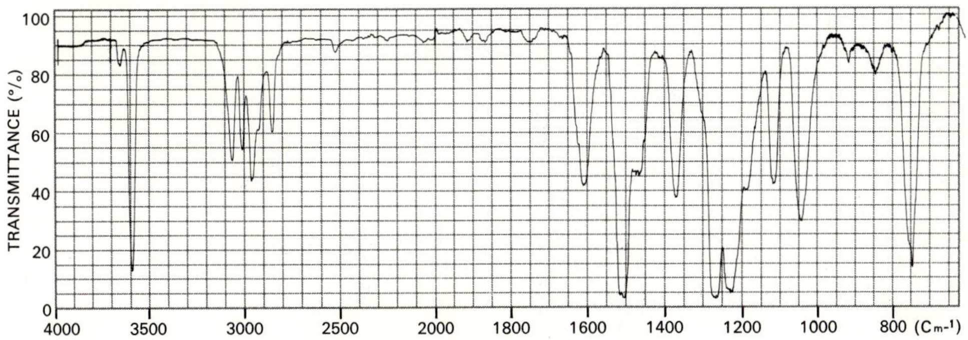

COMPOUND	MOL. FORMULA	B. PT.	INDEX No.
2, 6-Dimethoxyphenol	$C_8H_{10}O_2$	263°C	**63**

CELL TEMP.	APPROX AMOUNT	CELL PATH	REMARKS
~160°C	A few mg	Length 9·2cm Volume 25ml	

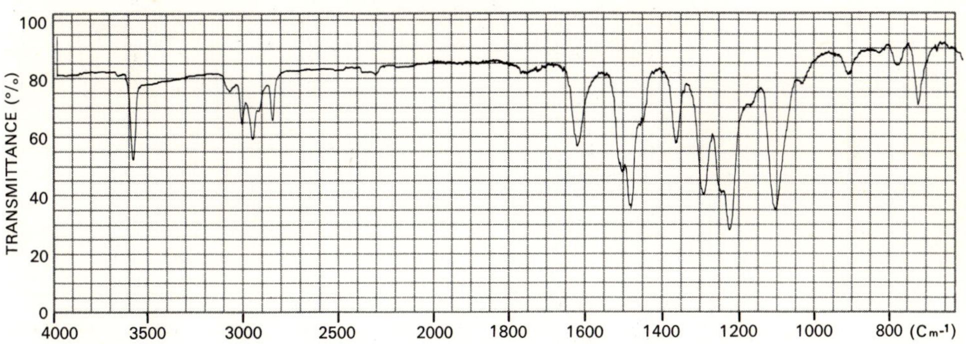

COMPOUND	MOL. FORMULA	B. PT.	INDEX No.
2-Chlorophenol	C_6H_5ClO	176°C	**64**

CELL TEMP.	APPROX AMOUNT	CELL PATH	REMARKS
~ 160°C	3μl	Length 9·2cm Volume 25ml	

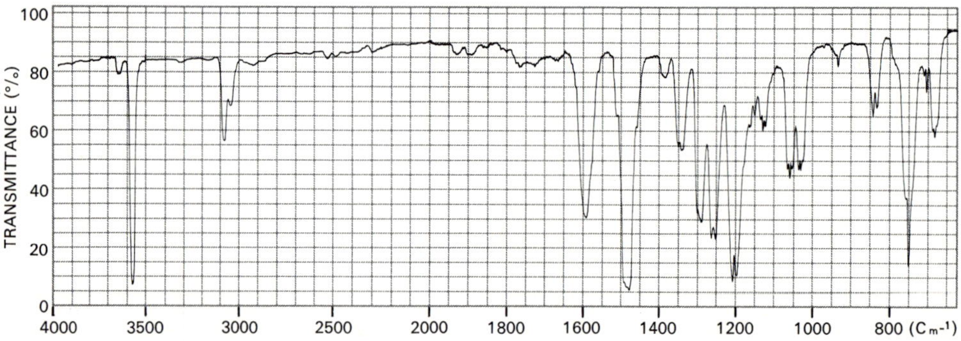

COMPOUND	MOL. FORMULA	B. PT.	INDEX No.
2-Bromophenol	C_6H_5BrO	195°C	**65**

CELL TEMP.	APPROX AMOUNT	CELL PATH	REMARKS
~ 145°C	3μl	Length 9·2cm Volume 25ml	

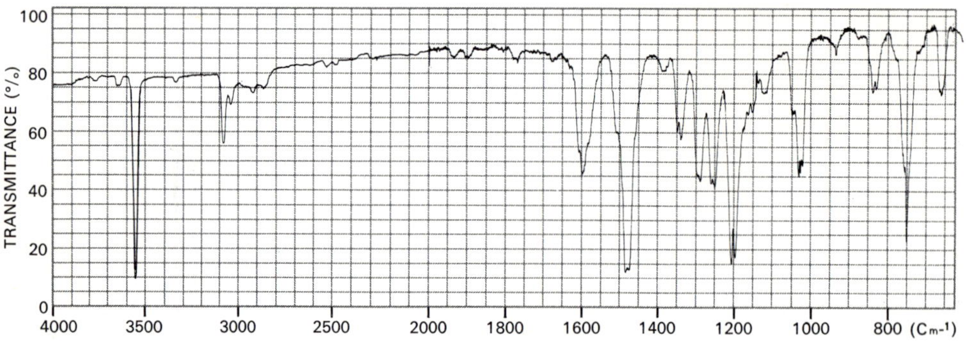

COMPOUND	MOL. FORMULA	B. PT.	INDEX No.
4-Chlorophenol	C_6H_5ClO	220°C	**66**

CELL TEMP.	APPROX AMOUNT	CELL PATH	REMARKS
~ 130°C	A few mg	Length 9·2cm Volume 25ml	

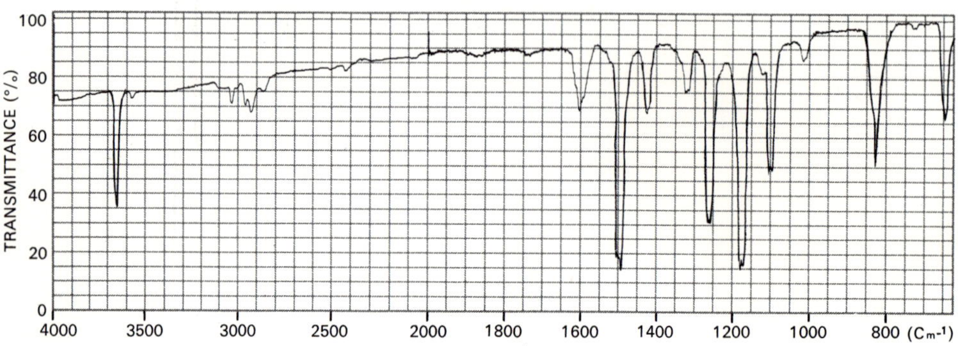

COMPOUND	MOL. FORMULA	B. PT.	INDEX No.
4-Bromophenol	C_6H_5BrO	238° C	**67**

CELL TEMP.	APPROX AMOUNT	CELL PATH	REMARKS
~150° C	A few mg	Length 9·2cm Volume 25ml	

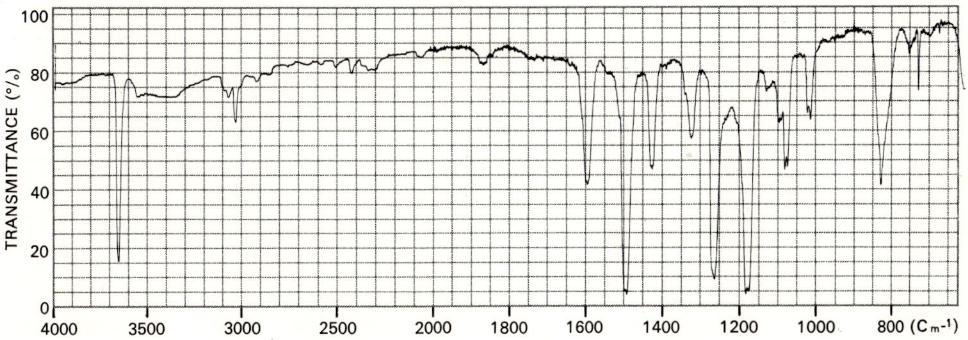

COMPOUND	MOL. FORMULA	B. PT.	INDEX No.
2-Isopropyl-5-methylphenol	$C_{10}H_{14}O$	233° C	**68**

CELL TEMP.	APPROX AMOUNT	CELL PATH	REMARKS
~150° C	2μl	Length 9·2cm Volume 25ml	

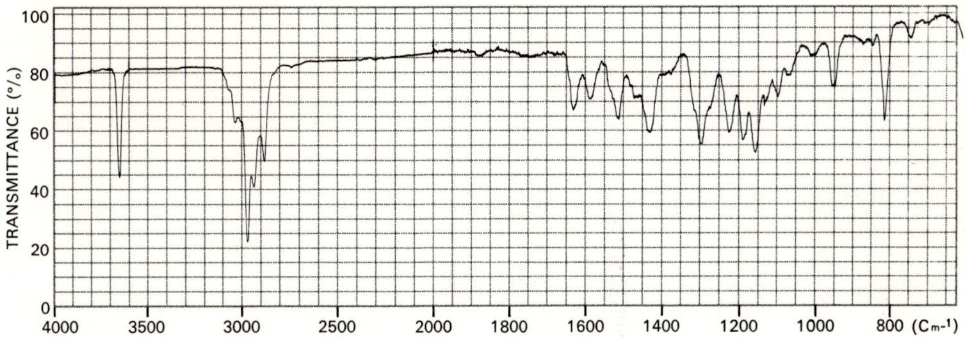

COMPOUND	MOL. FORMULA	B. PT.	INDEX No.
4 Allyl 2-methoxyphenol (Eugenol)	$C_{10}H_{12}O_2$	255° C	**69**

CELL TEMP.	APPROX AMOUNT	CELL PATH	REMARKS
~130° C	2μl	Length 9·2cm Volume 25ml	

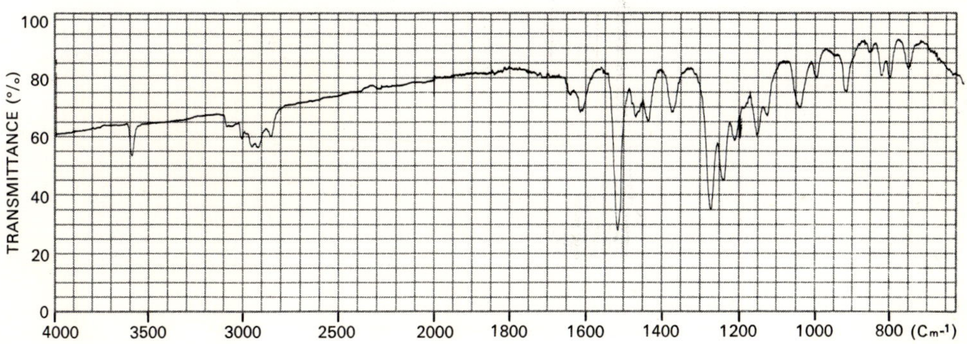

COMPOUND	MOL. FORMULA	B. PT.	INDEX No.
1, 2-Dihydroxybenzene (Catechol)	$C_6H_6O_2$	240° C	**70**

CELL TEMP.	APPROX AMOUNT	CELL PATH	REMARKS
~150° C	A few mg	Length 9·2cm Volume 25ml	

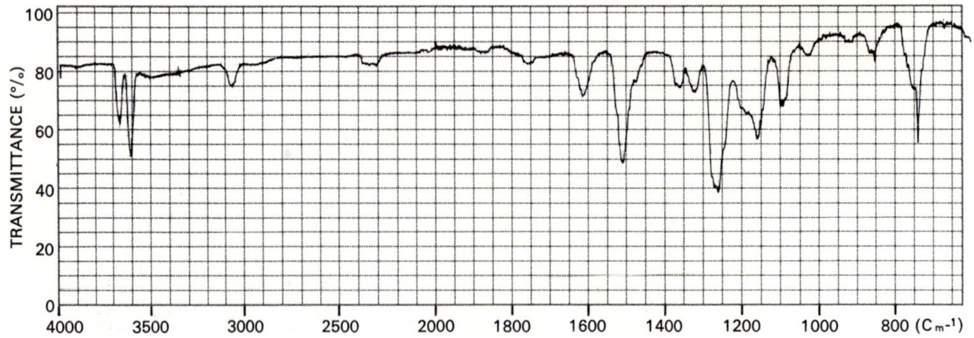

COMPOUND	MOL. FORMULA	B. PT.	INDEX No.
1, 2-Dihydroxy-3-methylbenzene	$C_7H_8O_2$	241° C	**71**

CELL TEMP.	APPROX AMOUNT	CELL PATH	REMARKS
~150° C	A few mg	Length 9·2cm Volume 25ml	

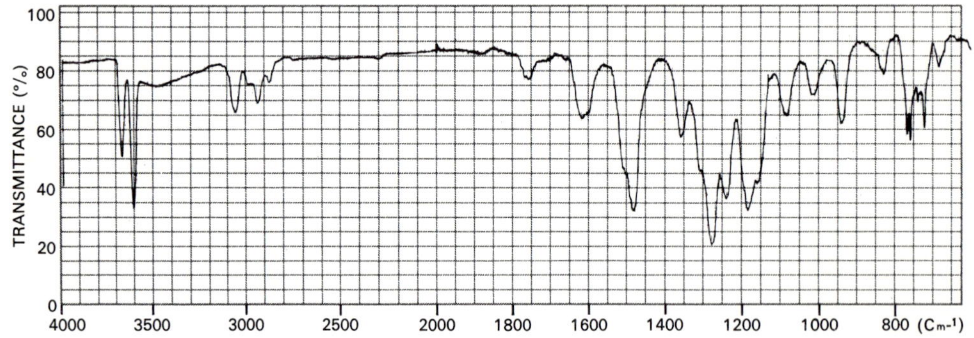

COMPOUND	MOL. FORMULA	B. PT.	INDEX No.
1, 2-Dihydroxy-4-methylbenzene	$C_7H_8O_2$	-	**72**

CELL TEMP.	APPROX AMOUNT	CELL PATH	REMARKS
~150° C	A few mg	Length 9·2cm Volume 25ml	

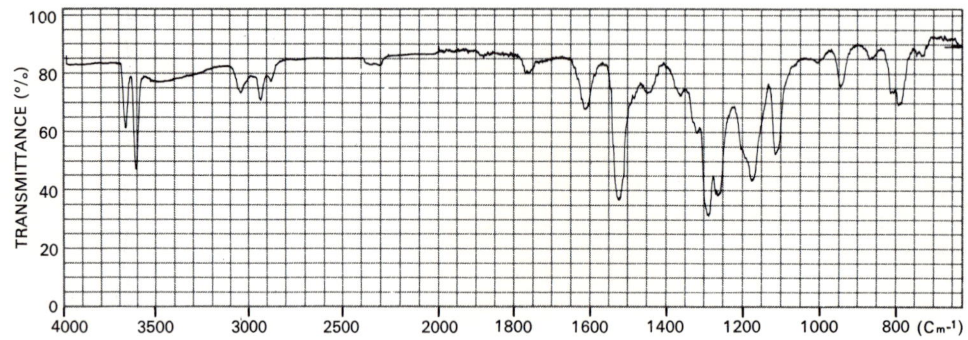

COMPOUND	MOL. FORMULA	B. PT.	INDEX No.
Formic acid	CH_2O_2	101°C	**73**

CELL TEMP.	APPROX AMOUNT	CELL PATH	REMARKS
~40°C	1·0μl	Length 9·2cm Volume 25ml	

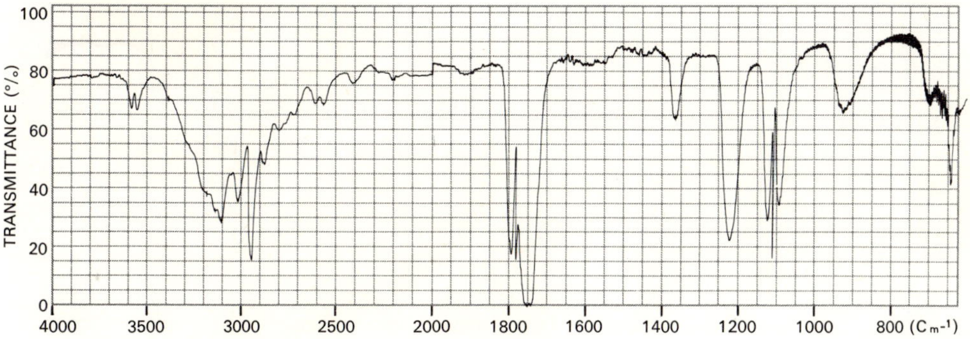

COMPOUND	MOL. FORMULA	B. PT.	INDEX No.
Formic acid	CH_2O_2	101°C	**74**

CELL TEMP.	APPROX AMOUNT	CELL PATH	REMARKS
~130°C	0·5μl	Length 9·2cm Volume 25ml	Mainly monomer

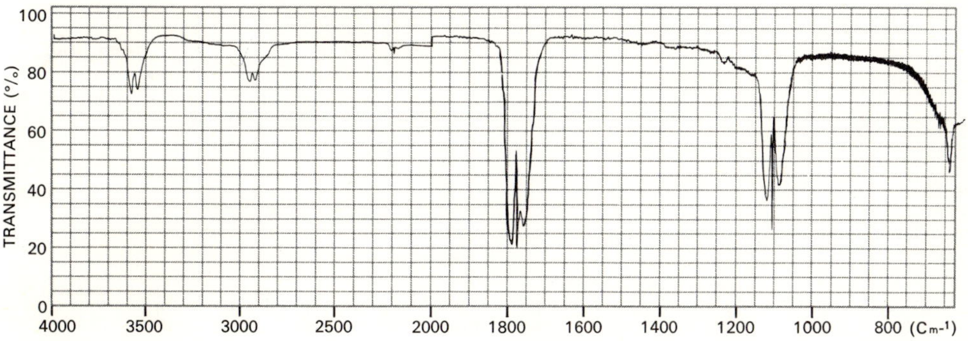

COMPOUND	MOL. FORMULA	B. PT.	INDEX No.
Acetic acid	$C_2H_4O_2$	119°C	**75**

CELL TEMP.	APPROX AMOUNT	CELL PATH	REMARKS
~35°C	1·0μl	Length 9·2cm Volume 25ml	Mainly dimer

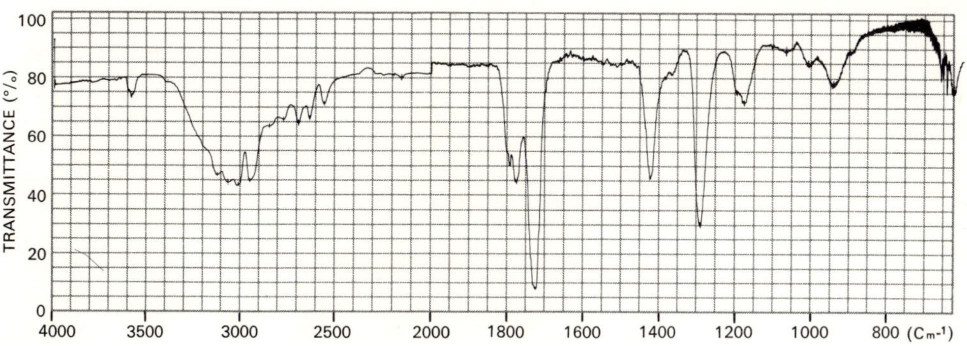

COMPOUND	MOL. FORMULA	B. PT.	INDEX No.
Acetic acid	$C_2H_4O_2$	119°C	**76**

CELL TEMP.	APPROX AMOUNT	CELL PATH	REMARKS
~130°C	0·5μl	Length 9·2cm Volume 25ml	Mainly monomer

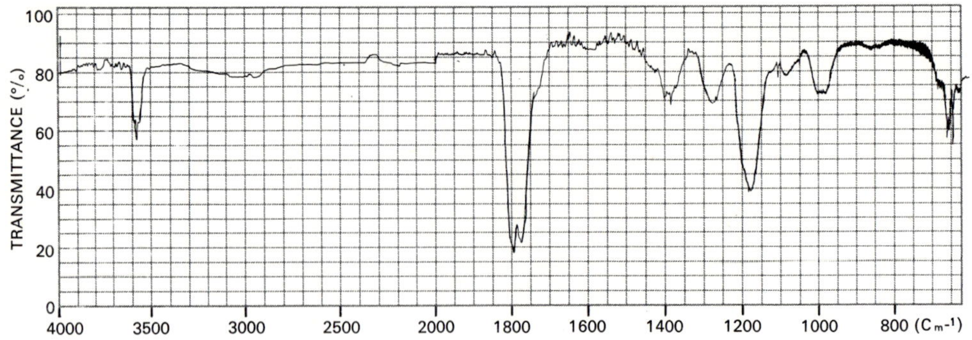

COMPOUND	MOL. FORMULA	B. PT.	INDEX No.
Propanoic acid	$C_3H_6O_2$	141°C	**77**

CELL TEMP.	APPROX AMOUNT	CELL PATH	REMARKS
~130°C	1μl	Length 9·2cm Volume 25ml	

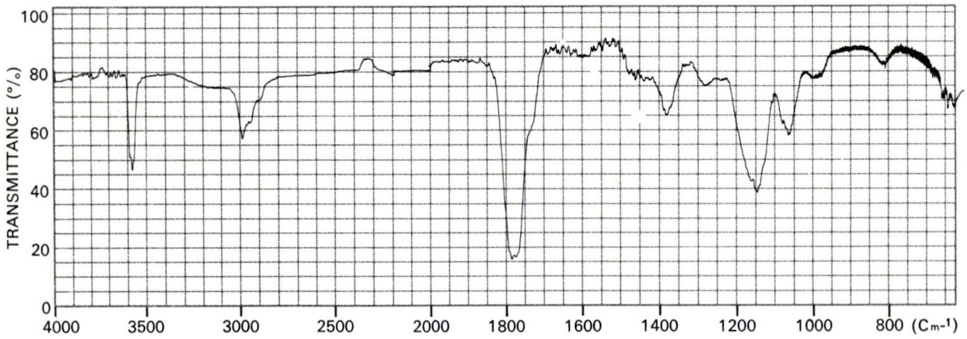

COMPOUND	MOL. FORMULA	B. PT.	INDEX No.
n-Butanoic acid	$C_4H_8O_2$	164°C	**78**

CELL TEMP.	APPROX AMOUNT	CELL PATH	REMARKS
~130°C	1μl	Length 9·2cm Volume 25ml	

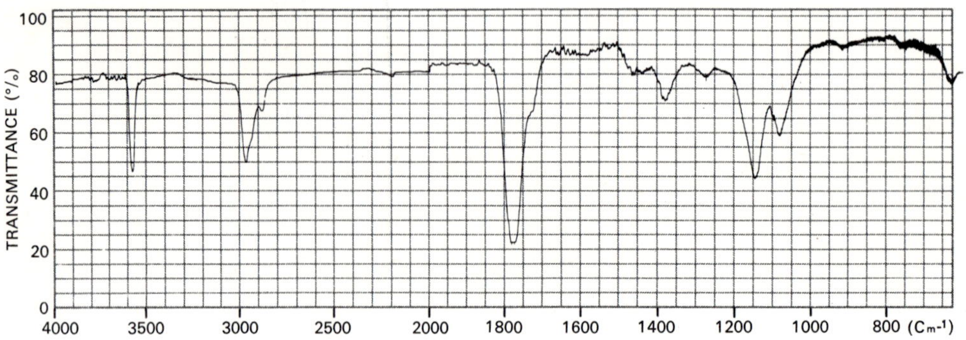

COMPOUND	MOL. FORMULA	B. PT.	INDEX No.
n-Pentanoic acid	$C_5H_{10}O_2$	186° C	**79**

CELL TEMP.	APPROX AMOUNT	CELL PATH	REMARKS
~130° C	2μl	Length 9·2cm Volume 25ml	

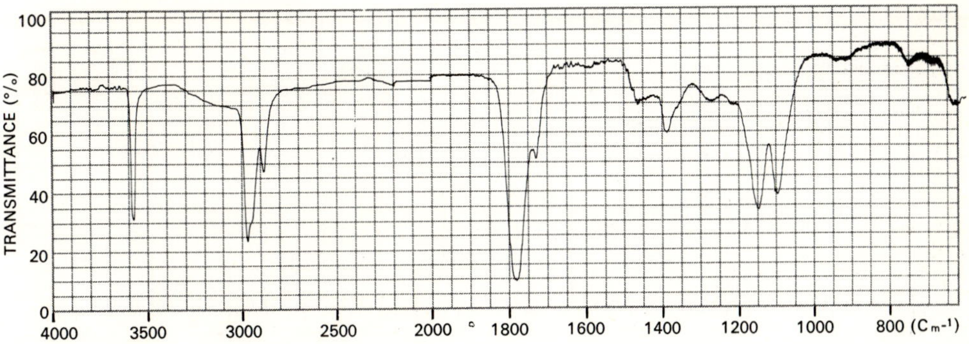

COMPOUND	MOL. FORMULA	B. PT.	INDEX No.
n-Hexanoic acid	$C_6H_{12}O_2$	205° C	**80**

CELL TEMP.	APPROX AMOUNT	CELL PATH	REMARKS
~130° C	2μl	Length 9·2cm Volume 25ml	

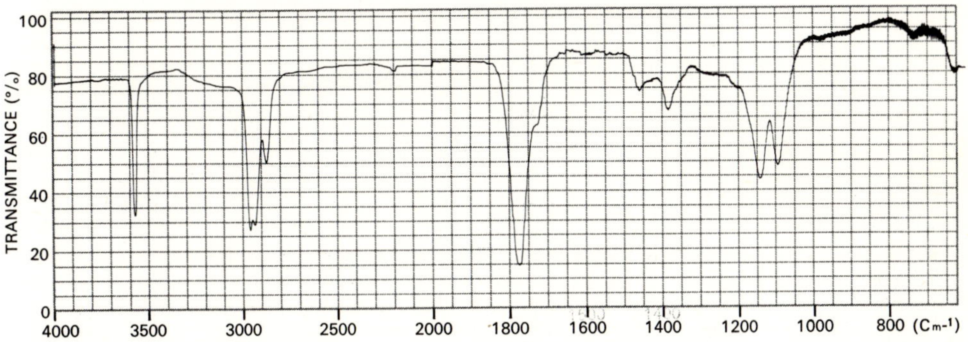

COMPOUND	MOL. FORMULA	B. PT.	INDEX No.
n-Heptanoic acid	$C_7H_{14}O_2$	223° C	**81**

CELL TEMP.	APPROX AMOUNT	CELL PATH	REMARKS
~160° C	2μl	Length 9·2cm Volume 25ml	

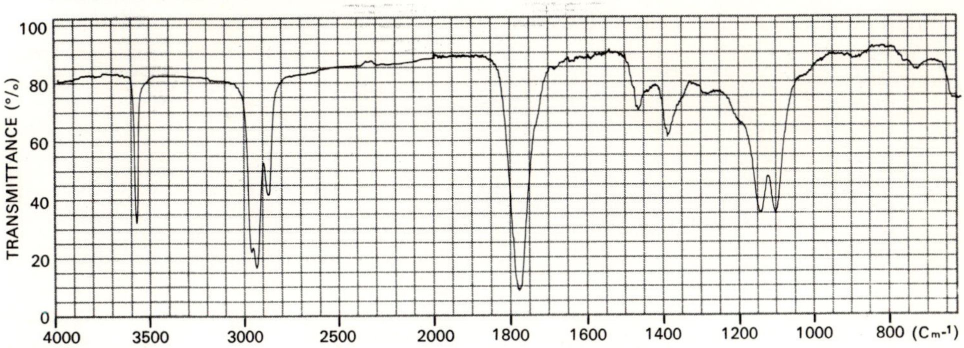

COMPOUND	MOL. FORMULA	B. PT.	INDEX No.
n-Octanoic acid	$C_8H_{16}O_2$	238°C	**82**
CELL TEMP. ~150°C	APPROX AMOUNT 2μl	CELL PATH Length 9·2cm Volume 25ml	REMARKS

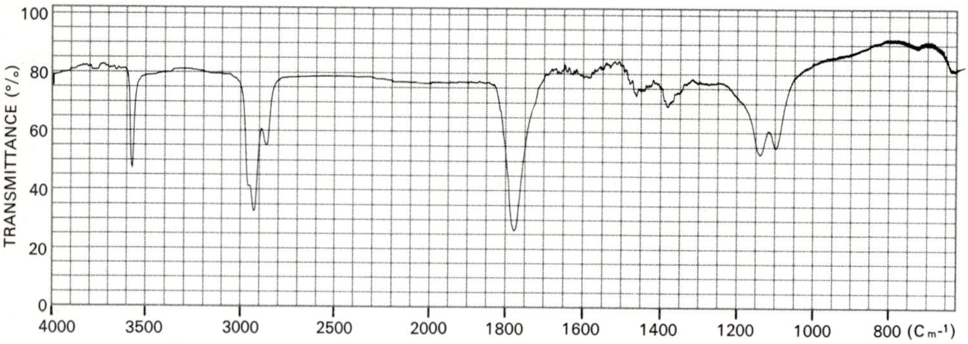

COMPOUND	MOL. FORMULA	B. PT.	INDEX No.
n-Nonanoic acid	$C_9H_{18}O_2$	255°C	**83**
CELL TEMP. ~150°C	APPROX AMOUNT 2μl	CELL PATH Length 9·2cm Volume 25ml	REMARKS

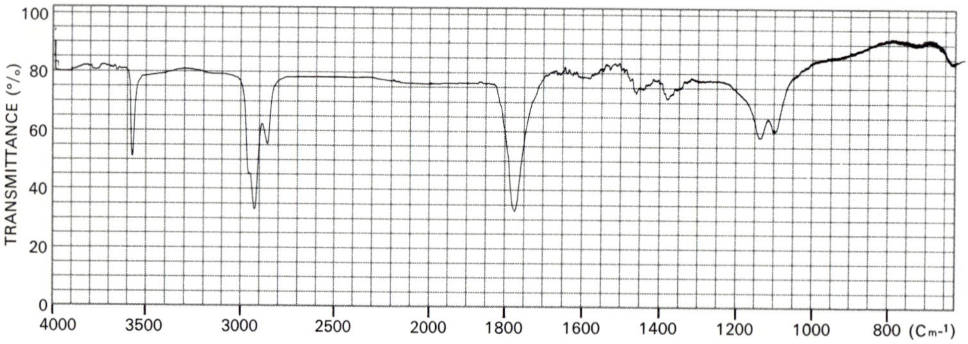

COMPOUND	MOL. FORMULA	B. PT.	INDEX No.
2-Methylpropanoic acid	$C_4H_8O_2$	154°C	**84**
CELL TEMP. ~150°C	APPROX AMOUNT 2μl	CELL PATH Length 9·2cm Volume 25ml	REMARKS

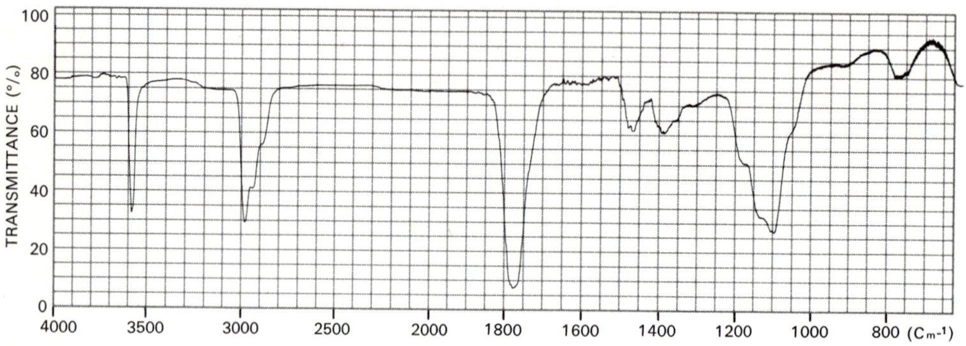

COMPOUND	MOL. FORMULA	B. PT.	INDEX No.
3-Methylbutanoic acid	$C_5H_{10}O_2$	177°C	**85**

CELL TEMP.	APPROX AMOUNT	CELL PATH	REMARKS
~150°C	2μl	Length 9·2cm Volume 25ml	

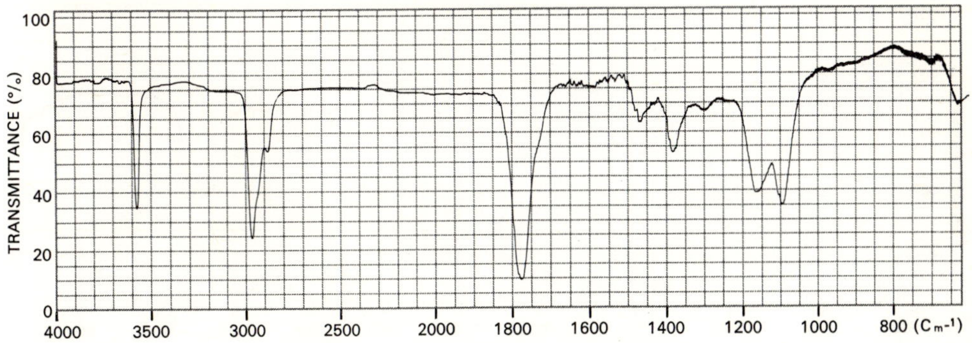

COMPOUND	MOL. FORMULA	B. PT.	INDEX No.
2-Methylbutanoic acid	$C_5H_{10}O_2$	201°C	**86**

CELL TEMP.	APPROX AMOUNT	CELL PATH	REMARKS
~150°C	2μl	Length 9·2cm Volume 25ml	

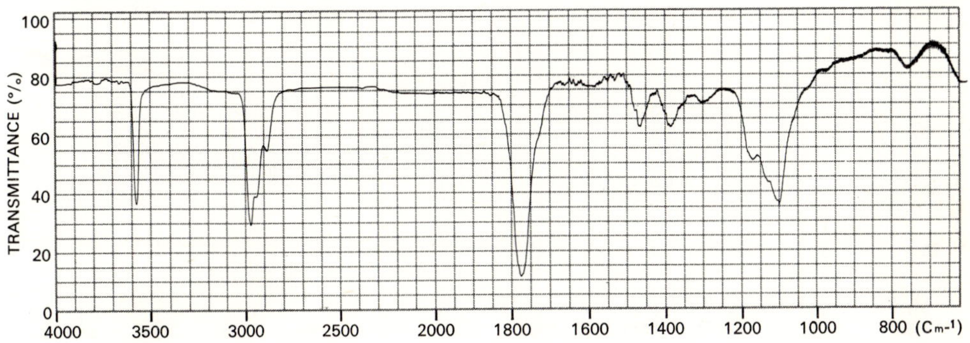

COMPOUND	MOL. FORMULA	B. PT.	INDEX No.
4-Methylpentanoic acid	$C_6H_{12}O_2$	200°C	**87**

CELL TEMP.	APPROX AMOUNT	CELL PATH	REMARKS
~150°C	2μl	Length 9·2cm Volume 25ml	

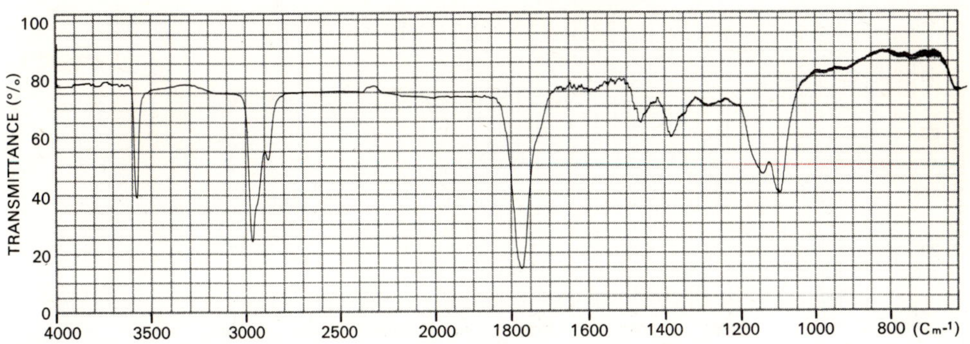

COMPOUND	MOL. FORMULA	B. PT.	INDEX No.
3-Methylpentanoic acid	$C_6H_{12}O_2$	198°C	**88**

CELL TEMP.	APPROX AMOUNT	CELL PATH	REMARKS
~150°C	2μl	Length 9·2cm Volume 25ml	

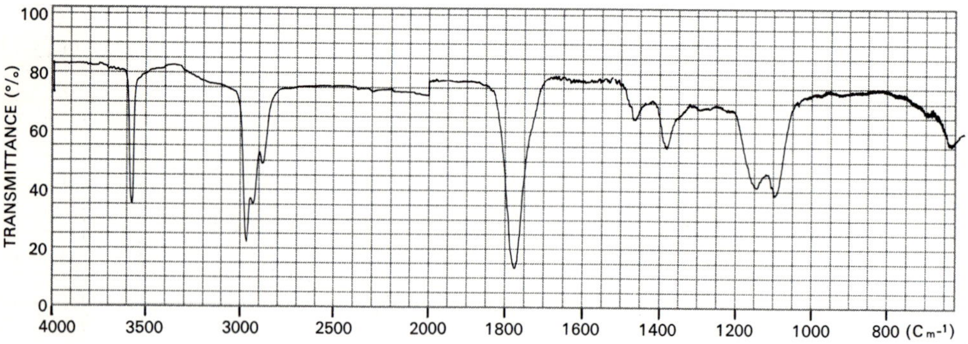

COMPOUND	MOL. FORMULA	B. PT.	INDEX No.
2-Ethylbutanoic acid	$C_6H_{12}O_2$	194°C	**89**

CELL TEMP.	APPROX AMOUNT	CELL PATH	REMARKS
~150°C	2μl	Length 9·2cm Volume 25ml	

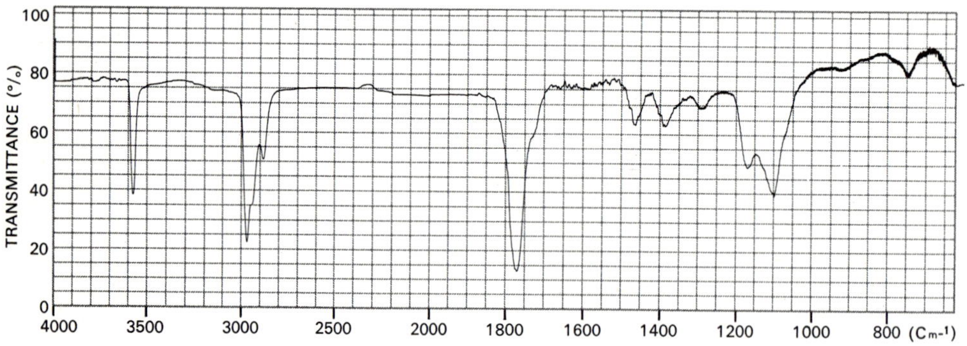

COMPOUND	MOL. FORMULA	B. PT.	INDEX No.
2-Ethylhexanoic acid	$C_8H_{16}O_2$	228°C	**90**

CELL TEMP.	APPROX AMOUNT	CELL PATH	REMARKS
~150°C	2μl	Length 9·2cm Volume 25ml	

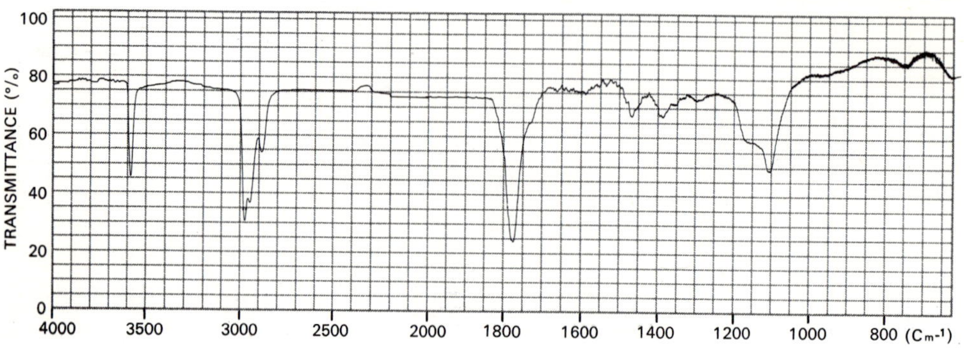

COMPOUND	MOL. FORMULA	B. PT.	INDEX No.
But-2-enoic acid	$C_4H_6O_2$	185°C	**91**

CELL TEMP.	APPROX AMOUNT	CELL PATH	REMARKS
~140°C	A few mg	Length 9·2cm Volume 25ml	

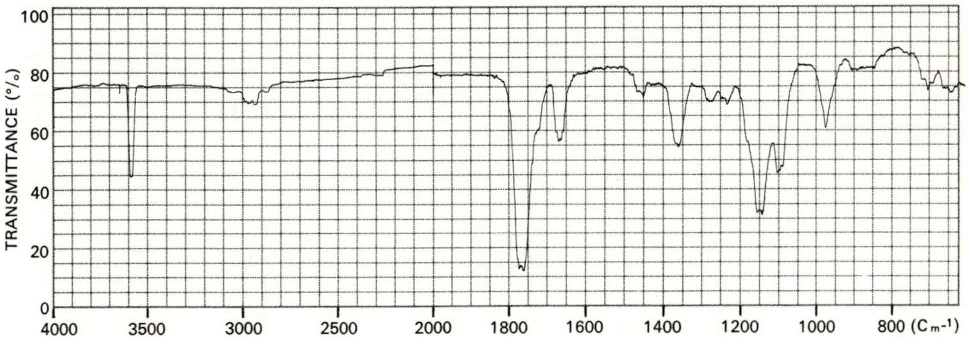

COMPOUND	MOL. FORMULA	B. PT.	INDEX No.
Hex-2-enoic acid	$C_6H_{10}O_2$	-	**92**

CELL TEMP.	APPROX AMOUNT	CELL PATH	REMARKS
~150°C	2μl	Length 9·2cm Volume 25ml	

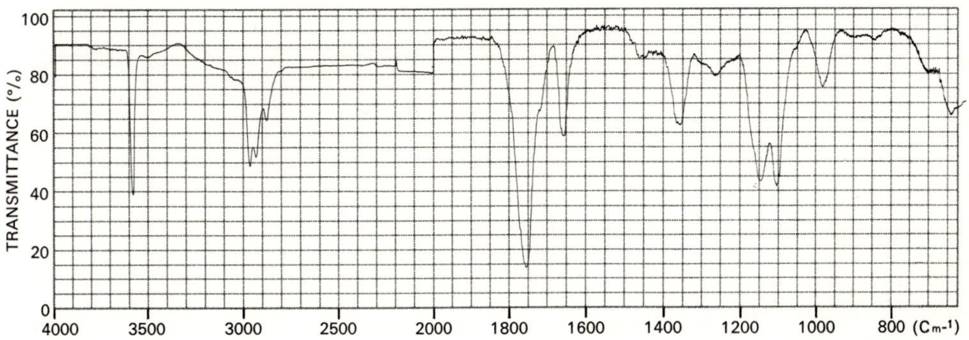

COMPOUND	MOL. FORMULA	B. PT.	INDEX No.
Hept-2-enoic acid	$C_7H_{12}O_2$	-	**93**

CELL TEMP.	APPROX AMOUNT	CELL PATH	REMARKS
~150°C	2μl	Length 9·2cm Volume 25ml	

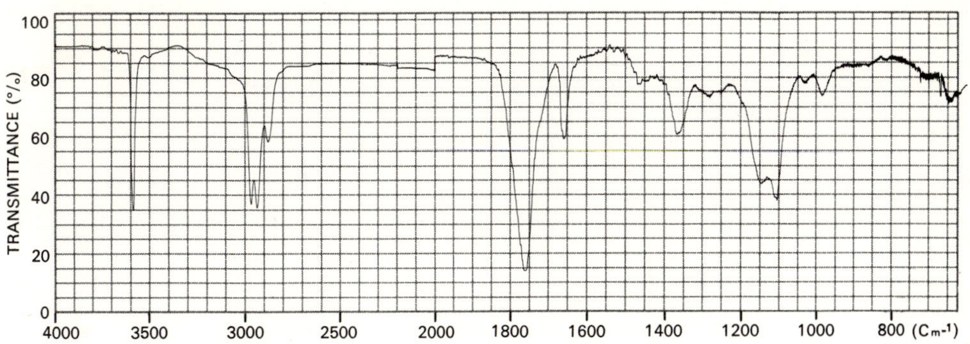

COMPOUND	MOL. FORMULA	B. PT.	INDEX No.
Oct-2-enoic acid	$C_8H_{14}O_2$	135-136° C/12mm	**94**

CELL TEMP.	APPROX AMOUNT	CELL PATH	REMARKS
~160°C	2μl	Length 9·2cm Volume 25ml	

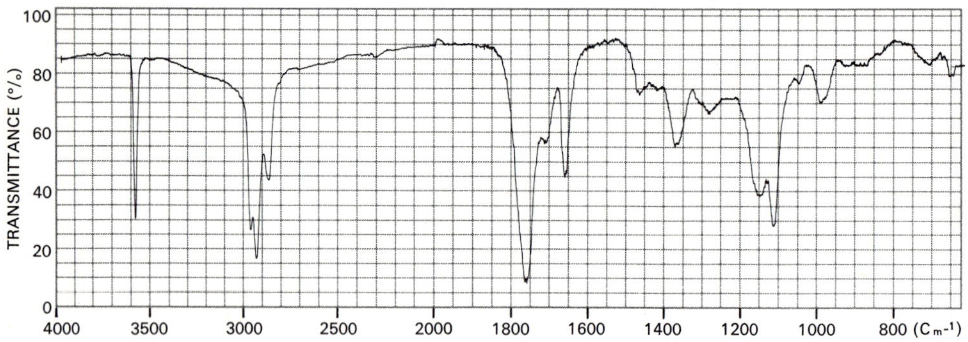

COMPOUND	MOL. FORMULA	B. PT.	INDEX No.
Non-2-enoic acid	$C_9H_{16}O_2$	136-139° C/5mm	**95**

CELL TEMP.	APPROX AMOUNT	CELL PATH	REMARKS
~150°C	2μl	Length 9·2cm Volume 25ml	

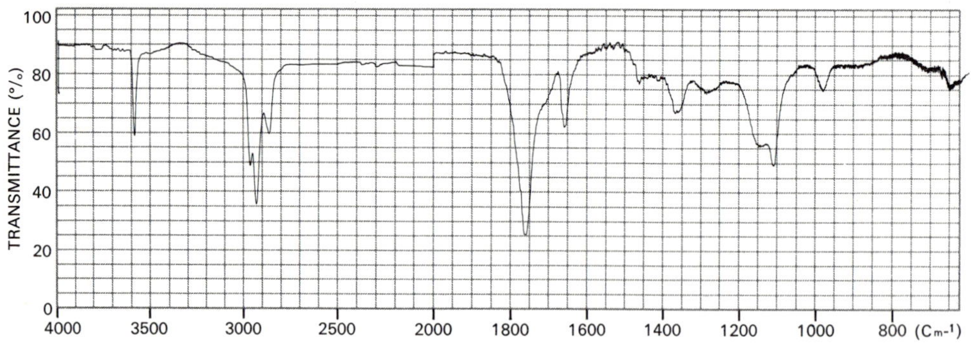

COMPOUND	MOL. FORMULA	B. PT.	INDEX No.
Dec-2-enoic acid	$C_{10}H_{18}O_2$	-	**96**

CELL TEMP.	APPROX AMOUNT	CELL PATH	REMARKS
~150°C	2μl	Length 9·2cm Volume 25ml	

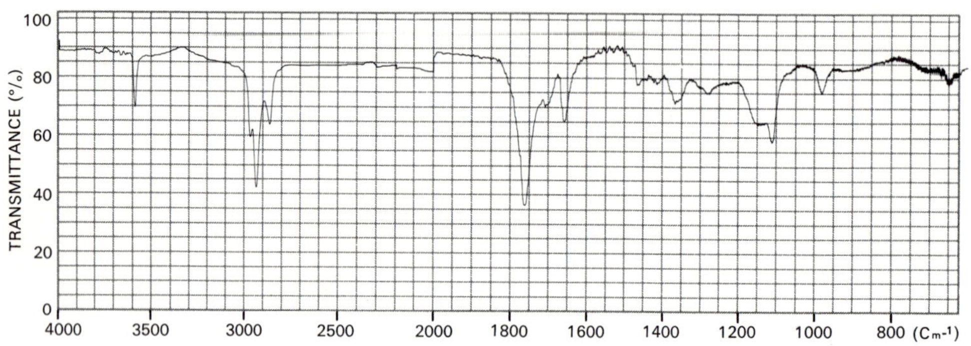

COMPOUND	MOL. FORMULA	B. PT.	INDEX No.
Hexa-2, 4-dienoic acid	$C_6H_8O_2$	-	**97**

CELL TEMP.	APPROX AMOUNT	CELL PATH	REMARKS
~145°C	A few mg	Length 9·2cm Volume 25ml	

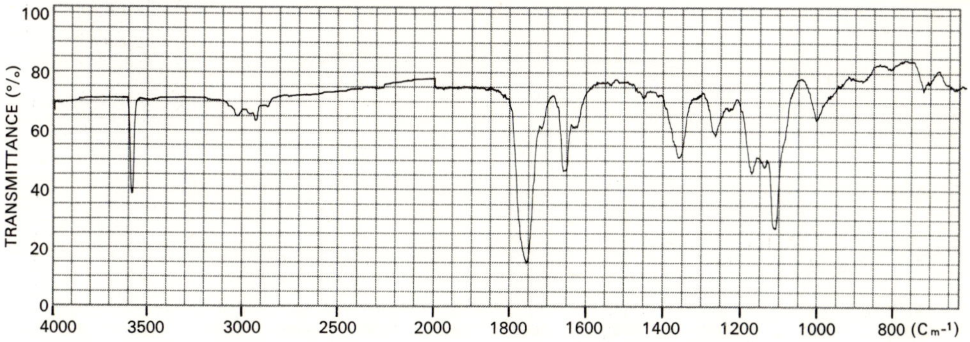

COMPOUND	MOL. FORMULA	B. PT.	INDEX No.
2-Methylprop-2-enoic acid(Methacrylic acid)	$C_4H_6O_2$	162°C	**98**

CELL TEMP.	APPROX AMOUNT	CELL PATH	REMARKS
~140°C	0·5μl	Length 9·2cm Volume 25ml	

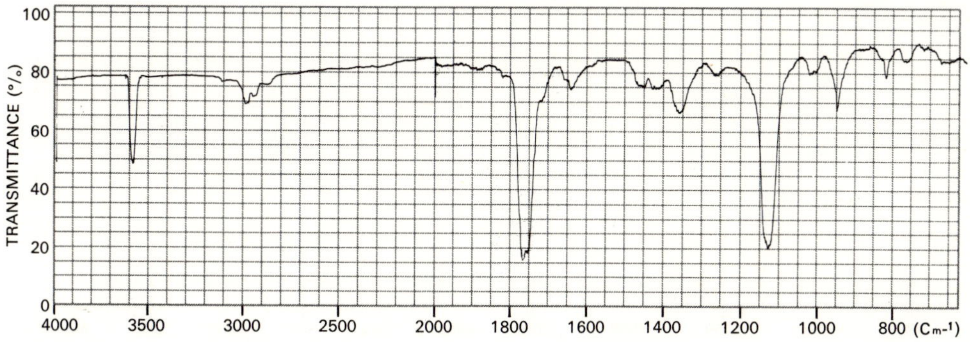

COMPOUND	MOL. FORMULA	B. PT.	INDEX No.
3-Methylbut-2-enoic acid	$C_5H_8O_2$	-	**99**

CELL TEMP.	APPROX AMOUNT	CELL PATH	REMARKS
~130°C	A few mg	Length 9·2cm Volume 25ml	

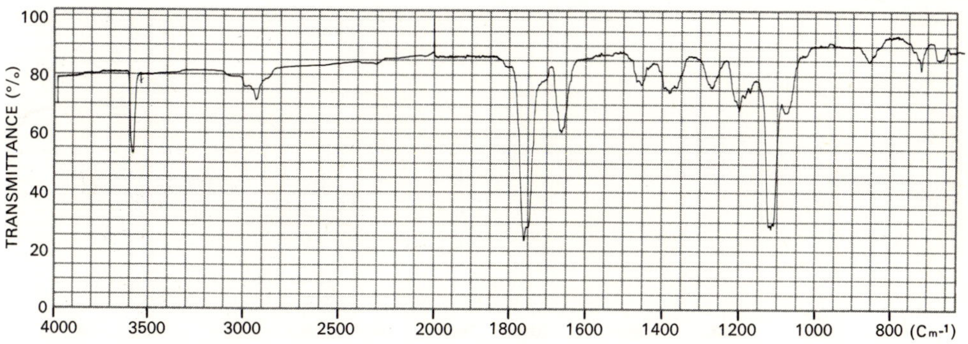

COMPOUND	MOL. FORMULA	B. PT.	INDEX No.
2-Methylbut-2-enoic acid (Tiglic acid)	$C_5H_8O_2$	199°C	**100**

CELL TEMP.	APPROX AMOUNT	CELL PATH	REMARKS
~140°C	A few mg	Length 9·2cm Volume 25ml	

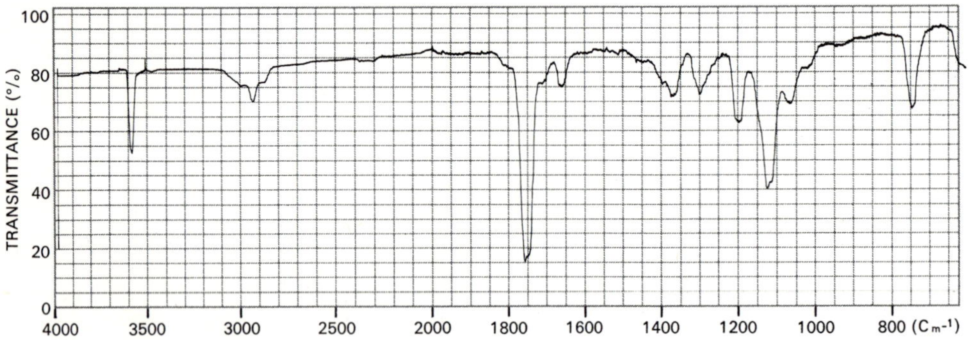

COMPOUND	MOL. FORMULA	B. PT.	INDEX No.
But-3-enoic acid	$C_4H_6O_2$	169°C	**101**

CELL TEMP.	APPROX AMOUNT	CELL PATH	REMARKS
~150°C	2μl	Length 9·2cm Volume 25ml	

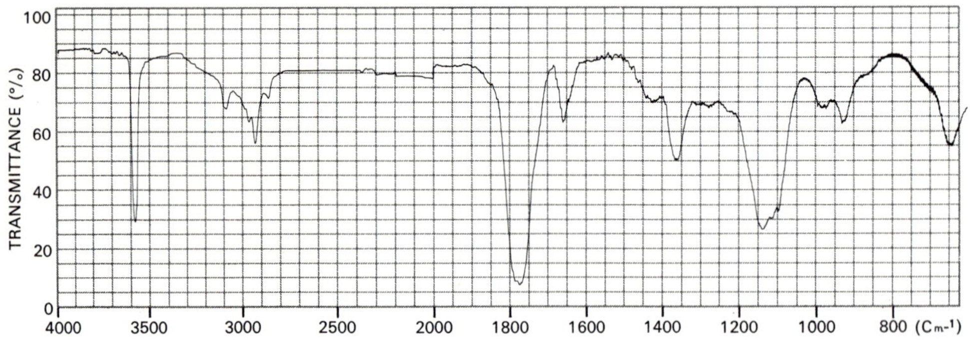

COMPOUND	MOL. FORMULA	B. PT.	INDEX No.
Hex-3-enoic acid	$C_6H_{10}O_2$	208°C	**102**

CELL TEMP.	APPROX AMOUNT	CELL PATH	REMARKS
~150°C	2μl	Length 9·2cm Volume 25ml	

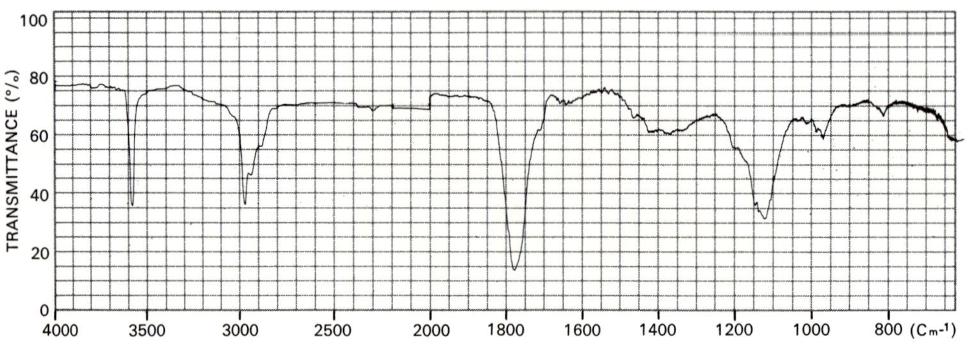

COMPOUND	MOL. FORMULA	B. PT.	INDEX No.
Oct-3-enoic acid	$C_8H_{14}O_2$	-	**103**

CELL TEMP.	APPROX AMOUNT	CELL PATH	REMARKS
~150°C	2µl	Length 9·2cm Volume 25ml	

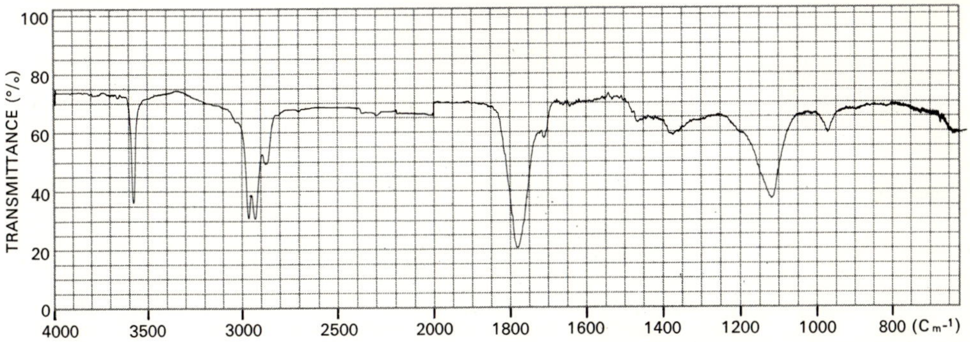

COMPOUND	MOL. FORMULA	B. PT.	INDEX No.
Undec-10-enoic acid	$C_{11}H_{20}O_2$	275° C	**104**

CELL TEMP.	APPROX AMOUNT	CELL PATH	REMARKS
155° C	2µl	Length 9·2cm Volume 25ml	

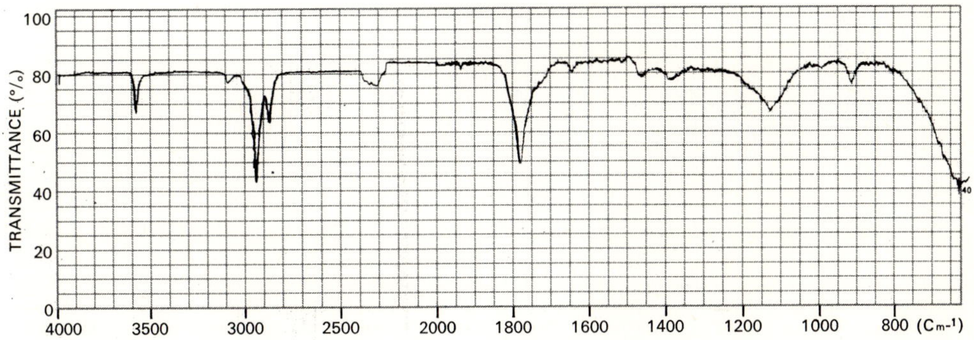

COMPOUND	MOL. FORMULA	B. PT.	INDEX No.
Benzoic acid	$C_7H_6O_2$	249° C	**105**

CELL TEMP.	APPROX AMOUNT	CELL PATH	REMARKS
~140°C	A few mg	Length 9·2cm Volume 25ml	

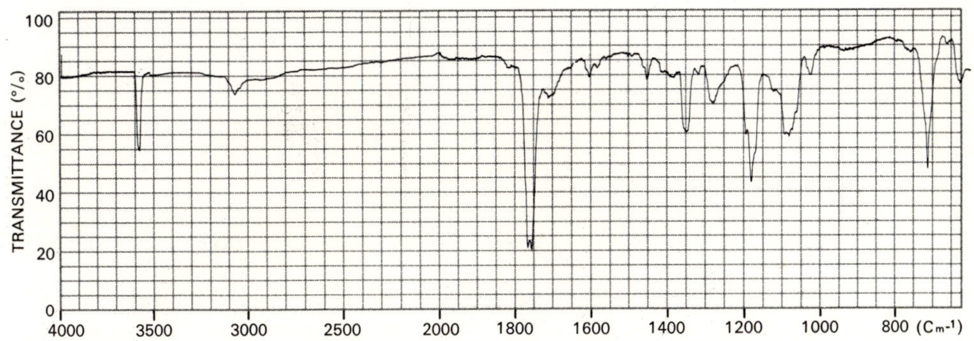

COMPOUND	MOL. FORMULA	B. PT.	INDEX No.
2-Oxo-propanoic acid (Pyruvic acid)	$C_3H_4O_3$	-	**106**

CELL TEMP.	APPROX AMOUNT	CELL PATH	REMARKS
~95°C	2μl	Length 9·2cm Volume 25ml	* Decomposing in cell (CO_2)

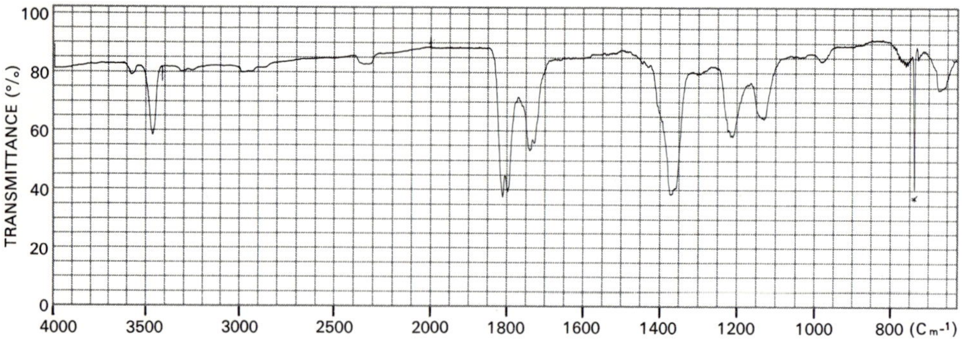

COMPOUND	MOL. FORMULA	B. PT.	INDEX No.
2-Hydroxypropanoic acid (Lactic acid)	$C_3H_6O_3$	-	**107**

CELL TEMP.	APPROX AMOUNT	CELL PATH	REMARKS
~145°C	1μl	Length 9·2cm Volume 25ml	

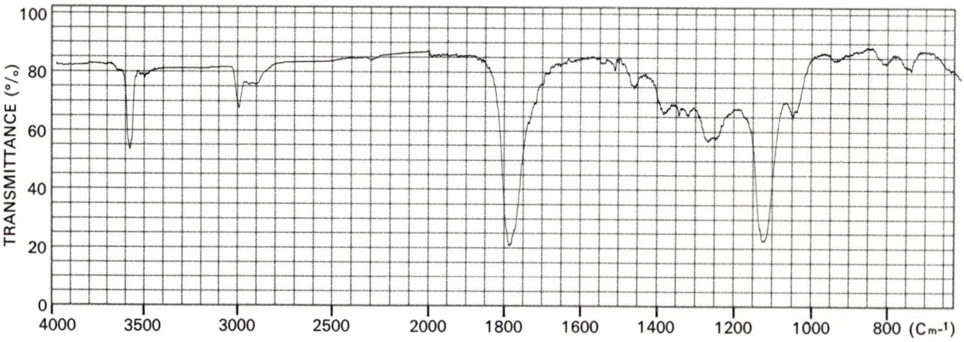

COMPOUND	MOL. FORMULA	B. PT.	INDEX No.
1-Aminoethane	C_2H_7N	17°C	**108**

CELL TEMP.	APPROX AMOUNT	CELL PATH	REMARKS
60°C	Unknown	Length 9·2cm Volume 25ml	

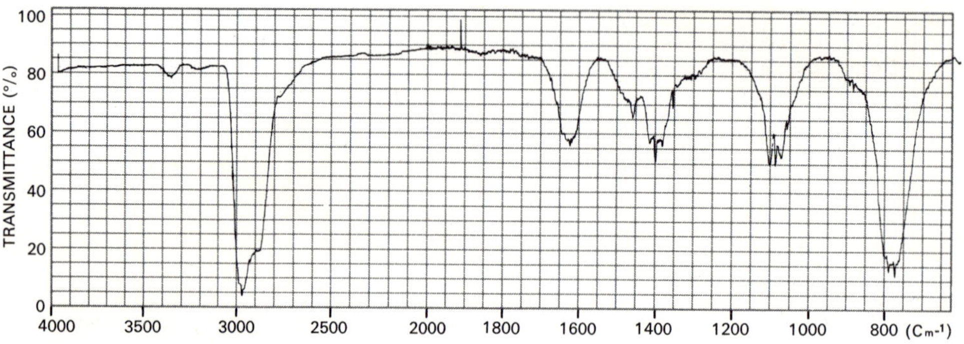

COMPOUND	MOL. FORMULA	B. PT.	INDEX No.
1-Aminopropane	C_3H_9N	49°C	**109**

CELL TEMP.	APPROX AMOUNT	CELL PATH	REMARKS
~130°C	4µl	Length 9·2cm Volume 25ml	

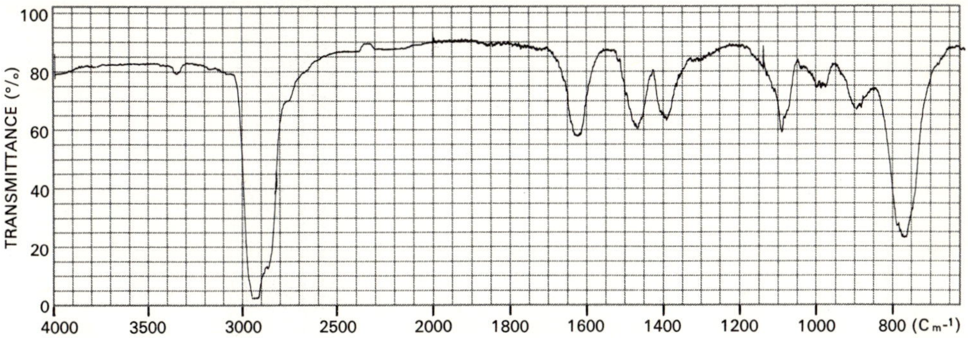

COMPOUND	MOL. FORMULA	B. PT.	INDEX No.
1-Aminobutane	$C_4H_{11}N$	78°C	**110**

CELL TEMP.	APPROX AMOUNT	CELL PATH	REMARKS
80°C	4µl	Length 9·2cm Volume 25ml	

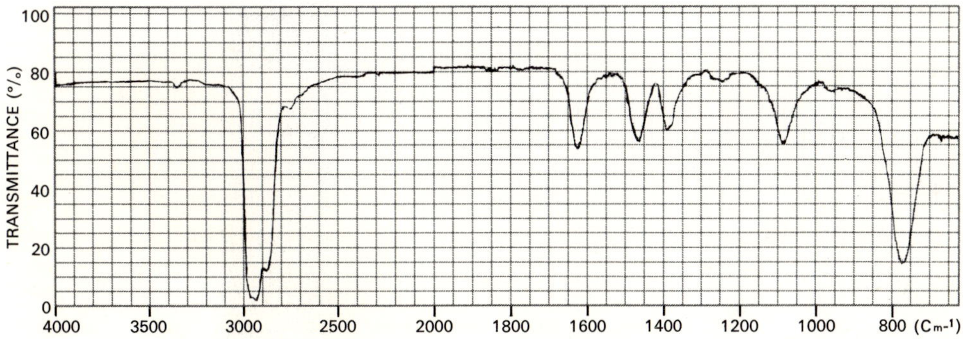

COMPOUND	MOL. FORMULA	B. PT.	INDEX No.
1-Aminopentane	$C_5H_{13}N$	103°C	**111**.

CELL TEMP.	APPROX AMOUNT	CELL PATH	REMARKS
~130°C	4µl	Length 9·2cm Volume 25ml	

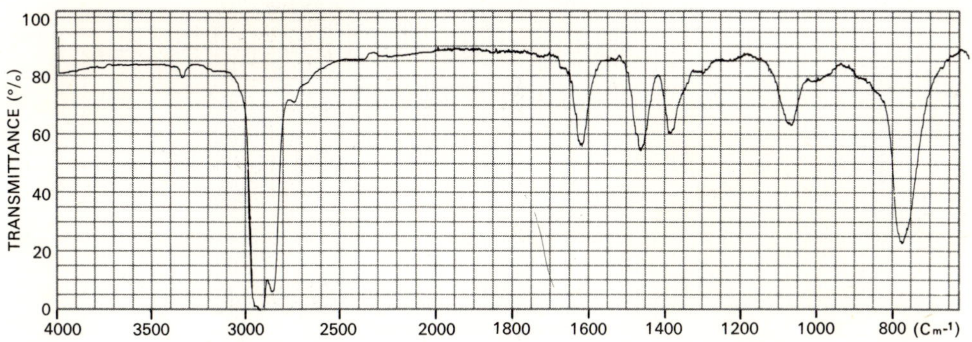

COMPOUND	MOL. FORMULA	B. PT.	INDEX No.
1-Aminohexane	$C_6H_{15}N$	130° C	**112**

CELL TEMP.	APPROX AMOUNT	CELL PATH	REMARKS
80° C	4μl	Length 9·2cm Volume 25ml	

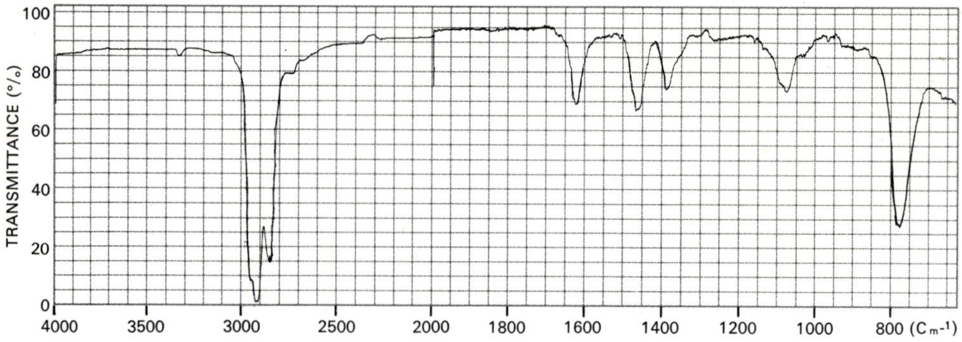

COMPOUND	MOL. FORMULA	B. PT.	INDEX No.
1-Aminoheptane	$C_7H_{17}N$	155° C	**113**

CELL TEMP.	APPROX AMOUNT	CELL PATH	REMARKS
~125°C	5μl	Length 9·2cm Volume 25ml	

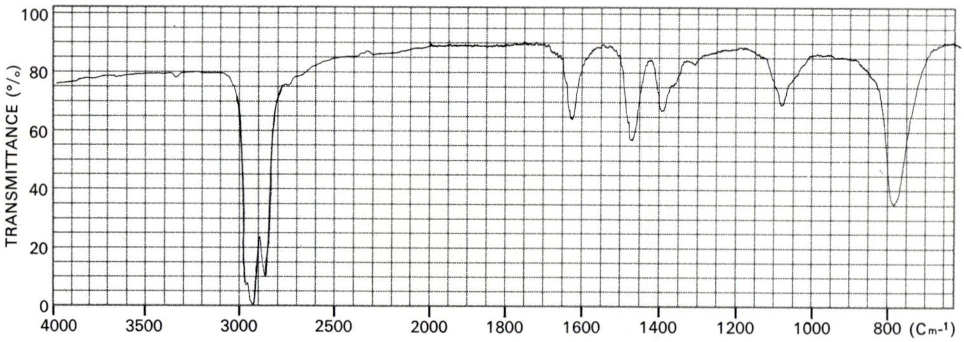

COMPOUND	MOL. FORMULA	B. PT.	INDEX No.
1-Aminooctane	$C_8H_{19}N$	180° C	**114**

CELL TEMP.	APPROX AMOUNT	CELL PATH	REMARKS
~125°C	4μl	Length 9·2cm Volume 25ml	

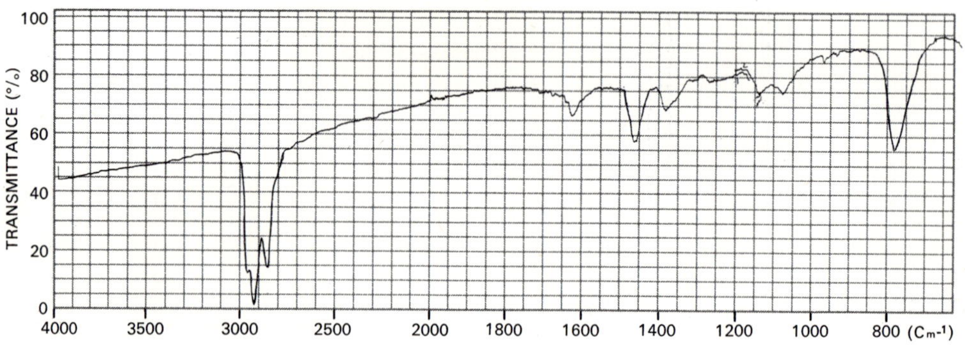

COMPOUND	MOL. FORMULA	B. PT.	INDEX No.
1, 1, 3, 3-Tetramethyl-1-aminobutane	$C_8H_{19}N$	137-143°C	**115**

CELL TEMP.	APPROX AMOUNT	CELL PATH	REMARKS
~130°C	2μl	Length 9·2cm Volume 25ml	

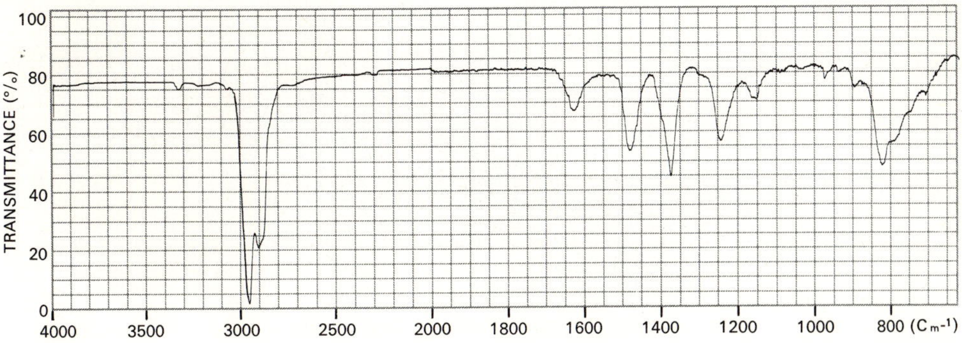

COMPOUND	MOL. FORMULA	B. PT.	INDEX No.
1-Aminononane	$C_9H_{21}N$	202°C	**116**

CELL TEMP.	APPROX AMOUNT	CELL PATH	REMARKS
~125°C	4μl	Length 9·2cm Volume 25ml	

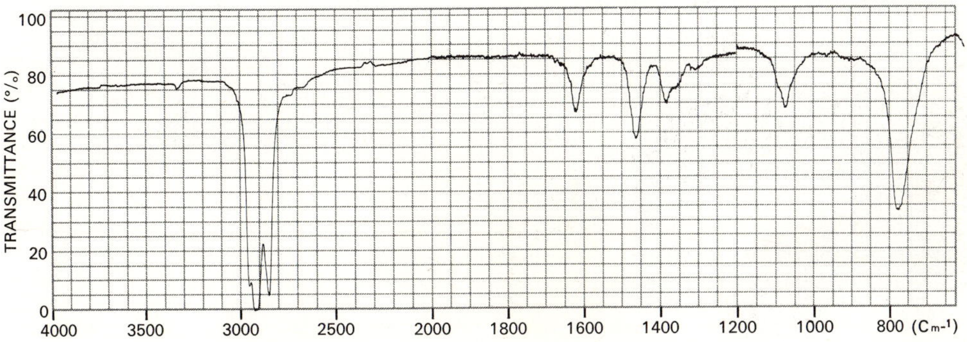

COMPOUND	MOL. FORMULA	B. PT.	INDEX No.
Piperidine	$C_5H_{11}N$	106°C	**117**

CELL TEMP.	APPROX AMOUNT	CELL PATH	REMARKS
~120°C	2μl	Length 9·2cm Volume 25ml	

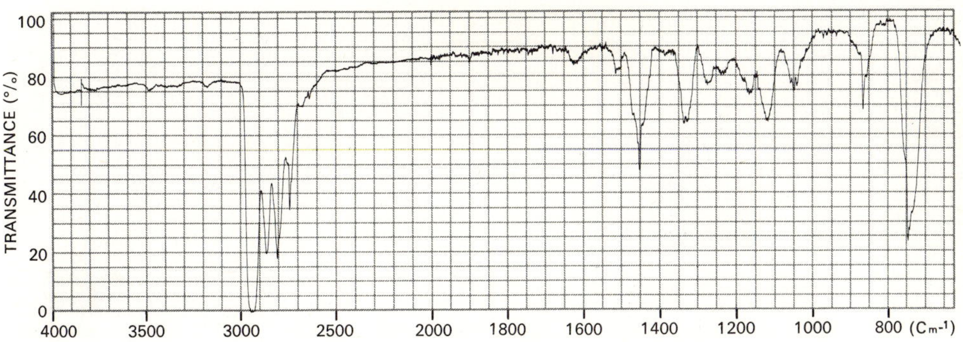

COMPOUND	MOL. FORMULA	B. PT.	INDEX No.
1-Aminoprop-2-ene	C_3H_7N	58° C	**118**

CELL TEMP.	APPROX AMOUNT	CELL PATH	REMARKS
~110°C	3μl	Length 9·2cm Volume 25ml	

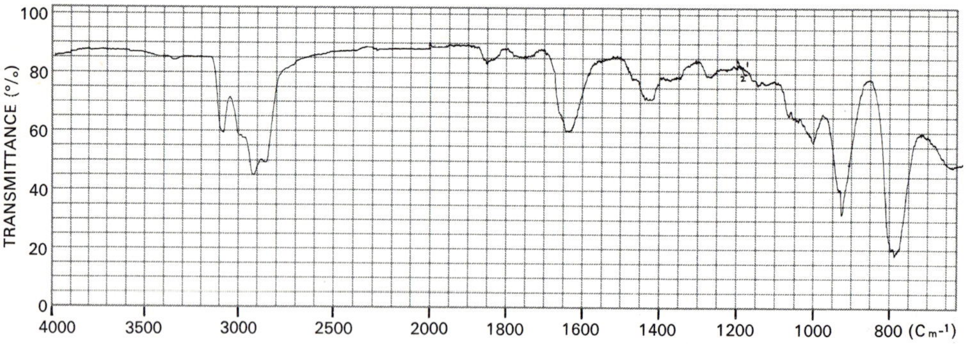

COMPOUND	MOL. FORMULA	B. PT.	INDEX No.
Benzylamine	C_7H_9N	185° C	**119**

CELL TEMP.	APPROX AMOUNT	CELL PATH	REMARKS
~110° C	4μl	Length 9·2cm Volume 25ml	

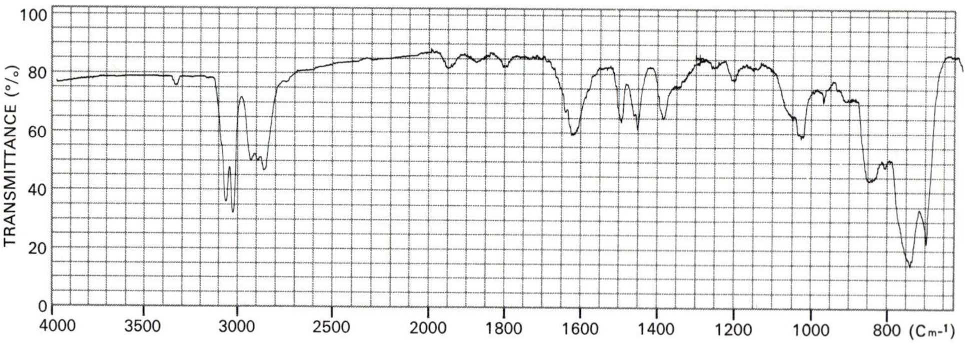

COMPOUND	MOL. FORMULA	B. PT.	INDEX No.
2-Aminopropane	C_3H_9N	33° C	**120**

CELL TEMP.	APPROX AMOUNT	CELL PATH	REMARKS
~75° C	4μl	Length 9·2cm Volume 25ml	

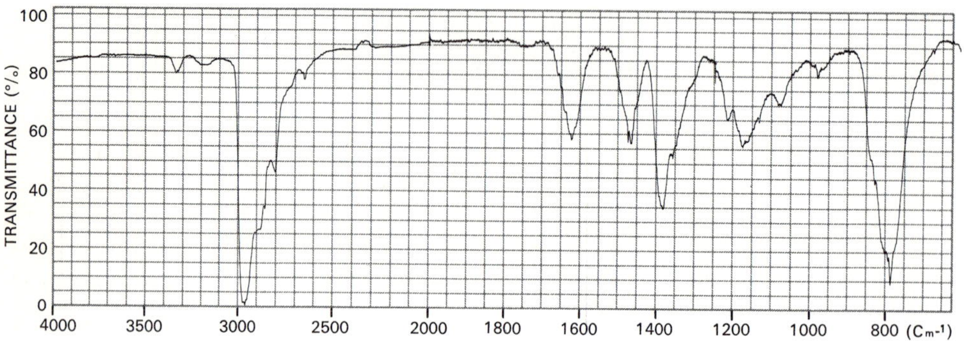

COMPOUND	MOL. FORMULA	B. PT.	INDEX No.
2-Aminobutane	$C_4H_{11}N$	67°C	**121**

CELL TEMP.	APPROX AMOUNT	CELL PATH	REMARKS
~130°C	4μl	Length 9·2cm Volume 25ml	

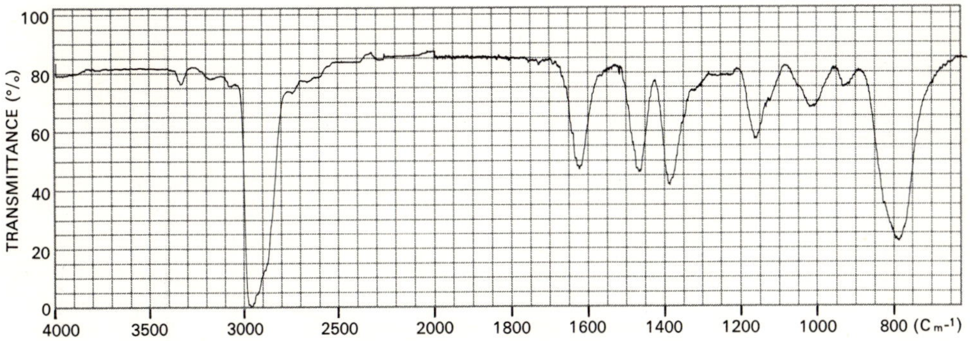

COMPOUND	MOL. FORMULA	B. PT.	INDEX No.
2-Aminopentane	$C_5H_{13}N$	-	**122**

CELL TEMP.	APPROX AMOUNT	CELL PATH	REMARKS
~130°C	4μl	Length 9·2cm Volume 25ml	

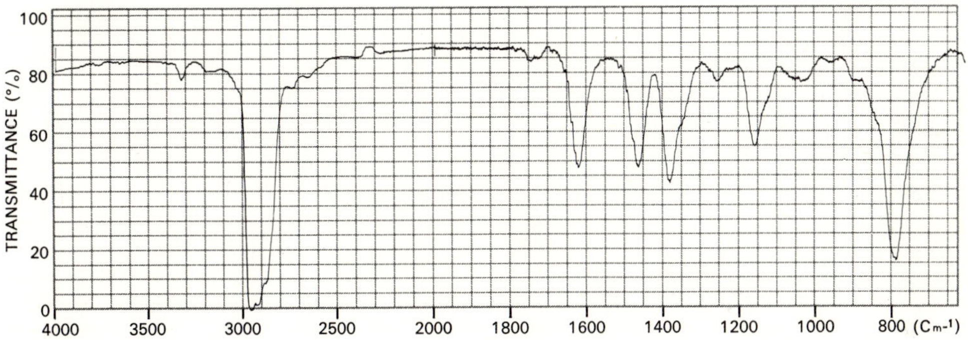

COMPOUND	MOL. FORMULA	B. PT.	INDEX No.
2-Aminooctane	$C_8H_{19}N$	164°C	**123**

CELL TEMP.	APPROX AMOUNT	CELL PATH	REMARKS
~125°C	3μl	Length 9·2cm Volume 25ml	

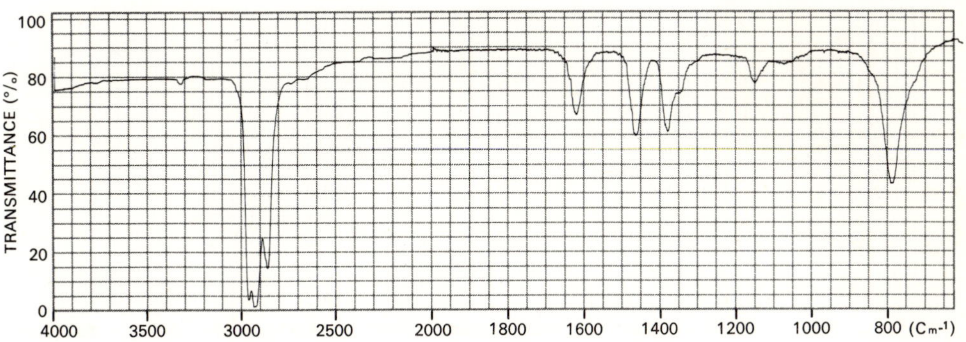

COMPOUND	MOL. FORMULA	B. PT.	INDEX No.
3-Aminopentane	$C_5H_{13}N$	90°C	**124**

CELL TEMP.	APPROX AMOUNT	CELL PATH	REMARKS
~140°C	4μl	Length 9·2cm Volume 25ml	

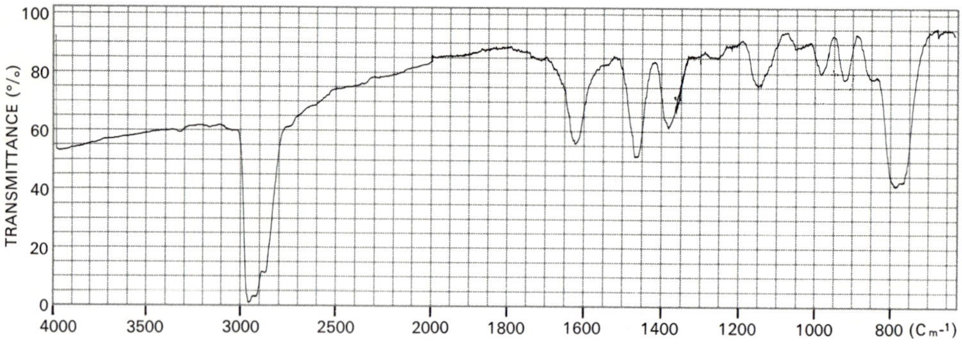

COMPOUND	MOL. FORMULA	B. PT.	INDEX No.
Diethylamine	$C_4H_{11}N$	56°C	**125**

CELL TEMP.	APPROX AMOUNT	CELL PATH	REMARKS
~130°C	4μl	Length 9·2cm Volume 25ml	

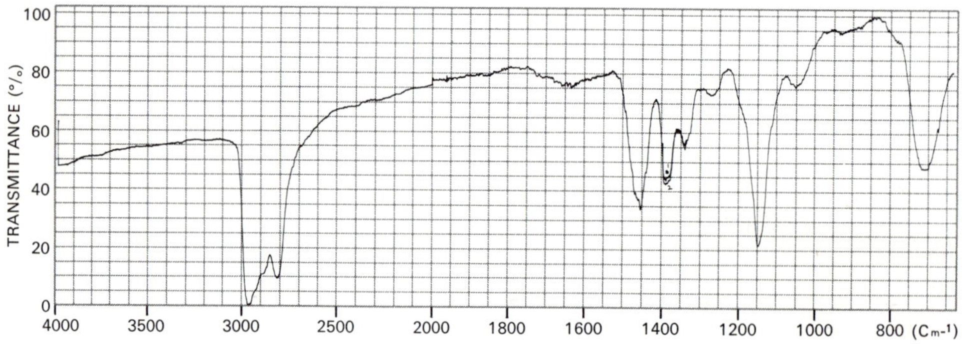

COMPOUND	MOL. FORMULA	B. PT.	INDEX No.
Dipropylamine	$C_6H_{15}N$	110°C	**126**

CELL TEMP.	APPROX AMOUNT	CELL PATH	REMARKS
~135°C	4μl	Length 9·2cm Volume 25ml	

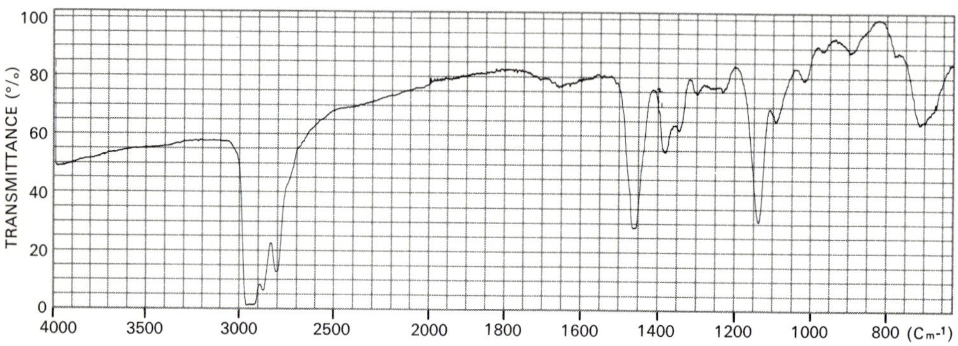

COMPOUND	MOL. FORMULA	B. PT.	INDEX No.
Dibutylamine	$C_8H_{19}N$	159°C	**127**

CELL TEMP.	APPROX AMOUNT	CELL PATH	REMARKS
~135°C	4µl	Length 9·2cm Volume 25ml	

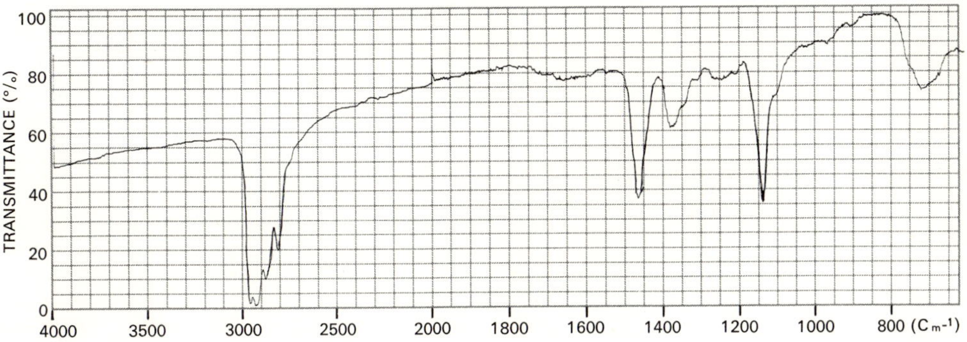

COMPOUND	MOL. FORMULA	B. PT.	INDEX No.
Dipentylamine	$C_{10}H_{23}N$	203°C	**128**

CELL TEMP.	APPROX AMOUNT	CELL PATH	REMARKS
~145°C	4µl	Length 9·2cm Volume 25ml	

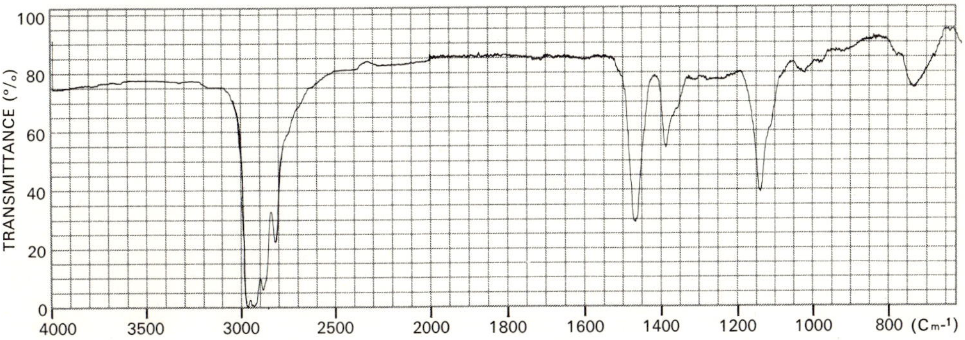

COMPOUND	MOL. FORMULA	B. PT.	INDEX No.
Dihexylamine	$C_{12}H_{27}N$	192-195°C	**129**

CELL TEMP.	APPROX AMOUNT	CELL PATH	REMARKS
~145°C	5µl	Length 9·2cm Volume 25ml	

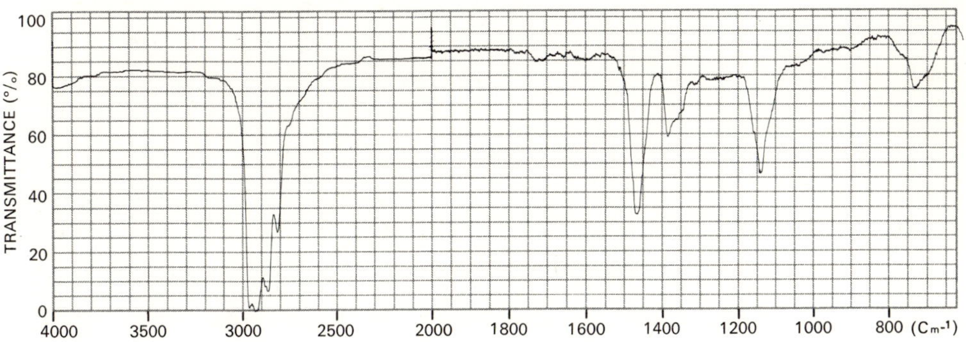

COMPOUND	MOL. FORMULA	B. PT.	INDEX No.
N-Methylbutylamine	$C_5H_{13}N$	-	**130**

CELL TEMP.	APPROX AMOUNT	CELL PATH	REMARKS
~140°C	2μl	Length 9·2cm Volume 25ml	

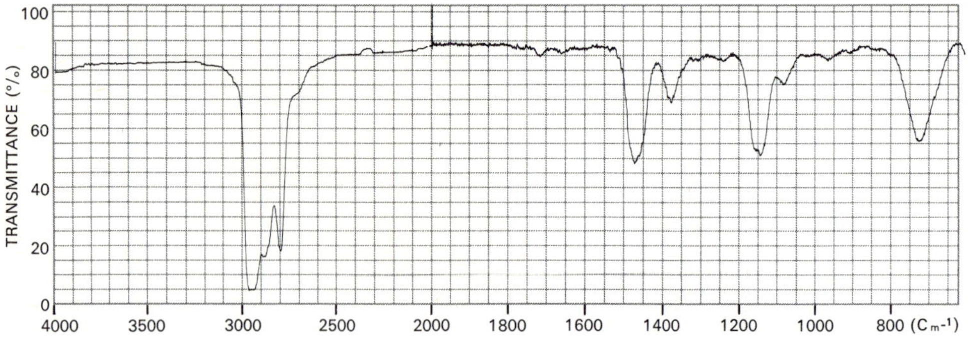

COMPOUND	MOL. FORMULA	B. PT.	INDEX No.
Tetrahydropyrrole (Pyrrolidine)	C_4H_9N	89° C	**131**

CELL TEMP.	APPROX AMOUNT	CELL PATH	REMARKS
~145°C	2μl	Length 9·2cm Volume 25ml	

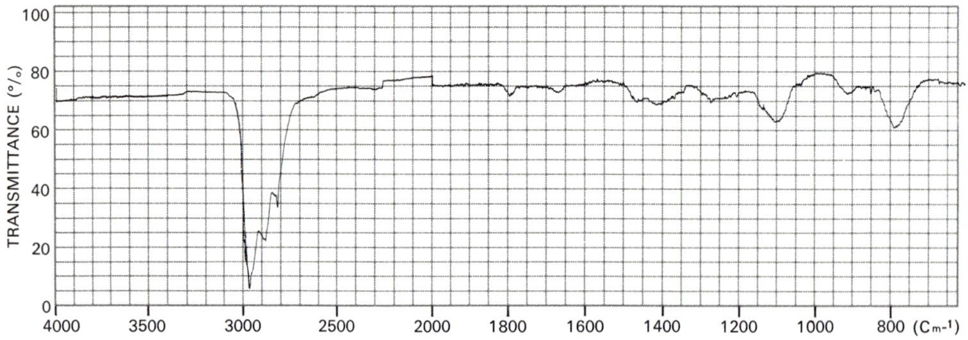

COMPOUND	MOL. FORMULA	B. PT.	INDEX No.
Aniline	C_6H_7N	184° C	**132**

CELL TEMP.	APPROX AMOUNT	CELL PATH	REMARKS
~100°C	2μl	Length 9·2cm Volume 25ml	

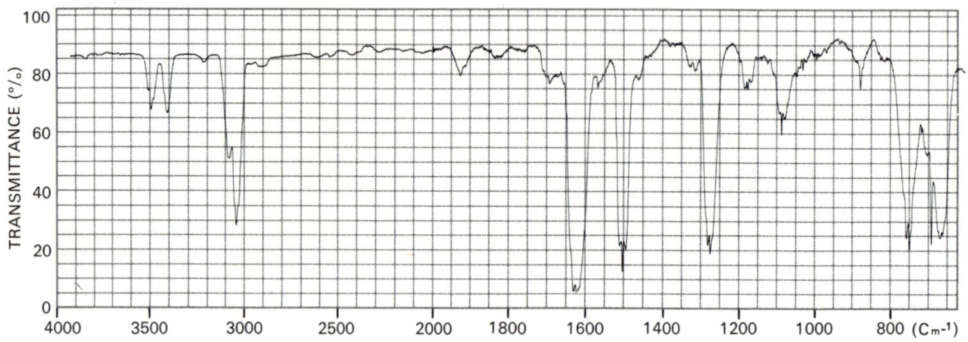

COMPOUND	MOL. FORMULA	B. PT.	INDEX No.
N-Methylaniline	C_7H_9N	196°C	**133**

CELL TEMP.	APPROX AMOUNT	CELL PATH	REMARKS
~100°C	2μl	Length 9·2cm Volume 25ml	

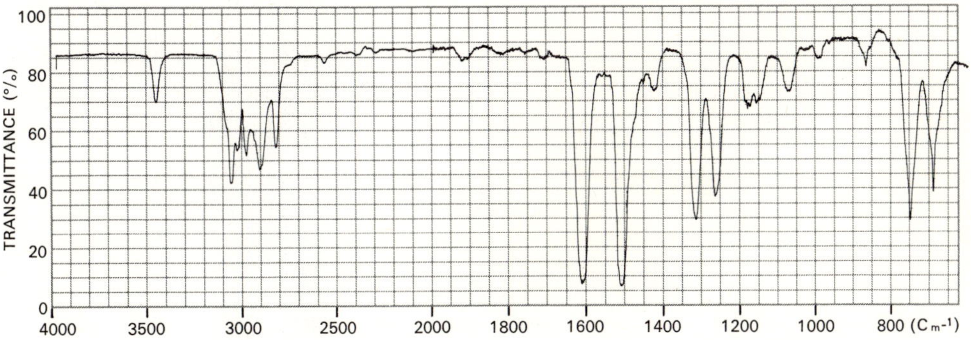

COMPOUND	MOL. FORMULA	B. PT.	INDEX No.
2-Methylaniline (o-Toluidine)	C_7H_9N	200°C	**134**

CELL TEMP.	APPROX AMOUNT	CELL PATH	REMARKS
~130°C	2μl	Length 9·2cm Volume 25ml	

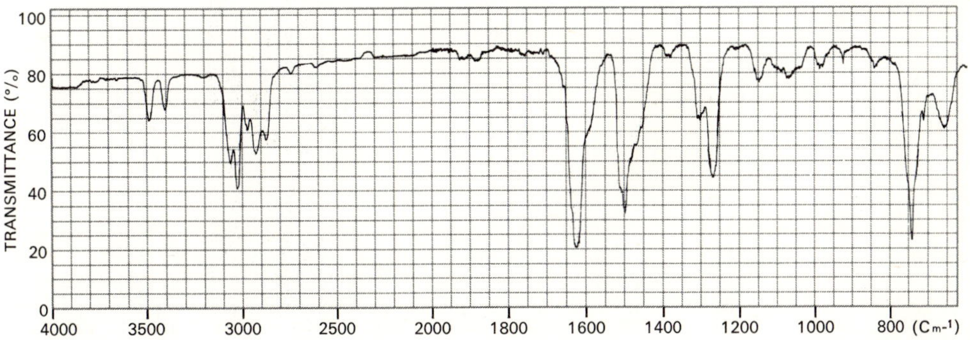

COMPOUND	MOL. FORMULA	B. PT.	INDEX No.
3-Methylaniline (m-Toluidine)	C_7H_9N	203°C	**135**

CELL TEMP.	APPROX AMOUNT	CELL PATH	REMARKS
~130°C	3μl	Length 9·2cm Volume 25ml	

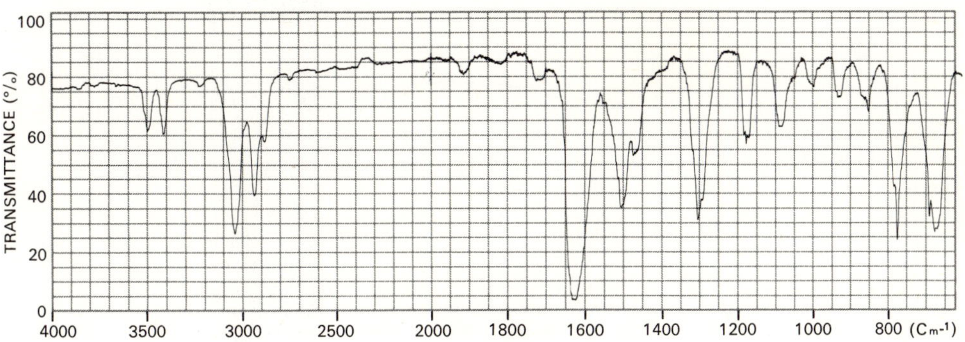

COMPOUND	MOL. FORMULA	B. PT.	INDEX No.
2,4-Dimethylaniline (2,4-Xylidene)	$C_8H_{11}N$	214°C	**136**

CELL TEMP.	APPROX AMOUNT	CELL PATH	REMARKS
~125°C	4μl	Length 9·2cm Volume 25ml	

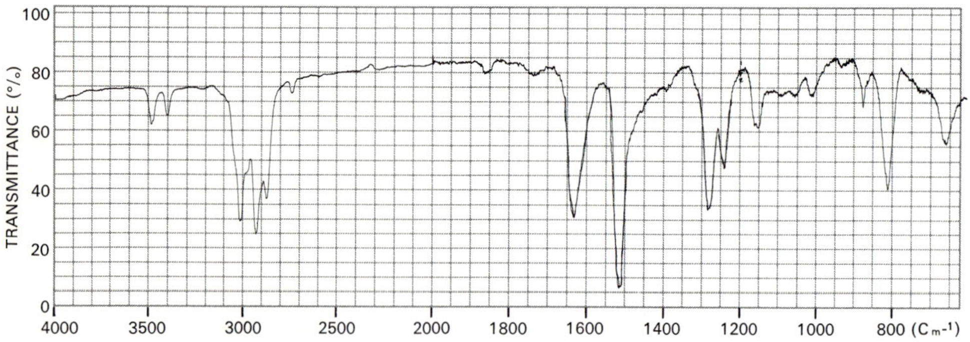

COMPOUND	MOL. FORMULA	B. PT.	INDEX No.
2-Methoxyaniline (o-Anisidine)	C_7H_9NO	224°C	**137**

CELL TEMP.	APPROX AMOUNT	CELL PATH	REMARKS
~120°C	2μl	Length 9·2cm Volume 25ml	

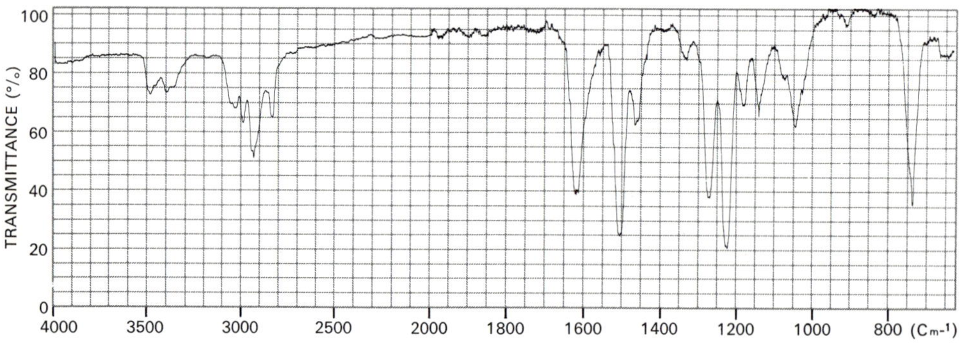

COMPOUND	MOL. FORMULA	B. PT.	INDEX No.
4-Ethoxyaniline (p-Phenetidine)	$C_8H_{11}NO$	250°C	**138**

CELL TEMP.	APPROX AMOUNT	CELL PATH	REMARKS
~135°C	3·5μl	Length 9·2cm Volume 25ml	

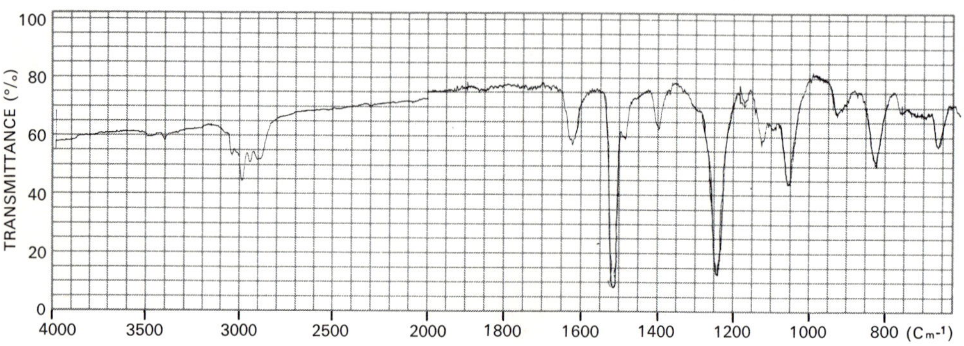

COMPOUND	MOL. FORMULA	B. PT.	INDEX No.
2-Chloroaniline	C_6H_6ClN	209°C	**139**

CELL TEMP.	APPROX AMOUNT	CELL PATH	REMARKS
~140°C	2μl	Length 9·2cm Volume 25ml	

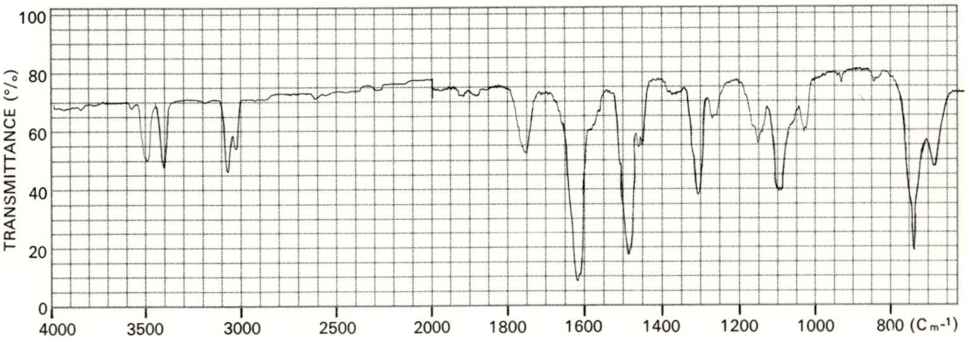

COMPOUND	MOL. FORMULA	B. PT.	INDEX No.
m-Chloroaniline	C_6H_6ClN	230°C	**140**

CELL TEMP.	APPROX AMOUNT	CELL PATH	REMARKS
~140°C	2μl	Length 9·2cm Volume 25ml	

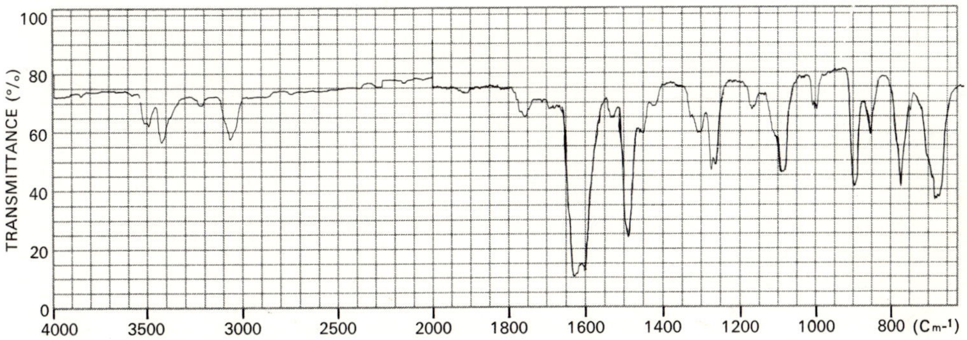

COMPOUND	MOL. FORMULA	B. PT.	INDEX No.
Pyrrole	C_4H_5N	131°C	**141**

CELL TEMP.	APPROX AMOUNT	CELL PATH	REMARKS
~150°C	3μl	Length 9·2cm Volume 25ml	

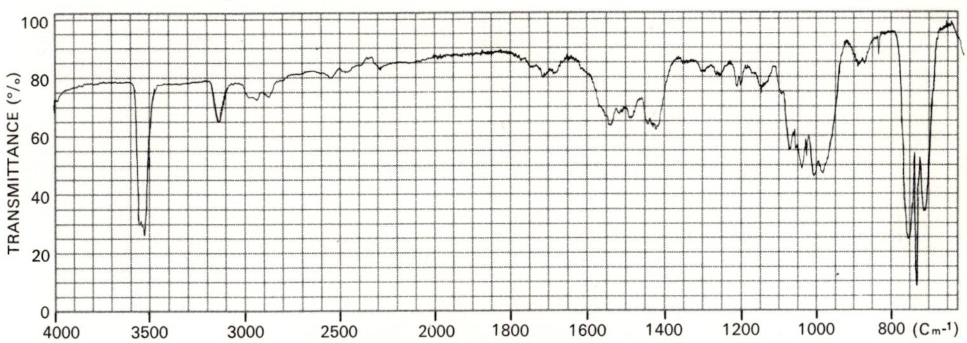

COMPOUND	MOL. FORMULA	B. PT.	INDEX No.
Acetamide	C_2H_5NO	221°C	**142**

CELL TEMP.	APPROX AMOUNT	CELL PATH	REMARKS
~140°C	1mg	Length 9·2cm Volume 25ml	

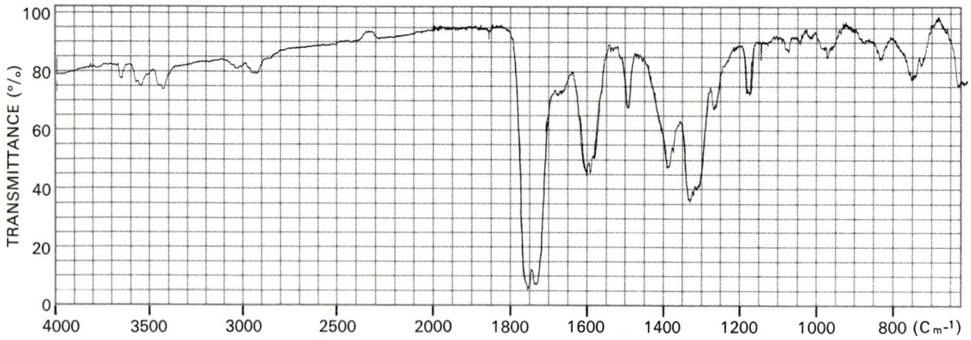

COMPOUND	MOL. FORMULA	B. PT.	INDEX No.
n-Propionamide	C_3H_7NO	213°C	**143**

CELL TEMP.	APPROX AMOUNT	CELL PATH	REMARKS
~130°C	A few mg	Length 9·2cm Volume 25ml	

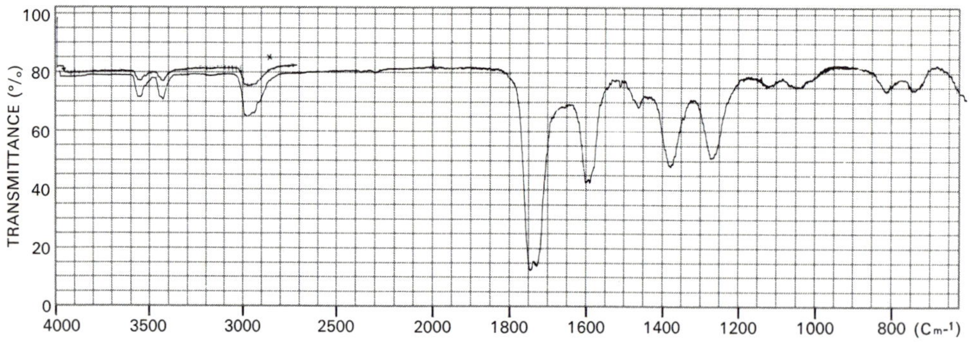

COMPOUND	MOL. FORMULA	B. PT.	INDEX No.
n-Butyramide	C_4H_9NO	216°C	**144**

CELL TEMP.	APPROX AMOUNT	CELL PATH	REMARKS
~140°C	A few mg	Length 9·2cm Volume 25ml	

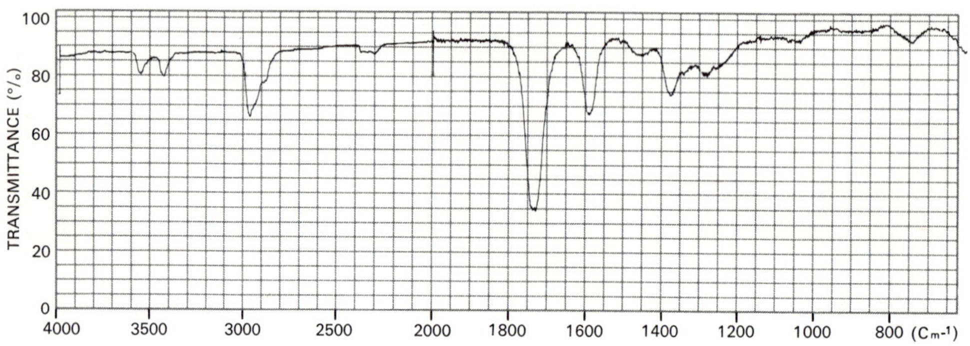

COMPOUND	MOL. FORMULA	B. PT.	INDEX No.
N, N-Diethylaniline	$C_{10}H_{15}N$	216°C	**145**

CELL TEMP.	APPROX AMOUNT	CELL PATH	REMARKS
~140°C	2μl	Length 9·2cm Volume 25ml	

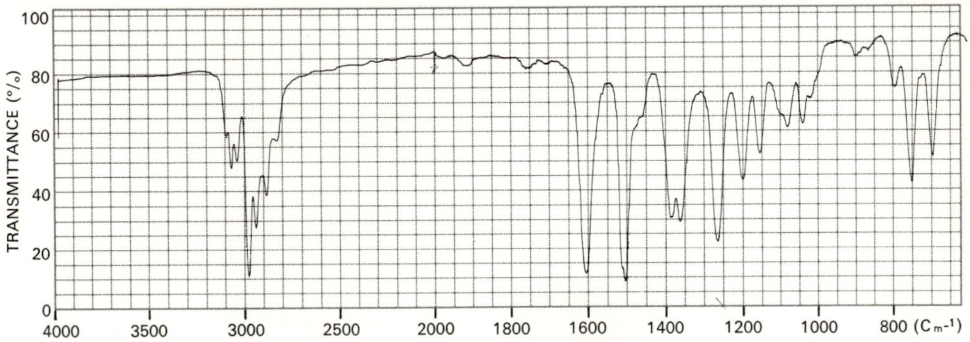

COMPOUND	MOL. FORMULA	B. PT.	INDEX No.
N, N-Dimethylformamide	C_3H_7NO	153°C	**146**

CELL TEMP.	APPROX AMOUNT	CELL PATH	REMARKS
~145°C	0·5μl	Length 9·2cm Volume 25ml	

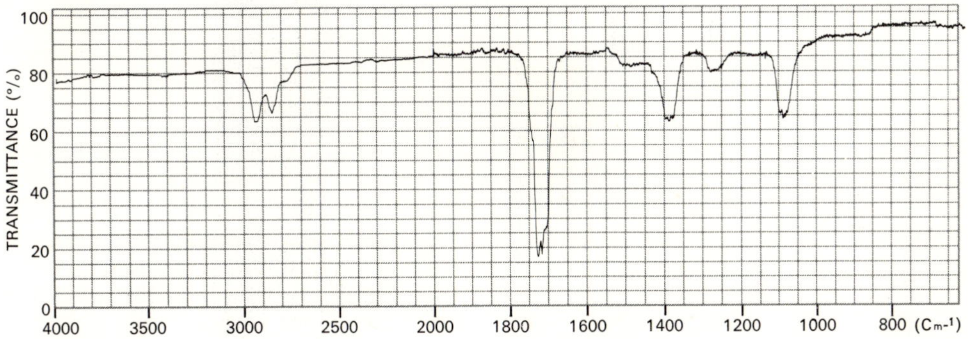

COMPOUND	MOL. FORMULA	B. PT.	INDEX No.
Pyridine	C_5H_5N	116°C	**147**

CELL TEMP.	APPROX AMOUNT	CELL PATH	REMARKS
~155°C	2μl	Length 9·2cm Volume 25ml	

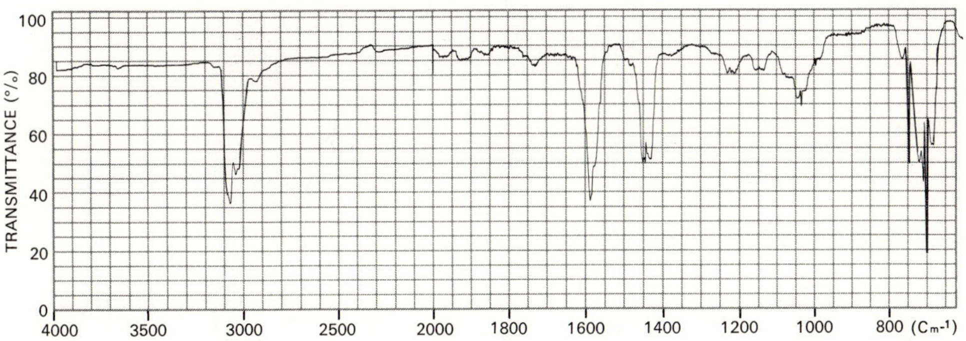

COMPOUND	MOL. FORMULA	B. PT.	INDEX No.
2-Methylpyridine	C_6H_7N	129° C	**148**

CELL TEMP.	APPROX AMOUNT	CELL PATH	REMARKS
~150° C	2μl	Length 9·2cm Volume 25ml	

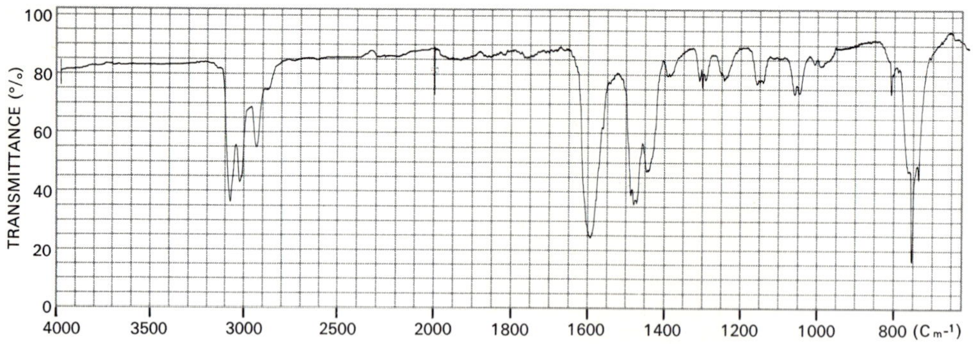

COMPOUND	MOL. FORMULA	B. PT.	INDEX No.
3-Methylpyridine	C_6H_7N	144° C	**149**

CELL TEMP.	APPROX AMOUNT	CELL PATH	REMARKS
~130° C	2μl	Length 9·2cm Volume 25ml	

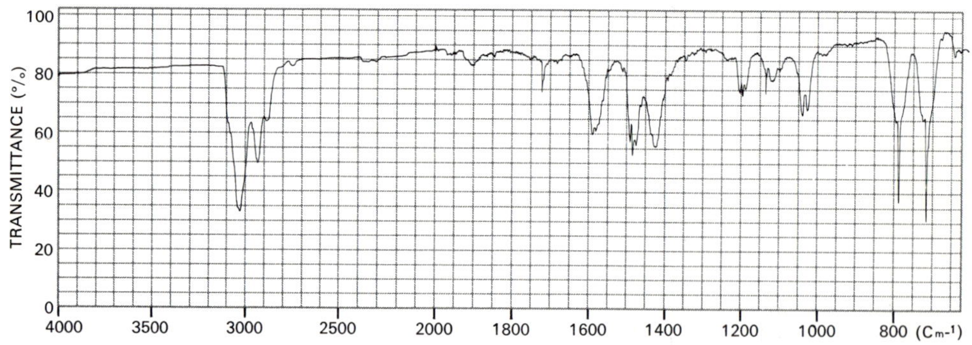

COMPOUND	MOL. FORMULA	B. PT.	INDEX No.
4-Methylpyridine	C_6H_7N	143° C	**150**

CELL TEMP.	APPROX AMOUNT	CELL PATH	REMARKS
~150° C	2μl	Length 9·2cm Volume 25ml	

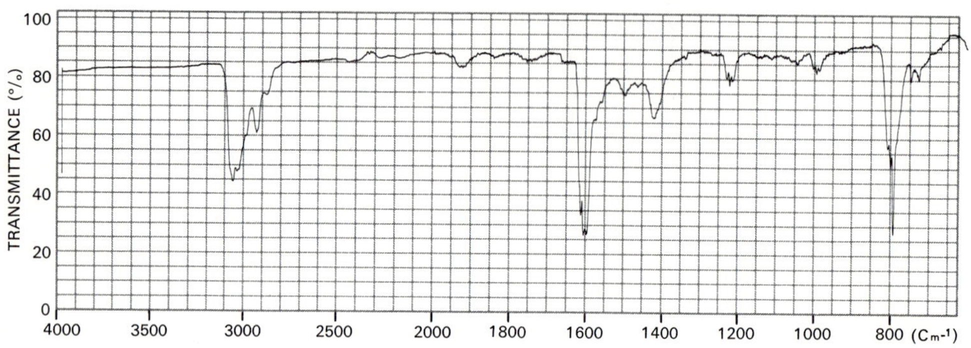

COMPOUND	MOL. FORMULA	B. PT.	INDEX No.
Quinoline	C_9H_7N	237°C	**151**

CELL TEMP.	APPROX AMOUNT	CELL PATH	REMARKS
~170°C	2·5μl	Length 9·2cm Volume 25ml	

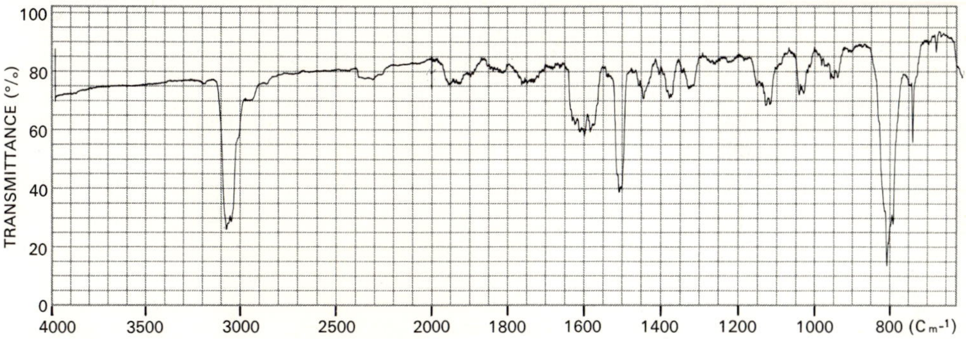

COMPOUND	MOL. FORMULA	B. PT.	INDEX No.
Acetaldehyde	C_2H_4O	21°C	**152**

CELL TEMP.	APPROX AMOUNT	CELL PATH	REMARKS
~130°C	1-2μl	Length 9·2cm Volume 25ml	

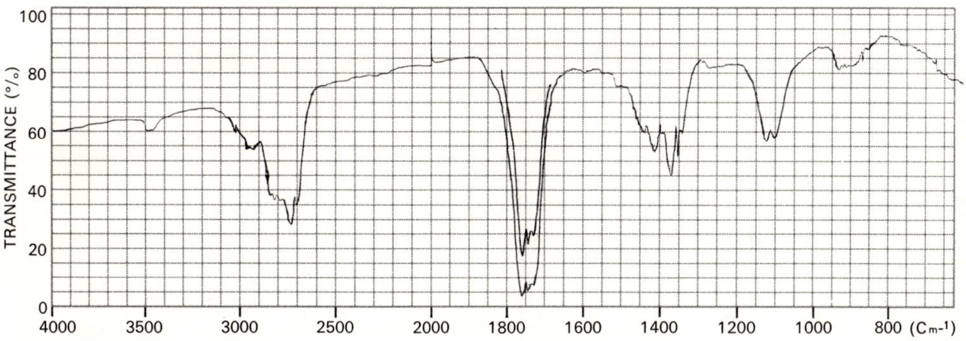

COMPOUND	MOL. FORMULA	B. PT.	INDEX No.
Propanal	C_3H_6O	49°C	**153**

CELL TEMP.	APPROX AMOUNT	CELL PATH	REMARKS
~130°C	1μl	Length 9·2cm Volume 25ml	Trace of acid

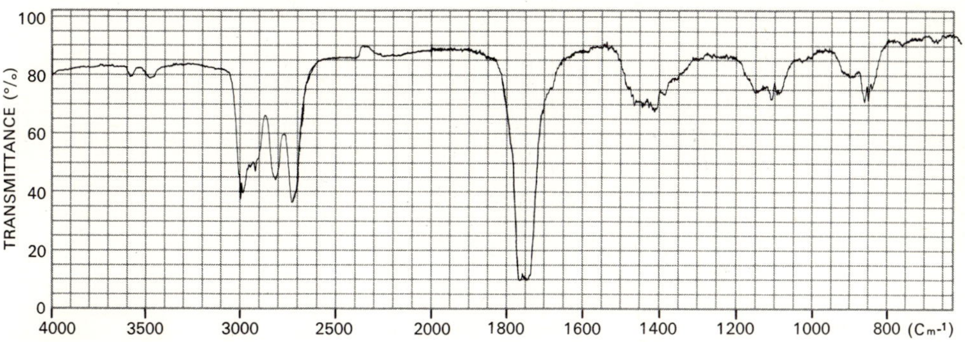

COMPOUND	MOL. FORMULA	B. PT.	INDEX No.
n-Butanal	C_4H_8O	76° C	**154**

CELL TEMP.	APPROX AMOUNT	CELL PATH	REMARKS
~100° C	2μl	Length 9·2cm Volume 25ml	Trace of acid

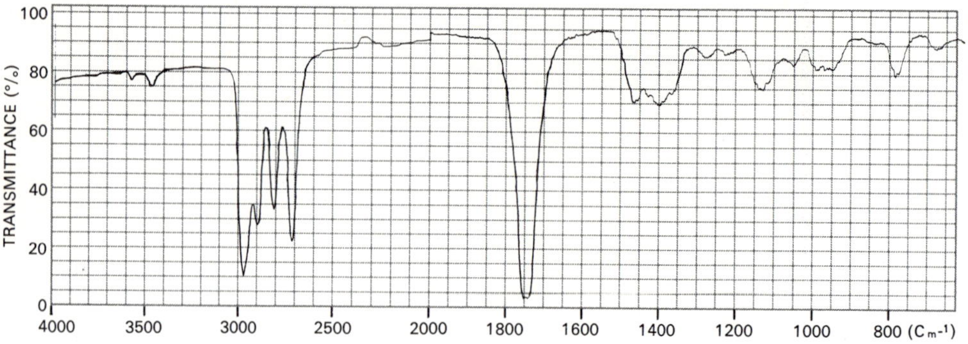

COMPOUND	MOL. FORMULA	B. PT.	INDEX No.
n-Pentanal	$C_5H_{10}O$	102° C	**155**

CELL TEMP.	APPROX AMOUNT	CELL PATH	REMARKS
~105° C	2μl	Length 9·2cm Volume 25ml	Trace of acid

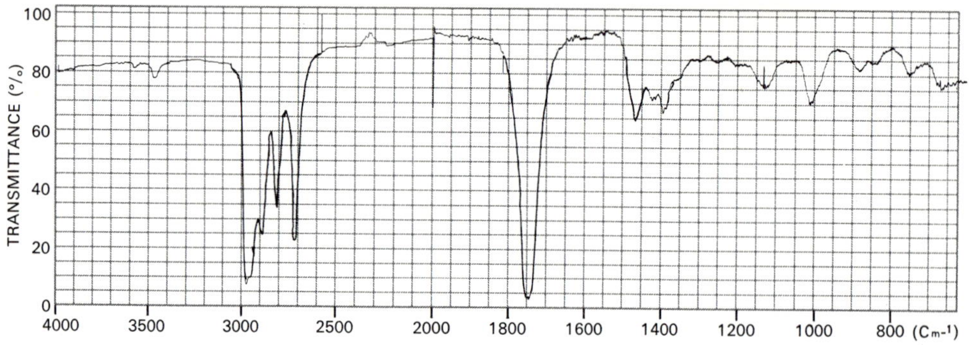

COMPOUND	MOL. FORMULA	B. PT.	INDEX No.
n-Hexanal	$C_6H_{12}O$	128° C	**156**

CELL TEMP.	APPROX AMOUNT	CELL PATH	REMARKS
~105° C	2μl	Length 9·2cm Volume 25ml	Trace of acid

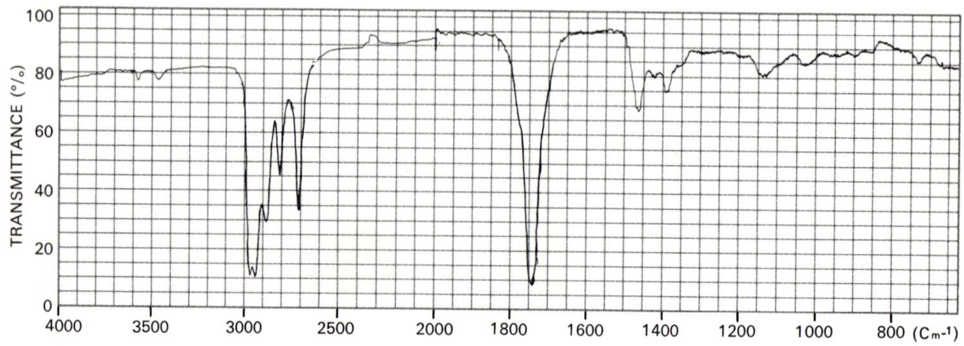

COMPOUND	MOL. FORMULA	B. PT.	INDEX No.
n-Heptanal	$C_7H_{14}O$	155°C	**157**

CELL TEMP.	APPROX AMOUNT	CELL PATH	REMARKS
~130°C	2μl	Length 9·2cm Volume 25ml	Slight trace of acid

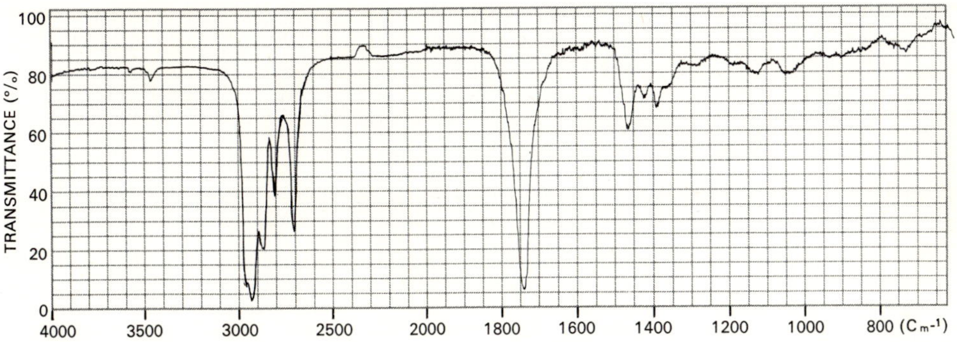

COMPOUND	MOL. FORMULA	B. PT.	INDEX No.
n-Octanal	$C_8H_{16}O$	167°C	**158**

CELL TEMP.	APPROX AMOUNT	CELL PATH	REMARKS
~105°C	2μl	Length 9·2cm Volume 25ml	

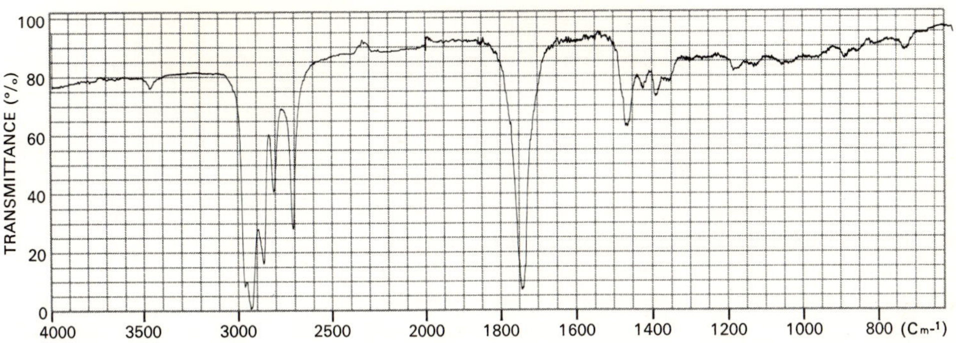

COMPOUND	MOL. FORMULA	B. PT.	INDEX No.
n-Nonanal	$C_9H_{18}O$	185°C	**159**

CELL TEMP.	APPROX AMOUNT	CELL PATH	REMARKS
~105°C	2μl	Length 9·2cm Volume 25ml	Trace of acid and liquid film

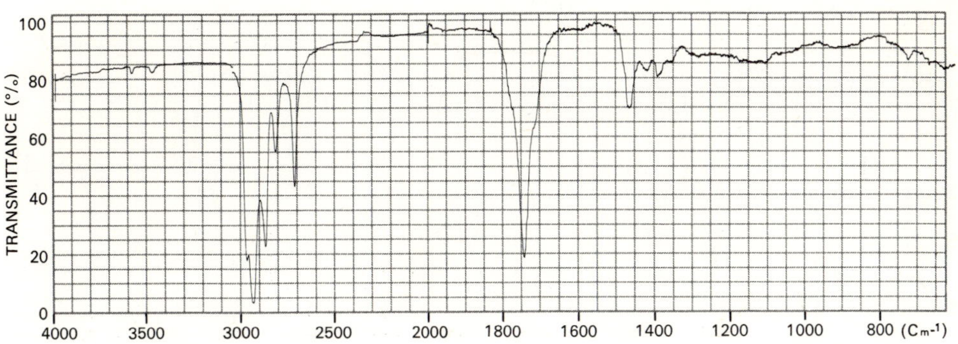

COMPOUND	MOL. FORMULA	B. PT.	INDEX No.
n-Decanal	$C_{10}H_{20}O$	208-209°C	**160**

CELL TEMP.	APPROX AMOUNT	CELL PATH	REMARKS
~ 120°C	2μl	Length 9·2cm Volume 25ml	Slight trace of acid and liquid film

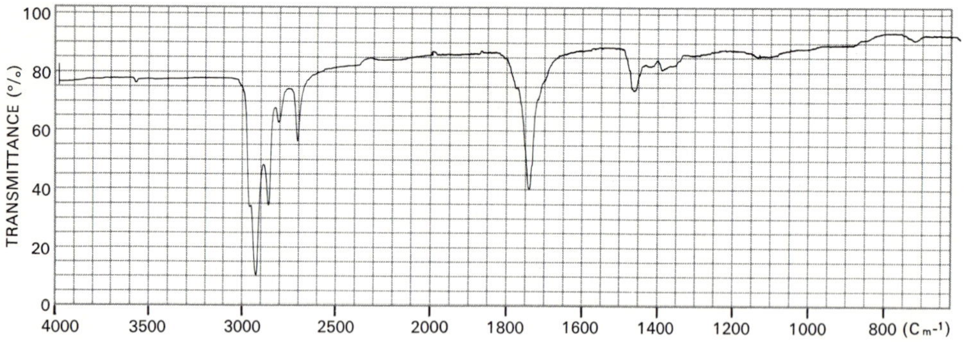

COMPOUND	MOL. FORMULA	B. PT.	INDEX No.
But-2-enal	C_4H_6O	104°C	**161**

CELL TEMP.	APPROX AMOUNT	CELL PATH	REMARKS
~130°C	2μl	Length 9·2cm Volume 25ml	Slight trace of acid

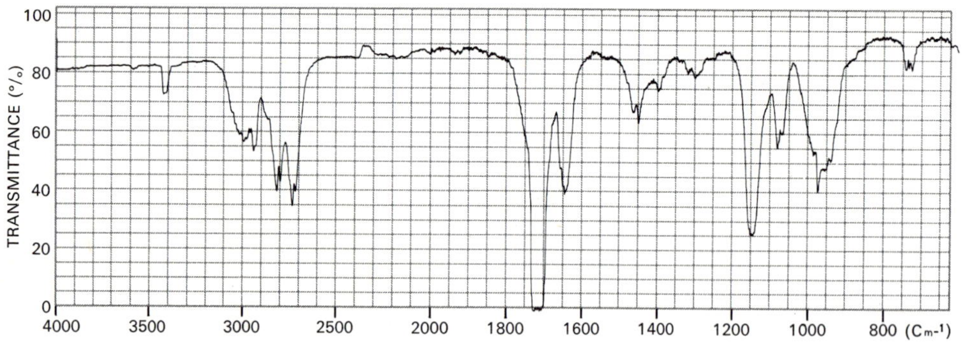

COMPOUND	MOL. FORMULA	B. PT.	INDEX No.
Hex-2-enal	$C_6H_{10}O$	47°C/17mm	**162**

CELL TEMP.	APPROX AMOUNT	CELL PATH	REMARKS
~130°C	2μl	Length 9·2cm Volume 25ml	Trace of acid

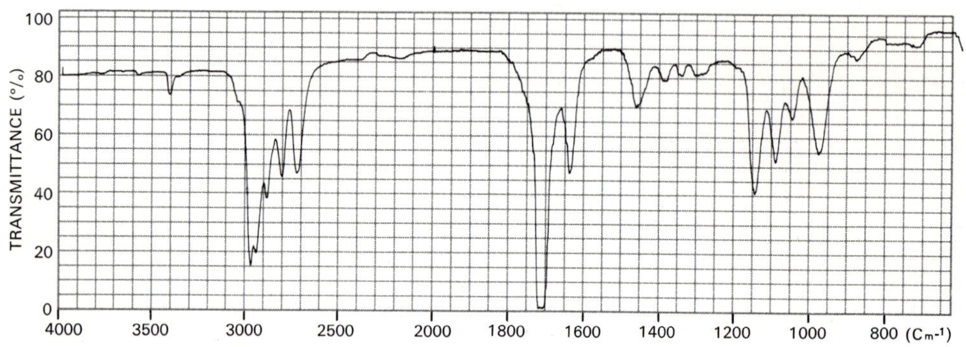

COMPOUND	MOL. FORMULA	B. PT.	INDEX No.
trans-Oct-2-enal	$C_8H_{14}O$	-	**163**

CELL TEMP.	APPROX AMOUNT	CELL PATH	REMARKS
~135°C	2μl	Length 9·2cm Volume 25ml	Slight trace of acid

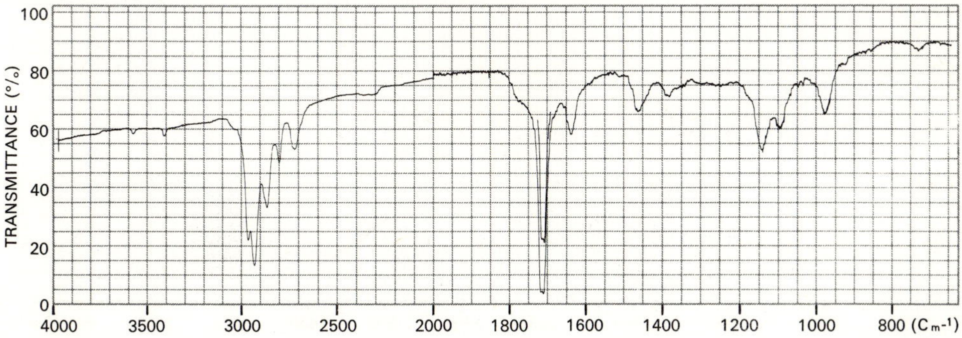

COMPOUND	MOL. FORMULA	B. PT.	INDEX No.
trans-Non-2-enal	$C_9H_{16}O$	-	**164**

CELL TEMP.	APPROX AMOUNT	CELL PATH	REMARKS
~130°C	2μl	Length 9·2cm Volume 25ml	

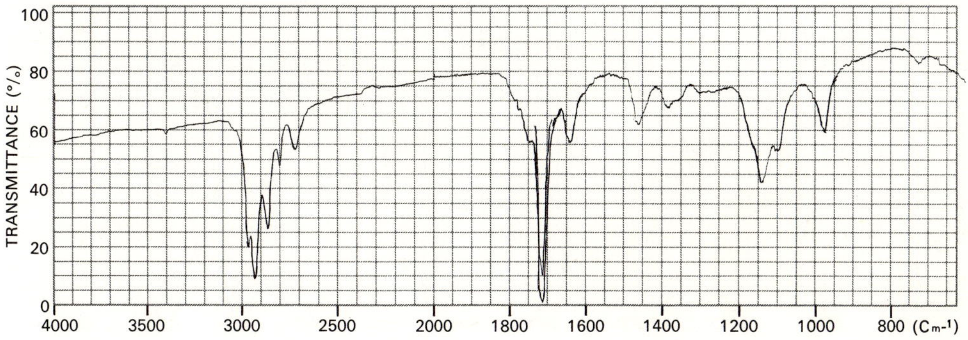

COMPOUND	MOL. FORMULA	B. PT.	INDEX No.
trans-Dec-2-enal	$C_{10}H_{18}O$	-	**165**

CELL TEMP.	APPROX AMOUNT	CELL PATH	REMARKS
~125°C	2μl	Length 9·2cm Volume 25ml	

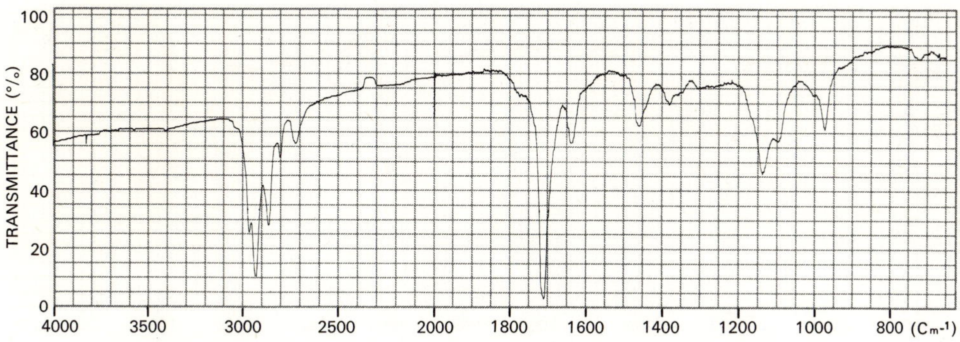

COMPOUND	MOL. FORMULA	B. PT.	INDEX No.
trans-Undec-2-enal	$C_{11}H_{20}O$	-	**166**

CELL TEMP.	APPROX AMOUNT	CELL PATH	REMARKS
~ 135° C	3μl	Length 9·2cm Volume 25ml	

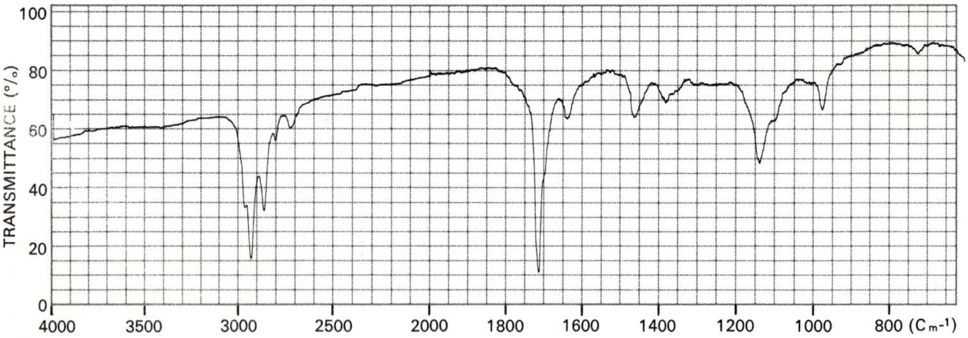

COMPOUND	MOL. FORMULA	B. PT.	INDEX No.
2,4-Hexadienal	C_6H_8O	76° C/30mm	**167**

CELL TEMP.	APPROX AMOUNT	CELL PATH	REMARKS
~ 130° C	2μl	Length 9·2cm Volume 25ml	

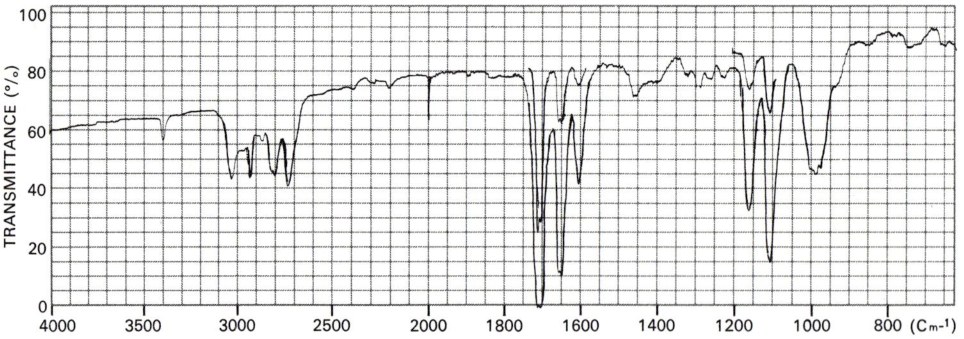

COMPOUND	MOL. FORMULA	B. PT.	INDEX No.
2 trans, 4 trans-Heptadienal	$C_7H_{10}O$	-	**168**

CELL TEMP.	APPROX AMOUNT	CELL PATH	REMARKS
~ 130° C	2μl	Length 9·2cm Volume 25ml	

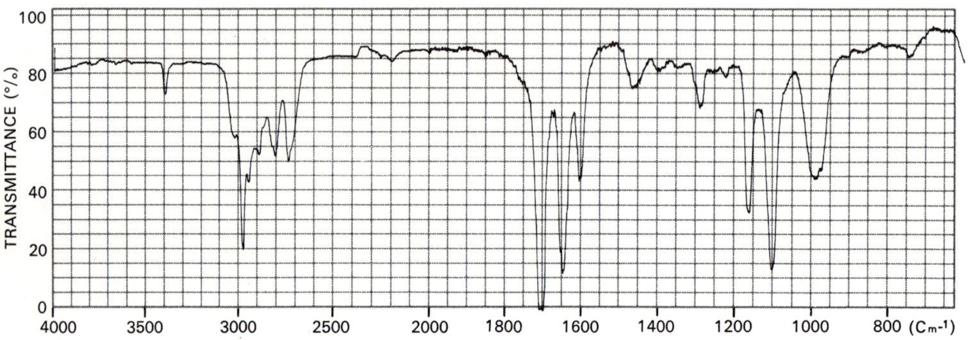

COMPOUND	MOL. FORMULA	B. PT.	INDEX No.
2 trans, 4 cis-Nonadienal	$C_9H_{14}O$	-	**169**
CELL TEMP. ~ 135°C	APPROX AMOUNT 2μl	CELL PATH Length 9·2cm Volume 25ml	REMARKS

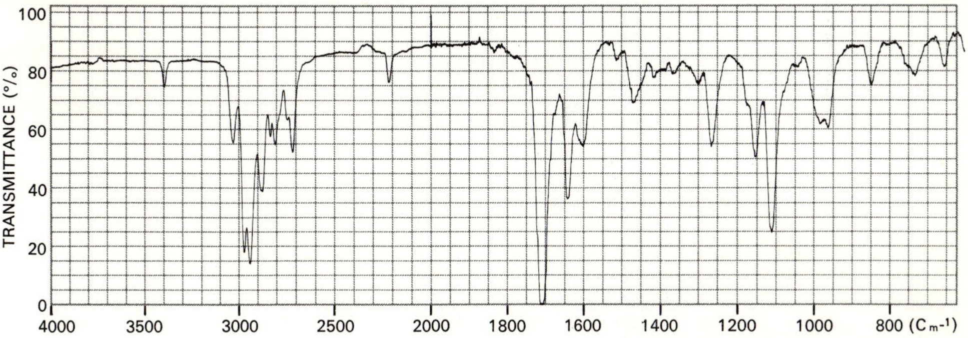

COMPOUND	MOL. FORMULA	B. PT.	INDEX No.
2 trans, 4 trans-Decadienal	$C_{10}H_{16}O$	-	**170**
CELL TEMP. ~ 125°C	APPROX AMOUNT 2μl	CELL PATH Length 9·2cm Volume 25ml	REMARKS

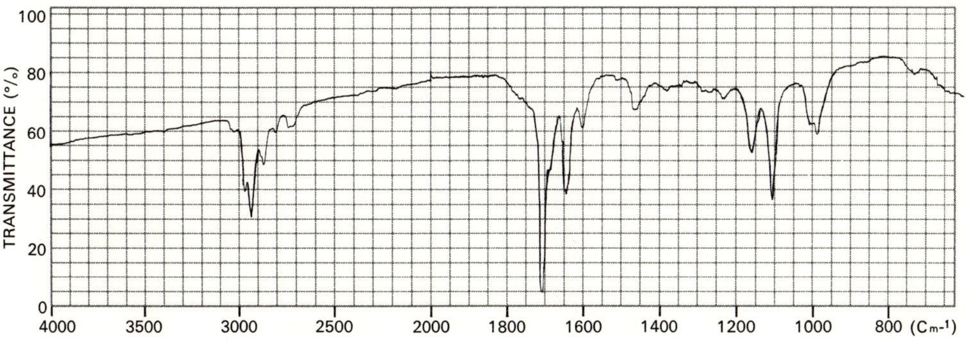

COMPOUND	MOL. FORMULA	B. PT.	INDEX No.
2 trans, 4 trans-Undecadienal	$C_{11}H_{18}O$	-	**171**
CELL TEMP. ~ 130°C	APPROX AMOUNT 2·5μl	CELL PATH Length 9·2cm Volume 25ml	REMARKS

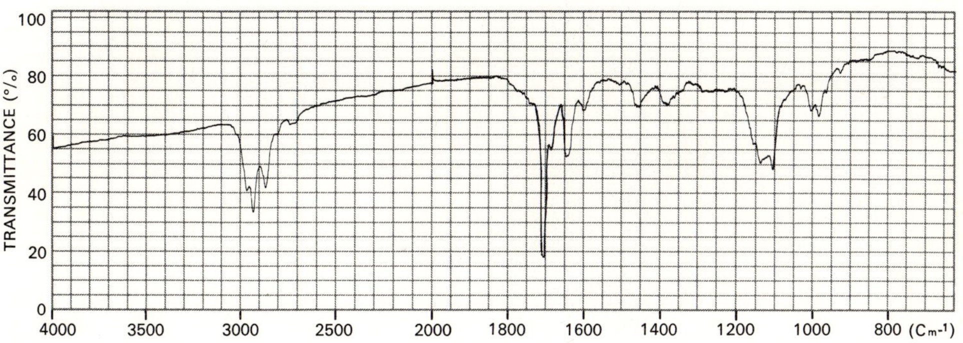

COMPOUND	MOL. FORMULA	B. PT.	INDEX No.
2 trans, 4 trans-Dodecadienal	$C_{12}H_{20}O$	-	**172**

CELL TEMP.	APPROX AMOUNT	CELL PATH	REMARKS
~ 135°C	3μl	Length 9·2cm Volume 25ml	

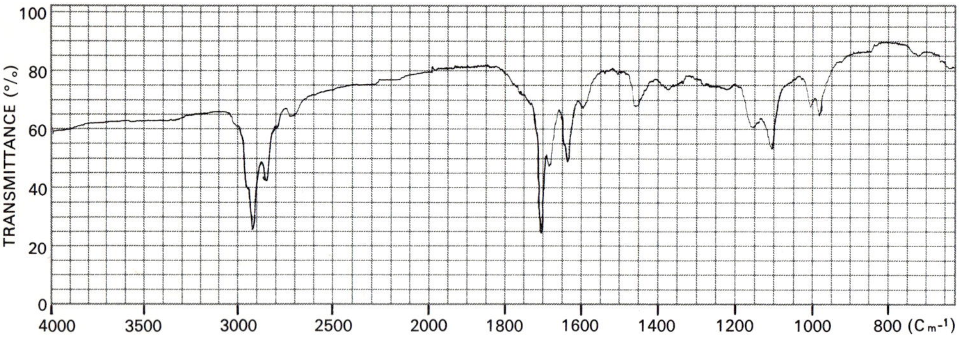

COMPOUND	MOL. FORMULA	B. PT.	INDEX No.
Benzaldehyde	C_7H_6O	178·1°C	**173**

CELL TEMP.	APPROX AMOUNT	CELL PATH	REMARKS
~ 130°C	2·5μl	Length 9·2cm Volume 25ml	Trace of acid

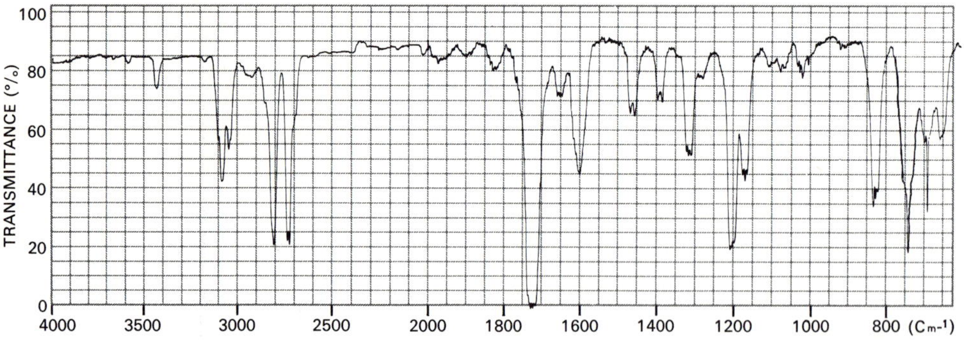

COMPOUND	MOL. FORMULA	B. PT.	INDEX No.
4-Isopropylbenzaldehyde	$C_{10}H_{12}O$	235-236°C	**174**

CELL TEMP.	APPROX AMOUNT	CELL PATH	REMARKS
~ 140°C	3μl	Length 9·2cm Volume 25ml	Slight trace of acid

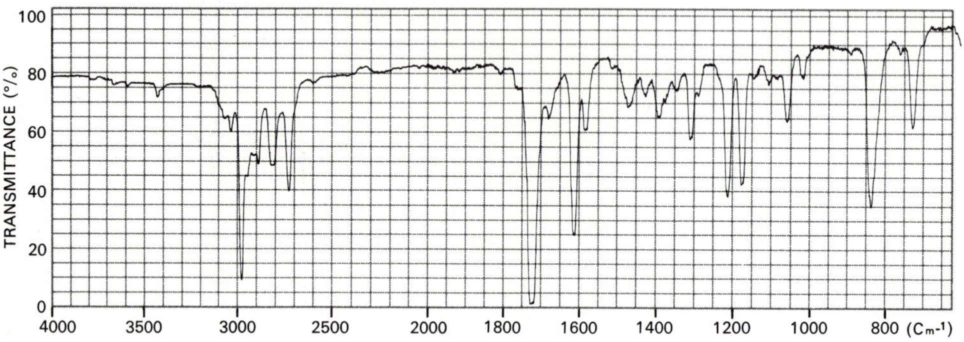

COMPOUND	MOL. FORMULA	B. PT.	INDEX No.
Furfural	$C_5H_4O_2$	161·7°C	**175**

CELL TEMP.	APPROX AMOUNT	CELL PATH	REMARKS
~ 130°C	1μl	Length 9·2cm Volume 25ml	

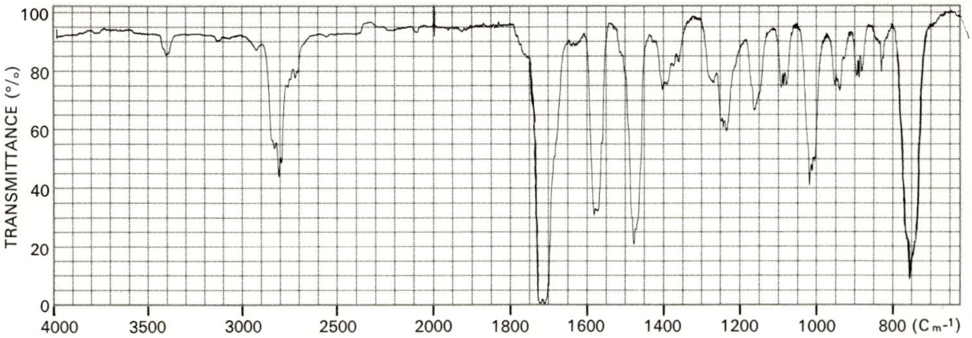

COMPOUND	MOL. FORMULA	B. PT.	INDEX No.
5-Methylfurfural	$C_6H_6O_2$	187°C	**176**

CELL TEMP.	APPROX AMOUNT	CELL PATH	REMARKS
~ 120°C	1μl	Length 9·2cm Volume 25ml	

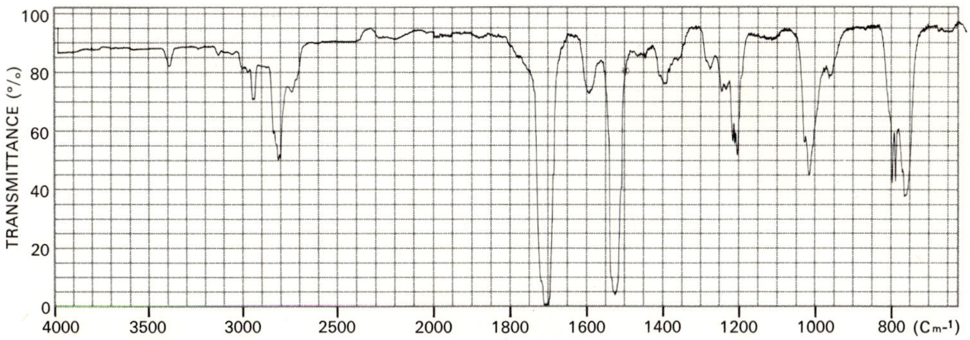

COMPOUND	MOL. FORMULA	B. PT.	INDEX No.
2-Methoxybenzaldehyde	$C_8H_8O_2$	243-244°C	**177**

CELL TEMP.	APPROX AMOUNT	CELL PATH	REMARKS
~ 125°C	A few mg	Length 9·2cm Volume 25ml	

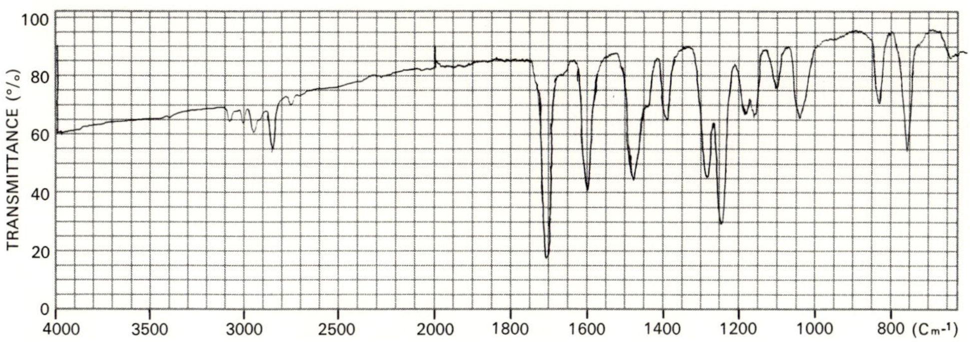

COMPOUND	MOL. FORMULA	B. PT.	INDEX No.
4-Methoxybenzaldehyde	$C_8H_8O_2$	249·5°C	**178**

CELL TEMP.	APPROX AMOUNT	CELL PATH	REMARKS
~ 135°C	2μl	Length 9·2cm Volume 25ml	

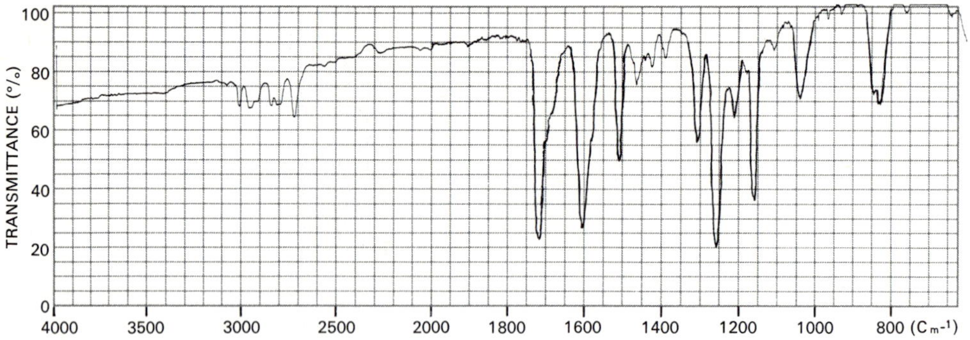

COMPOUND	MOL. FORMULA	B. PT.	INDEX No.
Salicylaldehyde	$C_7H_6O_2$	197°C	**179**

CELL TEMP.	APPROX AMOUNT	CELL PATH	REMARKS
~ 125°C	0·5μl	Length 9·2cm Volume 25ml	

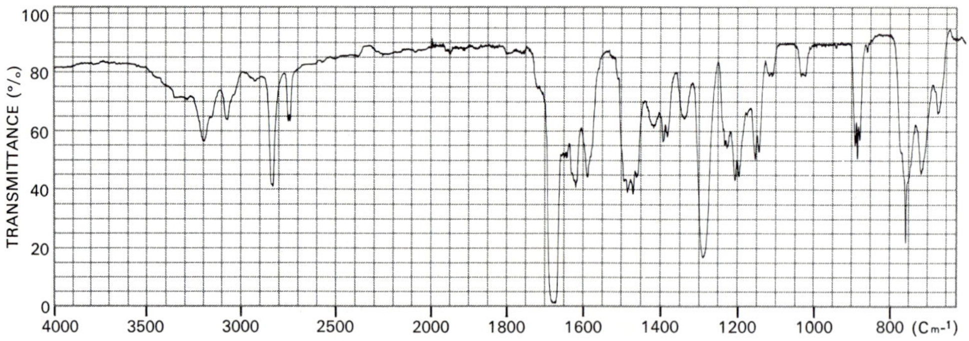

COMPOUND	MOL. FORMULA	B. PT.	INDEX No.
Propan-2-one	C_3H_6O	56°C	**180**

CELL TEMP.	APPROX AMOUNT	CELL PATH	REMARKS
~ 130°C	2μl	Length 9·2cm Volume 25ml	

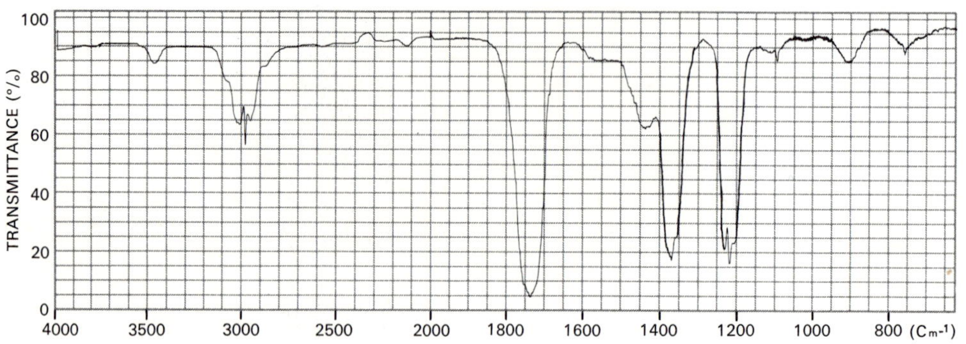

COMPOUND	MOL. FORMULA	B. PT.	INDEX No.
Butan-2-one	C_4H_8O	80°C	**181**

CELL TEMP.	APPROX AMOUNT	CELL PATH	REMARKS
~ 125°C	2µl	Length 9·2cm Volume 25ml	

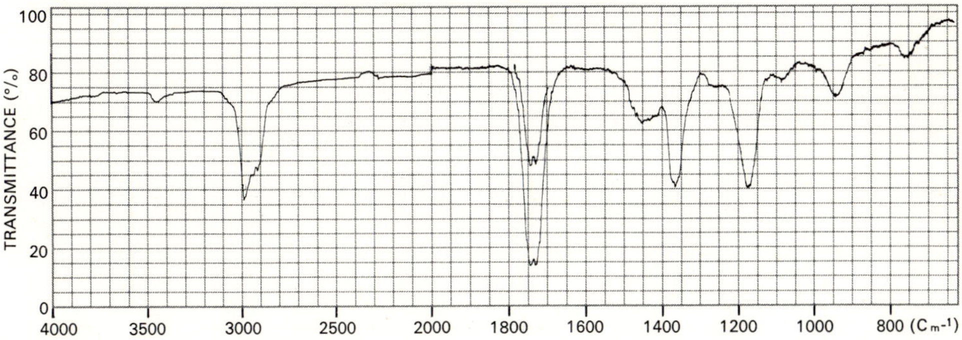

COMPOUND	MOL. FORMULA	B. PT.	INDEX No.
Pentan-2-one	$C_5H_{10}O$	102°C	**182**

CELL TEMP.	APPROX AMOUNT	CELL PATH	REMARKS
~ 130°C	2µl	Length 9·2cm Volume 25ml	

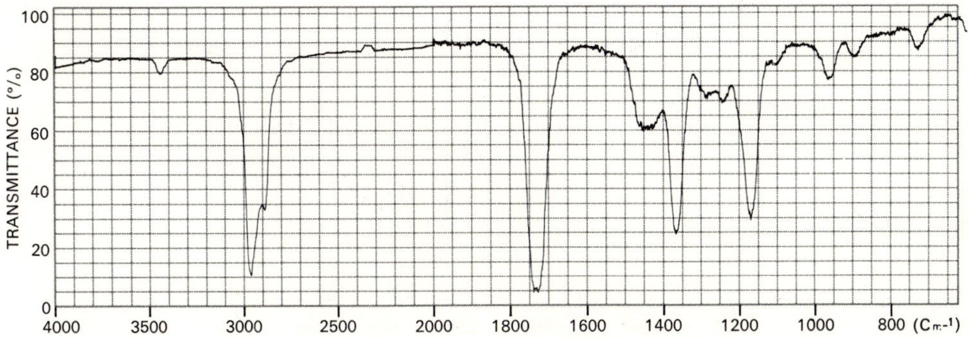

COMPOUND	MOL. FORMULA	B. PT.	INDEX No.
Hexan-2-one	$C_6H_{12}O$	126°C	**183**

CELL TEMP.	APPROX AMOUNT	CELL PATH	REMARKS
~ 130°C	2µl	Length 9·2cm Volume 25ml	

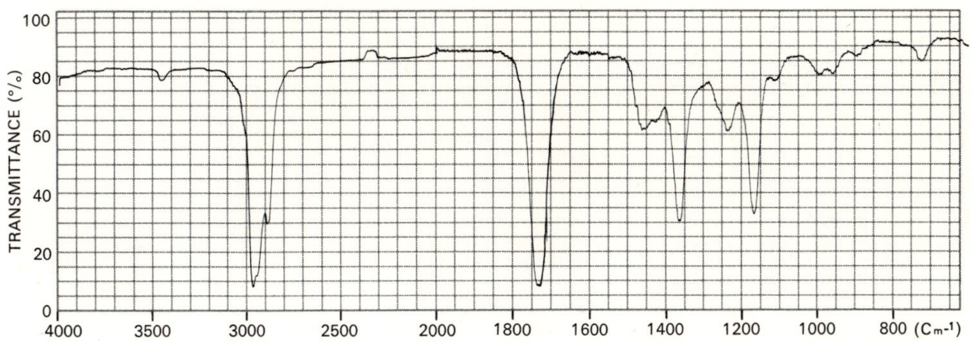

COMPOUND	MOL. FORMULA	B. PT.	INDEX No.
3-Methylpentan-2-one	$C_6H_{12}O$	118°C	**184**

CELL TEMP.	APPROX AMOUNT	CELL PATH	REMARKS
~130°C	2μl	Length 9·2cm Volume 25ml	

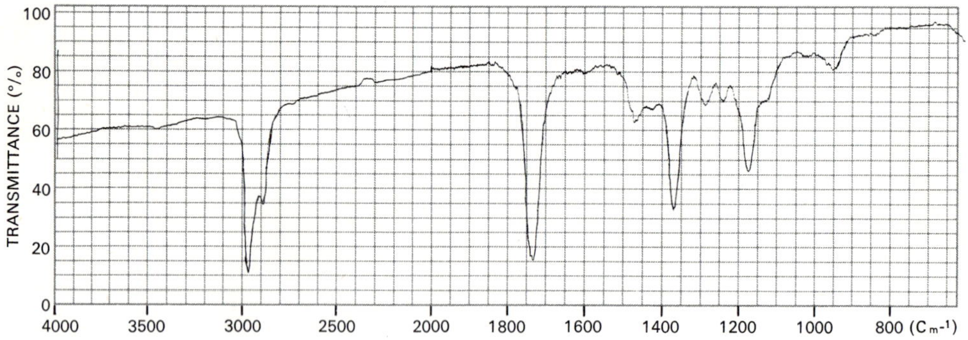

COMPOUND	MOL. FORMULA	B. PT.	INDEX No.
Heptan-2-one	$C_7H_{14}O$	151°C	**185**

CELL TEMP.	APPROX AMOUNT	CELL PATH	REMARKS
~135°C	2μl	Length 9·2cm Volume 25ml	

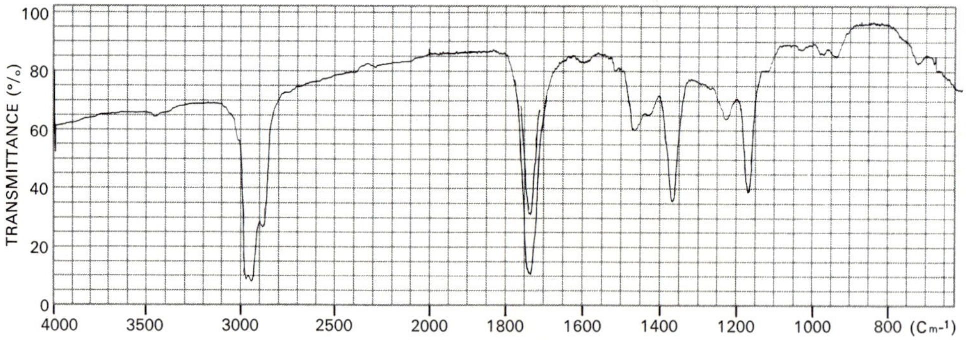

COMPOUND	MOL. FORMULA	B. PT.	INDEX No.
Octan-2-one	$C_8H_{16}O$	173°C	**186**

CELL TEMP.	APPROX AMOUNT	CELL PATH	REMARKS
~130°C	2μl	Length 9·2cm Volume 25ml	

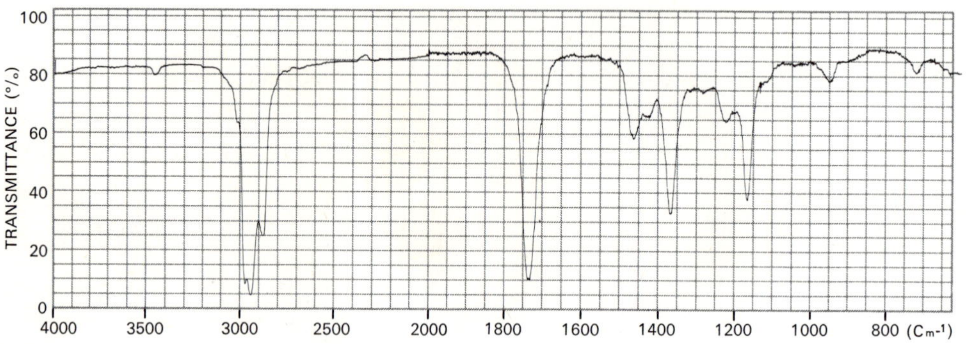

COMPOUND	MOL. FORMULA	B. PT.	INDEX No.
Nonan-2-one	$C_9H_{18}O$	193°C	**187**

CELL TEMP.	APPROX AMOUNT	CELL PATH	REMARKS
~ 135°C	2μl	Length 9·2cm Volume 25ml	

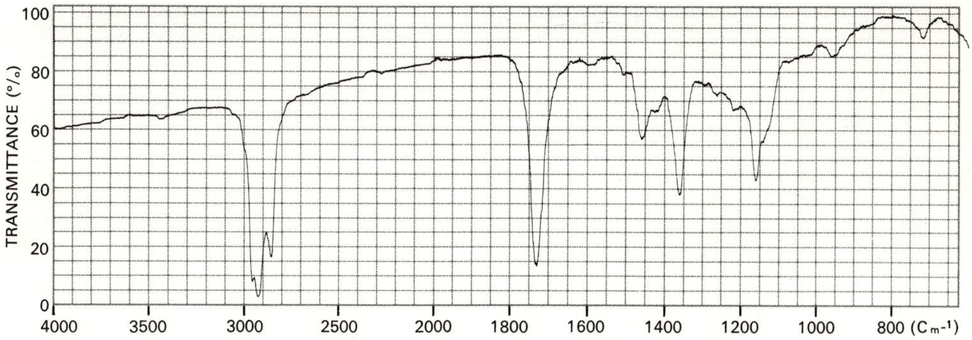

COMPOUND	MOL. FORMULA	B. PT.	INDEX No.
Decan-2-one	$C_{10}H_{20}O$	211°C	**188**

CELL TEMP.	APPROX AMOUNT	CELL PATH	REMARKS
~135°C	2μl	Length 9·2cm Volume 25ml	

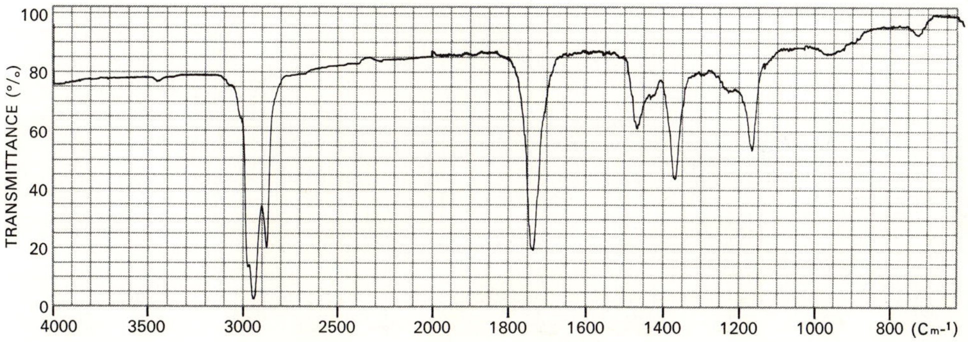

COMPOUND	MOL. FORMULA	B. PT.	INDEX No.
Undecan-2-one	$C_{11}H_{22}O$	230°C/748mm	**189**

CELL TEMP.	APPROX AMOUNT	CELL PATH	REMARKS
~135°C	2μl	Length 9·2cm Volume 25ml	

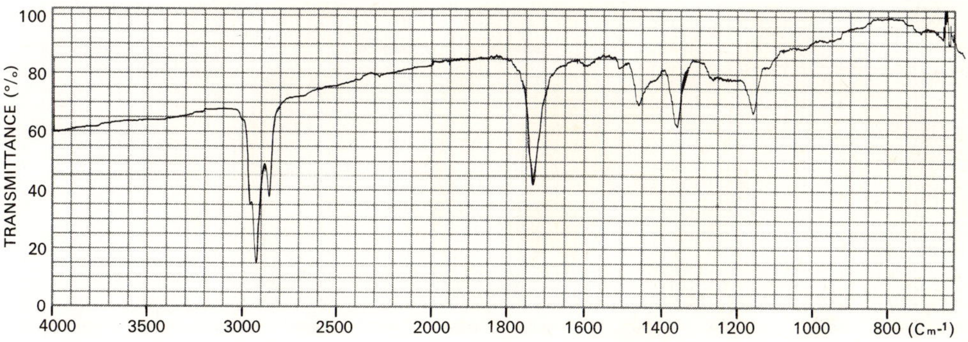

COMPOUND	MOL. FORMULA	B. PT.	INDEX No.
Pentan-3-one	$C_5H_{10}O$	103°C	**190**

CELL TEMP.	APPROX AMOUNT	CELL PATH	REMARKS
~ 140°C	2μl	Length 9·2cm Volume 25ml	

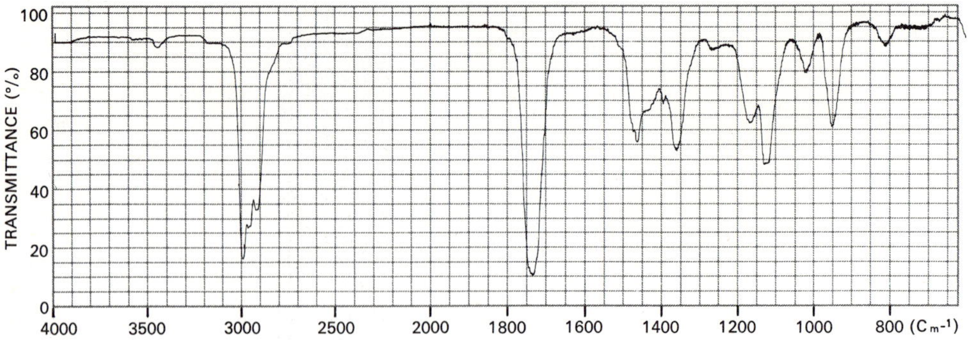

COMPOUND	MOL. FORMULA	B. PT.	INDEX No.
Octan-3-one	$C_8H_{16}O$	170°C/738mm	**191**

CELL TEMP.	APPROX AMOUNT	CELL PATH	REMARKS
~130°C	2μl	Length 9·2cm Volume 25ml	

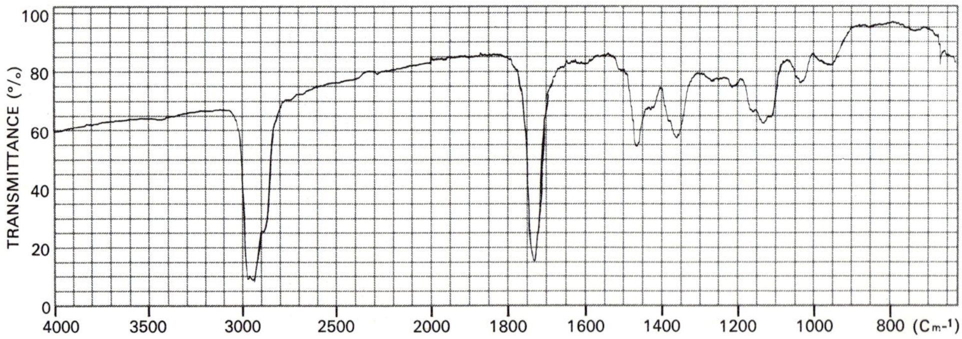

COMPOUND	MOL. FORMULA	B. PT.	INDEX No.
Undecan-5-one	$C_{11}H_{22}O$	227°C	**192**

CELL TEMP.	APPROX AMOUNT	CELL PATH	REMARKS
~ 130°C	2μl	Length 9·2cm Volume 25ml	

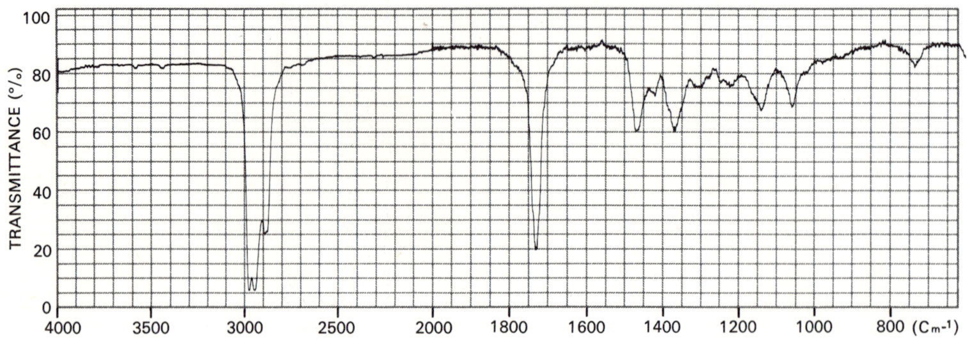

COMPOUND	MOL. FORMULA	B. PT.	INDEX No.
Butan-2,3-dione	$C_4H_6O_2$	90°C	**193**

CELL TEMP.	APPROX AMOUNT	CELL PATH	REMARKS
~135°C	1μl	Length 9·2cm Volume 25ml	

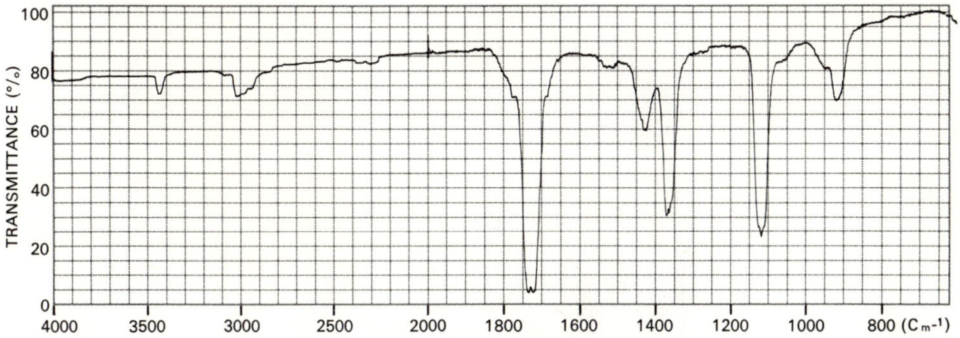

COMPOUND	MOL. FORMULA	B. PT.	INDEX No.
3-Hydroxybutan-2-one	$C_4H_8O_2$	143°C	**194**

CELL TEMP.	APPROX AMOUNT	CELL PATH	REMARKS
~150°C	A few mg	Length 9·2cm Volume 25ml	

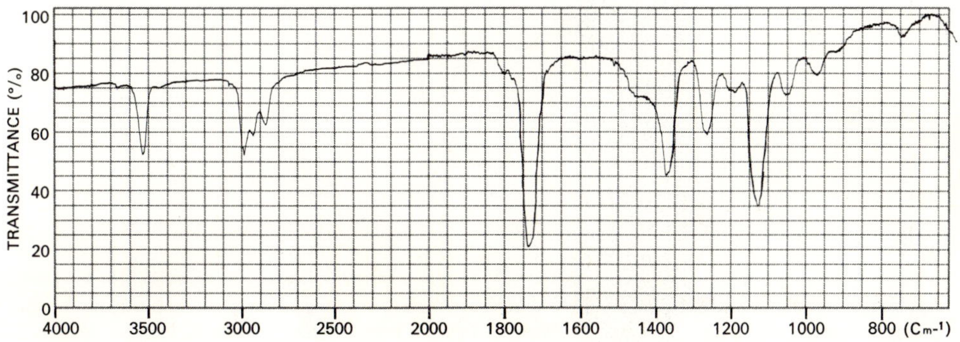

COMPOUND	MOL. FORMULA	B. PT.	INDEX No.
Penta-2,4-dione	$C_5H_8O_2$	139°C	**195**

CELL TEMP.	APPROX AMOUNT	CELL PATH	REMARKS
~130°C	1·5μl	Length 9·2cm Volume 25ml	

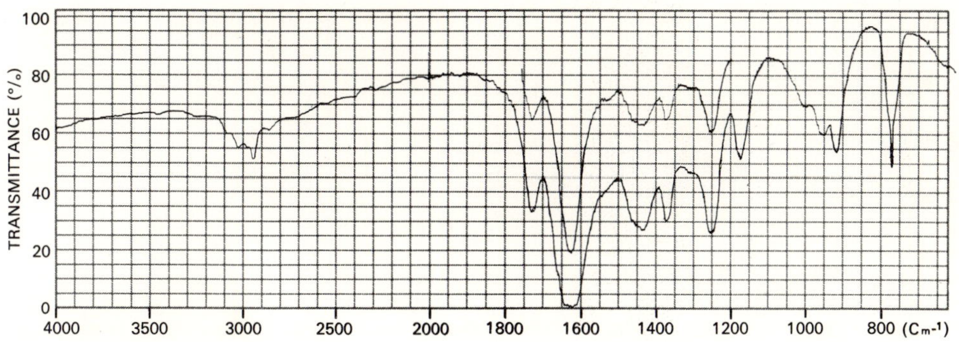

COMPOUND	MOL. FORMULA	B. PT.	INDEX No.
Cyclopentanone	C_5H_8O	131°C	**196**

CELL TEMP.	APPROX AMOUNT	CELL PATH	REMARKS
~130°C	2μl	Length 9·2cm Volume 25ml	

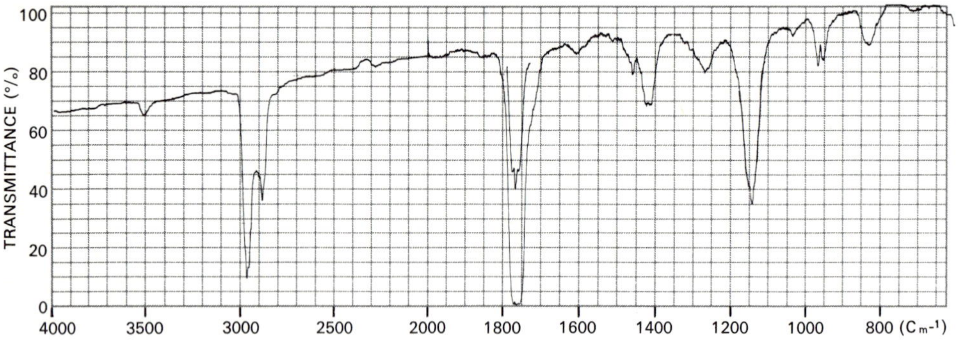

COMPOUND	MOL. FORMULA	B. PT.	INDEX No.
Cyclohexanone	$C_6H_{10}O$	156°C	**197**

CELL TEMP.	APPROX AMOUNT	CELL PATH	REMARKS
~130°C	2μl	Length 9·2cm Volume 25ml	

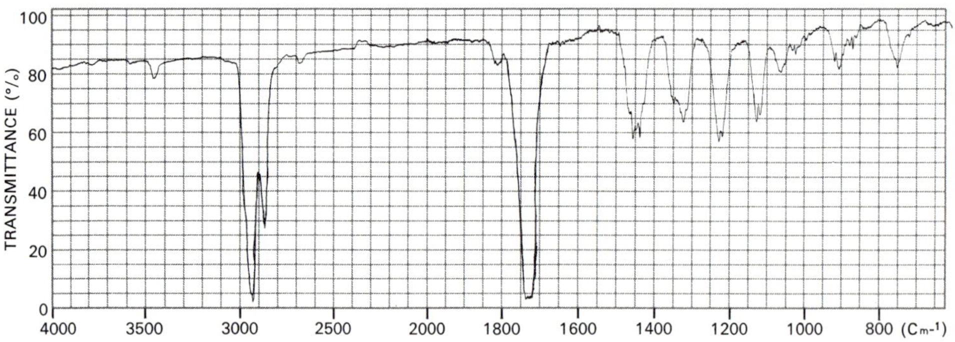

COMPOUND	MOL. FORMULA	B. PT.	INDEX No.
dl-Piperitone	$C_{10}H_{16}O$	233°C	**198**

CELL TEMP.	APPROX AMOUNT	CELL PATH	REMARKS
~120°C	2μl	Length 9·2cm Volume 25ml	

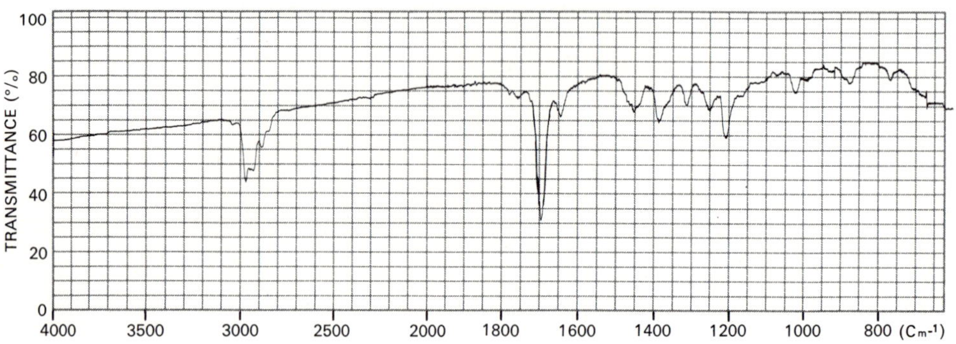

COMPOUND	MOL. FORMULA	B. PT.	INDEX No.
d-Carvone	$C_{10}H_{14}O$	231°C	**199**

CELL TEMP.	APPROX AMOUNT	CELL PATH	REMARKS
~130°C	2μl	Length 9·2cm Volume 25ml	

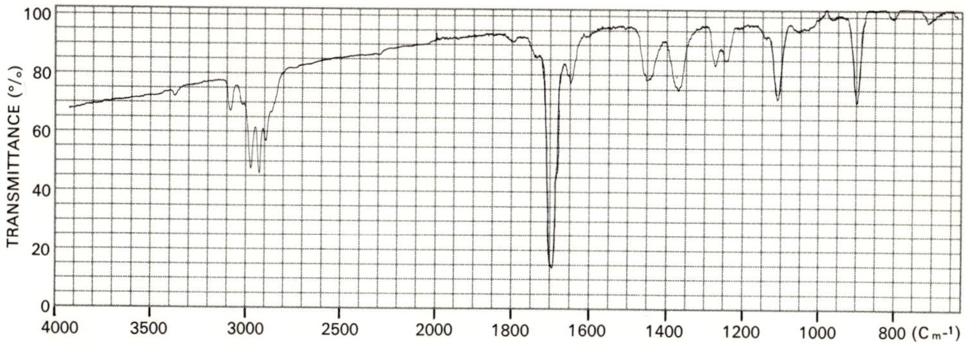

COMPOUND	MOL. FORMULA	B. PT.	INDEX No.
Pulegone	$C_{10}H_{16}O$	224°C	**200**

CELL TEMP.	APPROX AMOUNT	CELL PATH	REMARKS
~135°C	2μl	Length 9·2cm Volume 25ml	

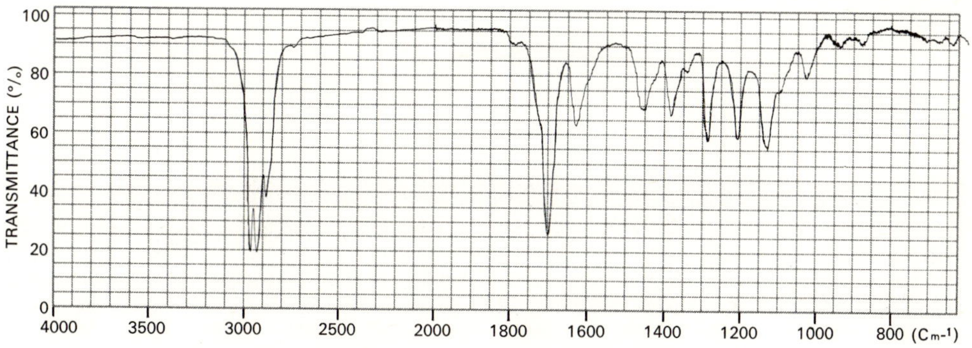

COMPOUND	MOL. FORMULA	B. PT.	INDEX No.
β-Ionone	$C_{13}H_{20}O$	140°C/18mm	**201**

CELL TEMP.	APPROX AMOUNT	CELL PATH	REMARKS
~140°C	2μl	Length 9·2cm Volume 25ml	

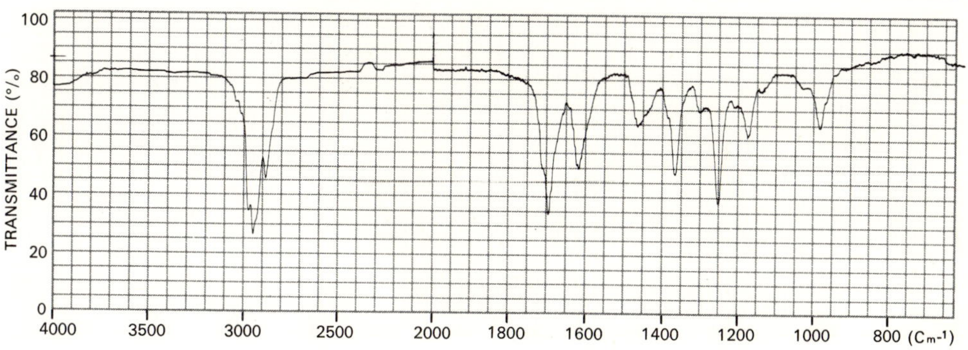

COMPOUND	MOL. FORMULA	B. PT.	INDEX No.
Acetophenone	C_8H_8O	202°C	**202**

CELL TEMP.	APPROX AMOUNT	CELL PATH	REMARKS
~ 130°C	2μl	Length 9·2cm Volume 25ml	

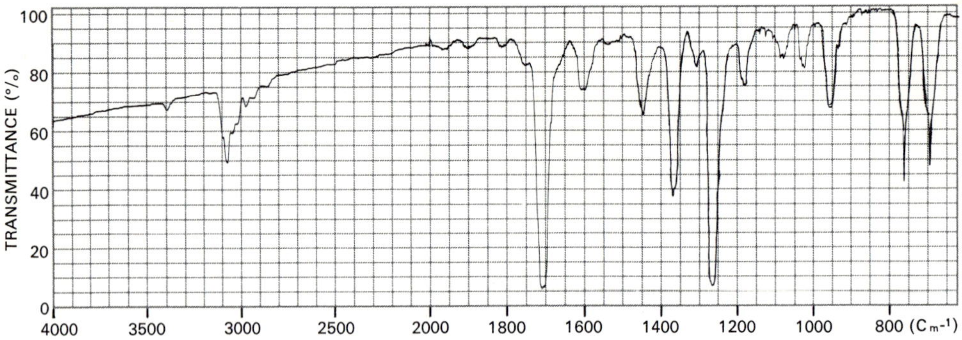

COMPOUND	MOL. FORMULA	B. PT.	INDEX No.
4-Methylacetophenone	$C_9H_{10}O$	226°C	**203**

CELL TEMP.	APPROX AMOUNT	CELL PATH	REMARKS
~ 130°C	2μl	Length 9·2cm Volume 25ml	

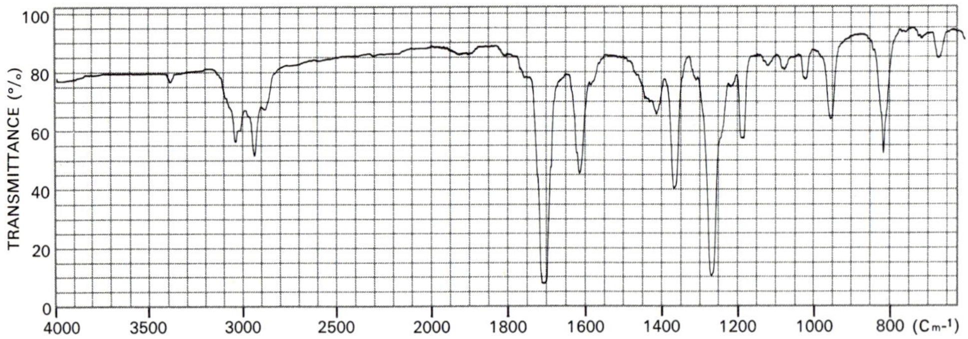

COMPOUND.	MOL. FORMULA	B. PT.	INDEX No.
Propiophenone	$C_9H_{10}O$	218°C	**204**

CELL TEMP.	APPROX AMOUNT	CELL PATH	REMARKS
~ 125°C	2μl	Length 9·2cm Volume 25ml	

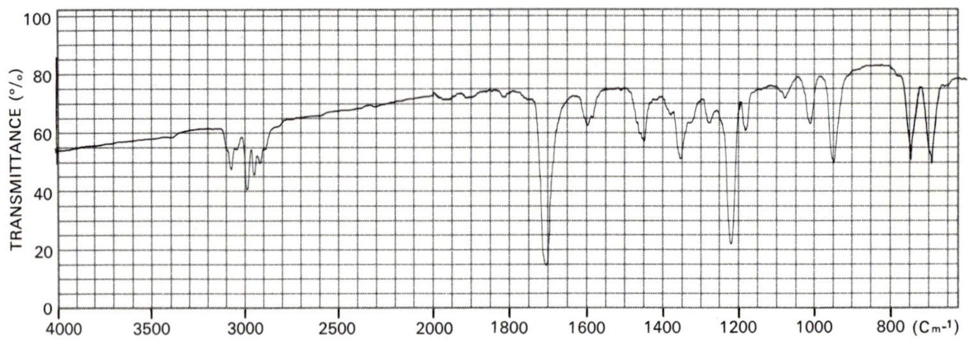

COMPOUND	MOL. FORMULA	B. PT.	INDEX No.
2-Acetylfuran	$C_6H_6O_2$	173°C	**205**

CELL TEMP.	APPROX AMOUNT	CELL PATH	REMARKS
~ 130°C	A few mg	Length 9·2cm Volume 25ml	

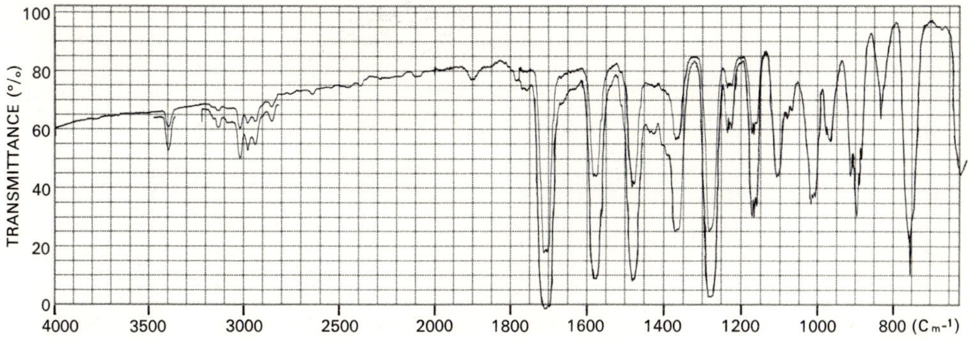

COMPOUND	MOL. FORMULA	B. PT.	INDEX No.
Methyl formate	$C_2H_4O_2$	32°C	**206**

CELL TEMP.	APPROX AMOUNT	CELL PATH	REMARKS
~ 135°C	1·5μl	Length 9·2cm Volume 25ml	

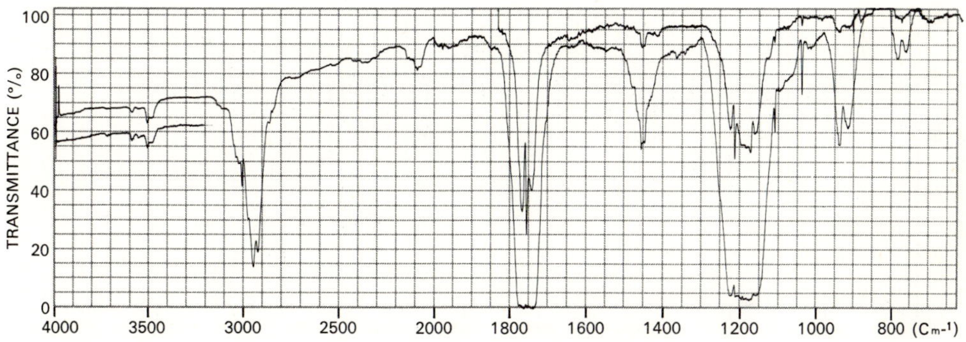

COMPOUND	MOL. FORMULA	B. PT.	INDEX No.
Ethyl formate	$C_3H_6O_2$	54°C	**207**

CELL TEMP.	APPROX AMOUNT	CELL PATH	REMARKS
~ 120°C	1μl	Length 9·2cm Volume 25ml	

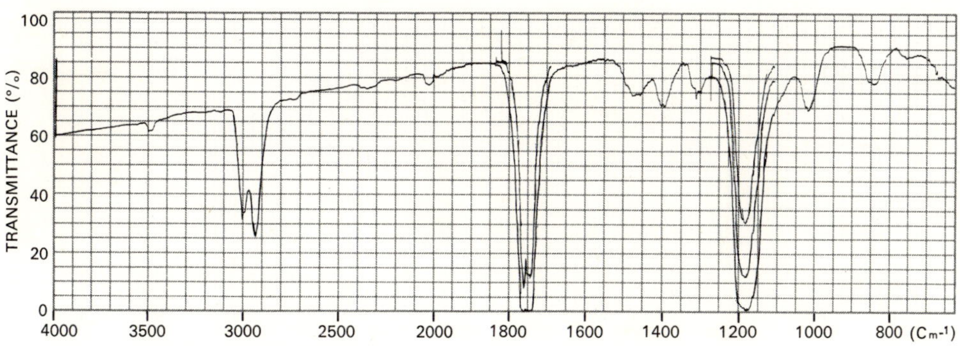

COMPOUND	MOL. FORMULA	B. PT.	INDEX No.
Propyl formate	$C_4H_8O_2$	81°C	**208**

CELL TEMP.	APPROX AMOUNT	CELL PATH	REMARKS
~ 135°C	1μl	Length 9·2cm Volume 25ml	

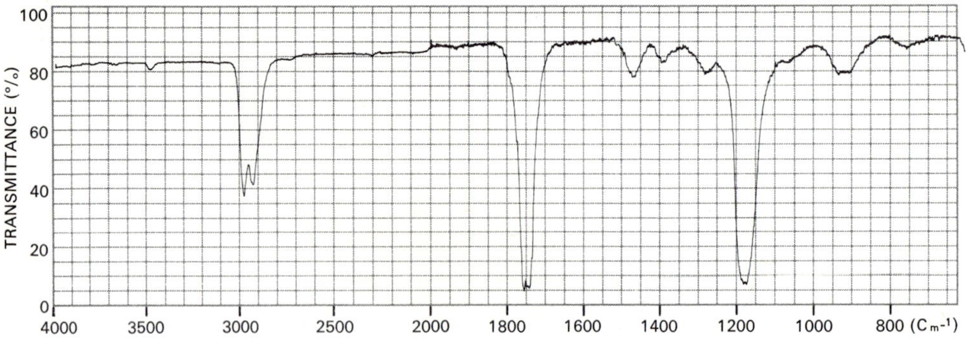

COMPOUND	MOL. FORMULA	B. PT.	INDEX No.
Butyl formate	$C_5H_{10}O_2$	107°C	**209**

CELL TEMP.	APPROX AMOUNT	CELL PATH	REMARKS
~ 130°C	2μl	Length 9·2cm Volume 25ml	

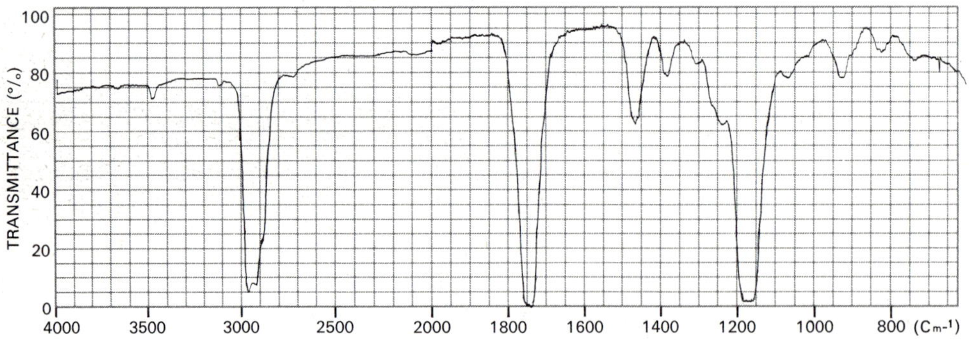

COMPOUND	MOL. FORMULA	B. PT.	INDEX No.
Amyl formate	$C_6H_{12}O_2$	132°C	**210**

CELL TEMP.	APPROX AMOUNT	CELL PATH	REMARKS
~ 130°C	1μl	Length 9·2cm Volume 25ml	Commercial. Mainly n-Pentyl

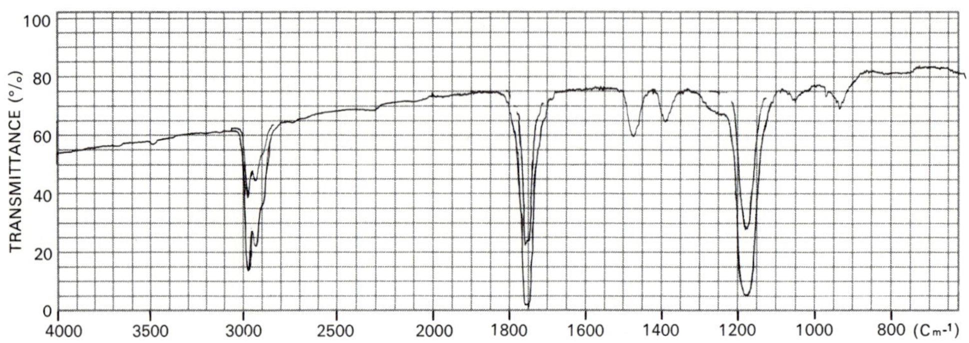

COMPOUND	MOL. FORMULA	B. PT.	INDEX No.
Methyl acetate	$C_3H_6O_2$	57°C	**211**

CELL TEMP.	APPROX AMOUNT	CELL PATH	REMARKS
~135°C	1μl	Length 9·2cm Volume 25ml	

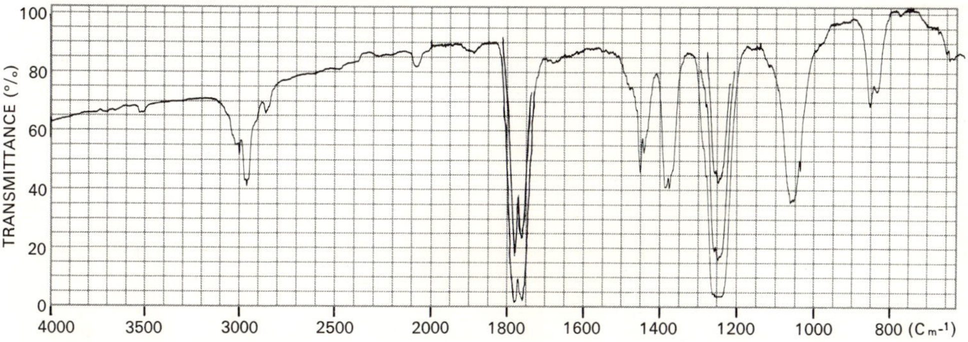

COMPOUND	MOL. FORMULA	B. PT.	INDEX No.
Ethyl acetate	$C_4H_8O_2$	77°C	**212**

CELL TEMP.	APPROX AMOUNT	CELL PATH	REMARKS
~140°C	0·5μl	Length 9·2cm Volume 25ml	

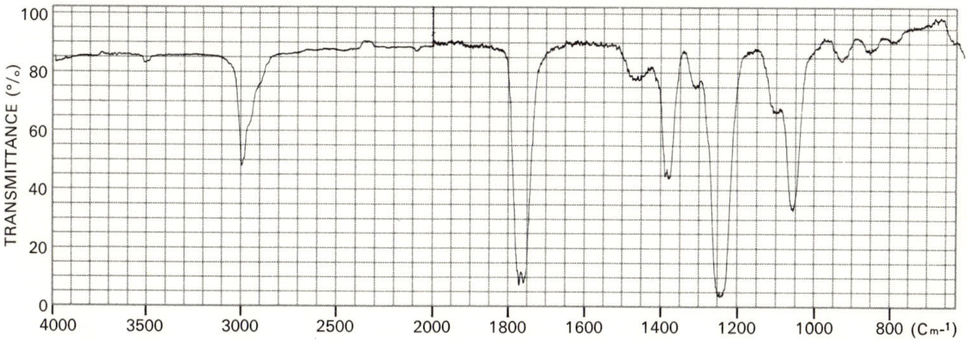

COMPOUND	MOL. FORMULA	B. PT.	INDEX No.
Propyl acetate	$C_5H_{10}O_2$	102°C	**213**

CELL TEMP.	APPROX AMOUNT	CELL PATH	REMARKS
~140°C	2μl	Length 9·2cm Volume 25ml	

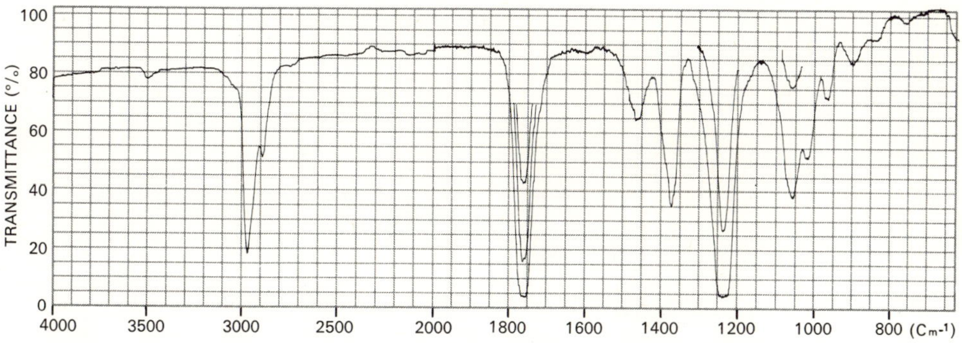

COMPOUND	MOL. FORMULA	B. PT.	INDEX No.
Butyl acetate	$C_6H_{12}O_2$	127°C	**214**

CELL TEMP.	APPROX AMOUNT	CELL PATH	REMARKS
~ 125°C	2μl	Length 9·2cm Volume 25ml	

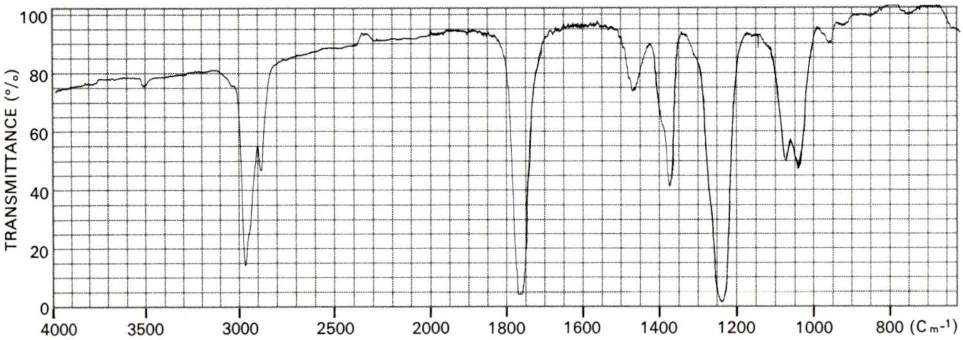

COMPOUND	MOL. FORMULA	B. PT.	INDEX No.
Amyl acetate	$C_7H_{14}O_2$	-	**215**

CELL TEMP.	APPROX AMOUNT	CELL PATH	REMARKS
~ 125°C	2μl	Length 9·2cm Volume 25ml	Commercial. Mainly Isopentyl

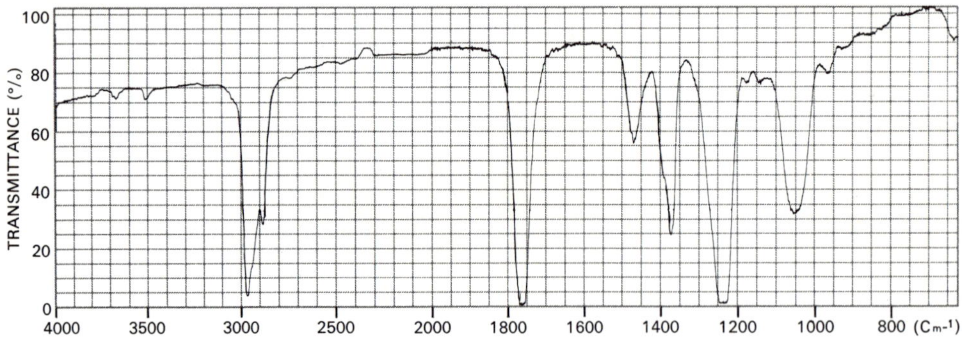

COMPOUND	MOL. FORMULA	B. PT.	INDEX No.
Hexyl acetate	$C_8H_{16}O_2$	169°C	**216**

CELL TEMP.	APPROX AMOUNT	CELL PATH	REMARKS
~ 135°C	2μl	Length 9·2cm Volume 25ml	

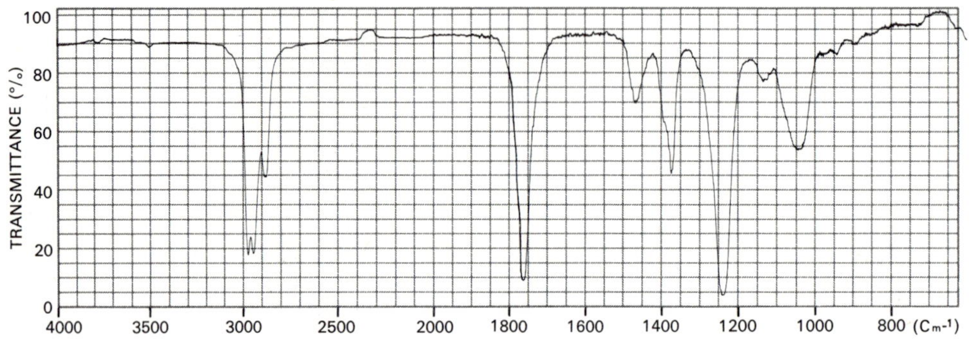

COMPOUND	MOL. FORMULA	B. PT.	INDEX No.
Octyl acetate	$C_{10}H_{20}O_2$	210°C	**217**

CELL TEMP.	APPROX AMOUNT	CELL PATH	REMARKS
~ 135°C	2μl	Length 9·2cm Volume 25ml	

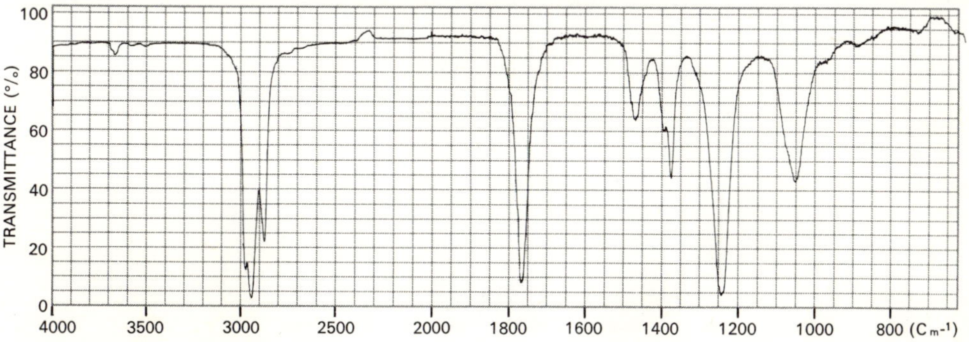

COMPOUND	MOL. FORMULA	B. PT.	INDEX No.
Ethyl propanoate	$C_5H_{10}O_2$	99°C	**218**

CELL TEMP.	APPROX AMOUNT	CELL PATH	REMARKS
~ 130°C	2μl	Length 9·2cm Volume 25ml	

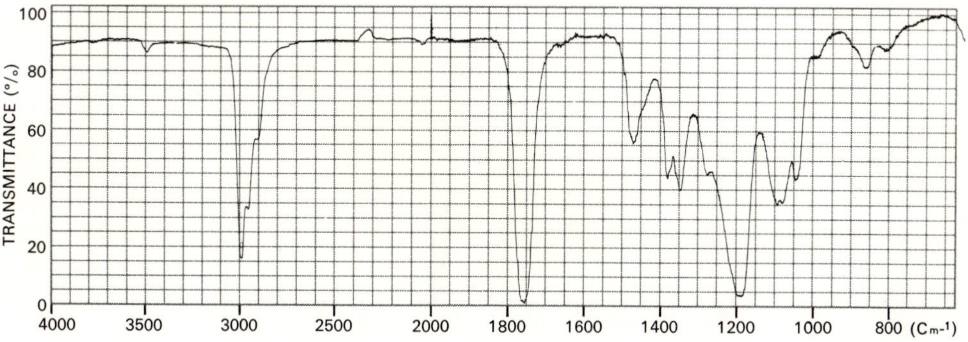

COMPOUND	MOL. FORMULA	B. PT.	INDEX No.
Propyl propanoate	$C_6H_{12}O_2$	123°C	**219**

CELL TEMP.	APPROX AMOUNT	CELL PATH	REMARKS
~ 140°C	2μl	Length 9·2cm Volume 25ml	

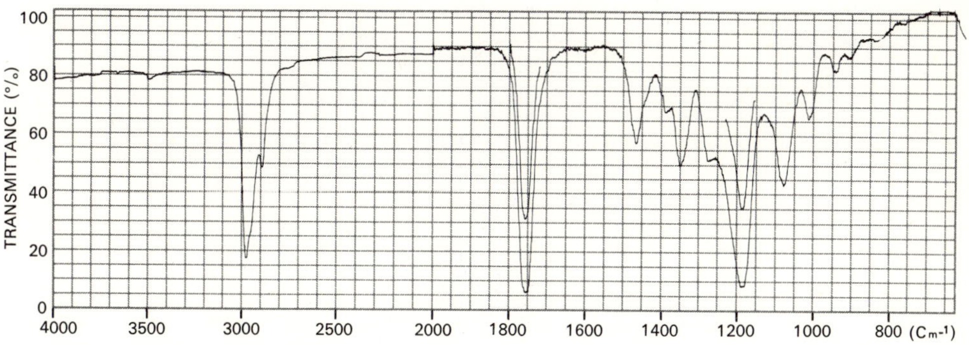

COMPOUND	MOL. FORMULA	B. PT.	INDEX No.
Butyl propanoate	$C_7H_{14}O_2$	146°C	**220**

CELL TEMP.	APPROX AMOUNT	CELL PATH	REMARKS
~ 125°C	2μl	Length 9·2cm Volume 25ml	

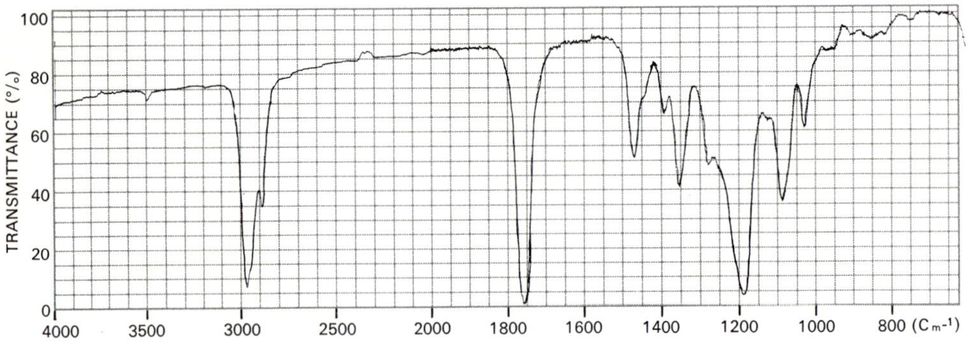

COMPOUND	MOL. FORMULA	B. PT.	INDEX No.
Amyl propanoate	$C_8H_{16}O_2$	-	**221**

CELL TEMP.	APPROX AMOUNT	CELL PATH	REMARKS
~ 120°C	2μl	Length 9·2cm Volume 25ml	Commercial. Probably Isopentyl

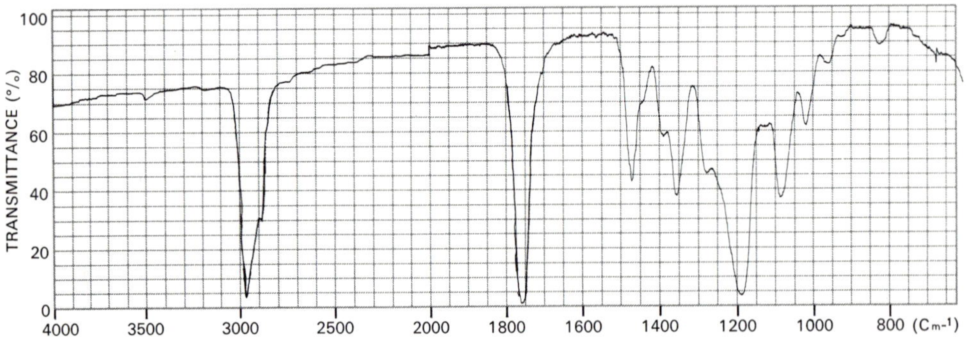

COMPOUND	MOL. FORMULA	B. PT.	INDEX No.
Allyl Propanoate	$C_6H_{10}O_2$	124°C	**222**

CELL TEMP.	APPROX AMOUNT	CELL PATH	REMARKS
~ 130°C	2μl	Length 9·2cm Volume 25ml	

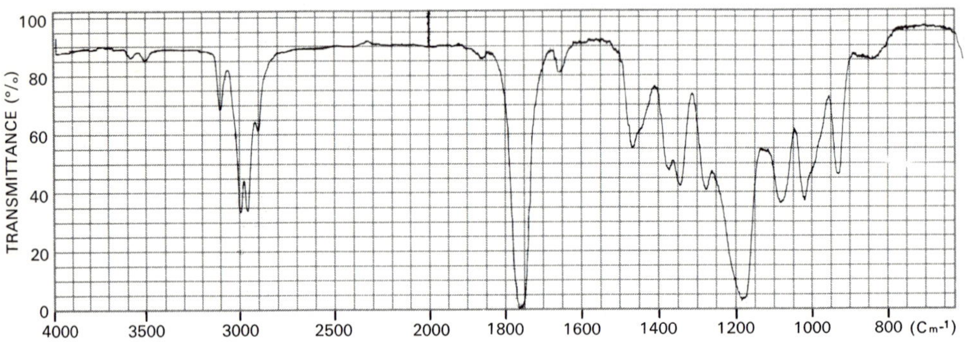

COMPOUND	MOL. FORMULA	B. PT.	INDEX No.
Methyl butanoate	$C_5H_{10}O_2$	103°C	**223**

CELL TEMP.	APPROX AMOUNT	CELL PATH	REMARKS
~130°C	2µl	Length 9·2cm Volume 25ml	

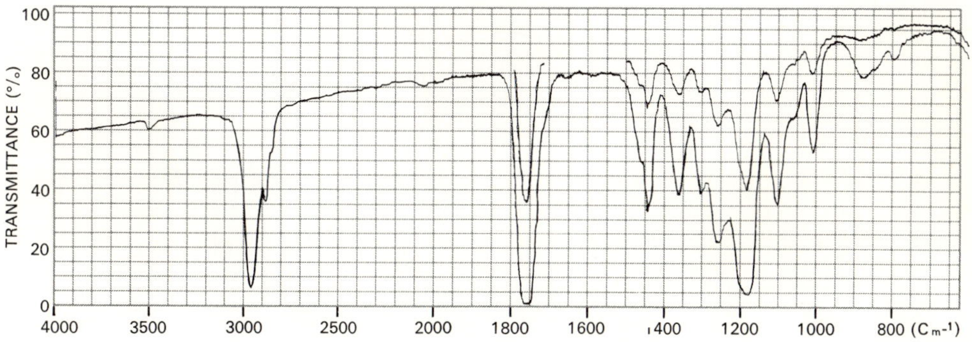

COMPOUND	MOL. FORMULA	B. PT.	INDEX No.
Ethyl butanoate	$C_6H_{12}O_2$	120°C	**224**

CELL TEMP.	APPROX AMOUNT	CELL PATH	REMARKS
~130°C	2µl	Length 9·2cm Volume 25ml	

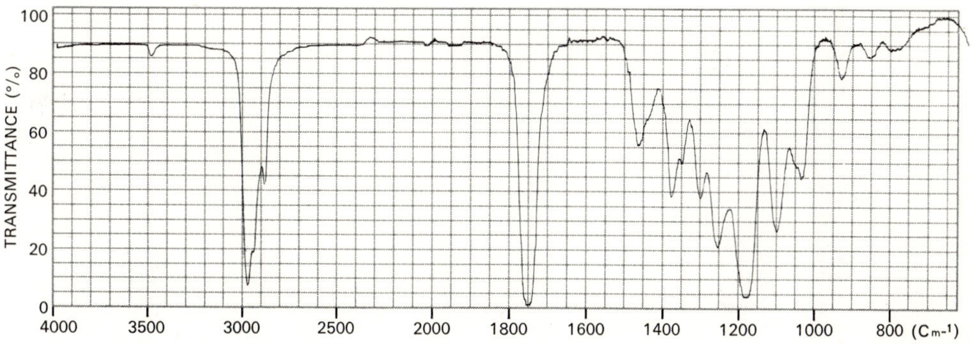

COMPOUND	MOL. FORMULA	B. PT.	INDEX No.
Propyl butanoate	$C_7H_{14}O_2$	143°C	**225**

CELL TEMP.	APPROX AMOUNT	CELL PATH	REMARKS
~145°C	2µl	Length 9·2cm Volume 25ml	

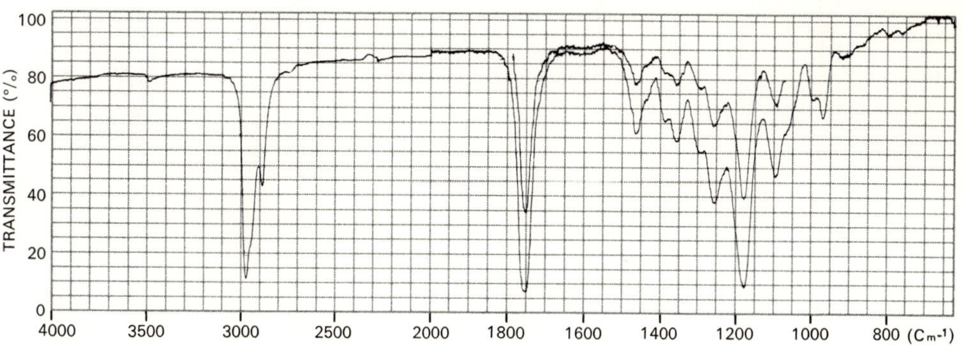

COMPOUND	MOL. FORMULA	B. PT.	INDEX No.
Butyl butanoate	$C_8H_{16}O_2$	167°C	**226**

CELL TEMP.	APPROX AMOUNT	CELL PATH	REMARKS
~125°C	2μl	Length 9·2cm Volume 25ml	

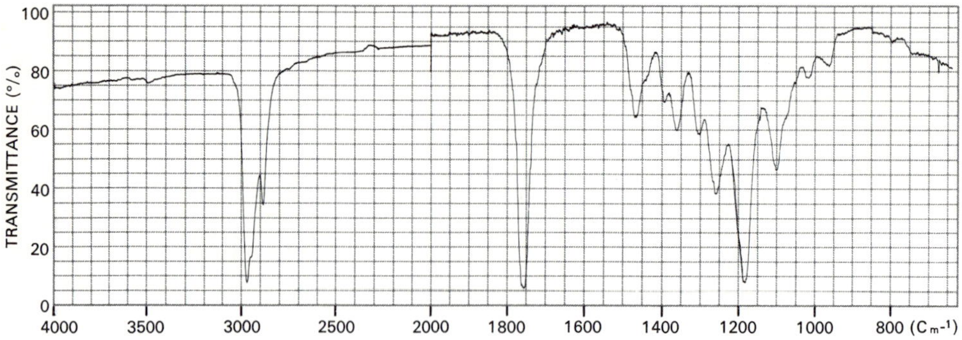

COMPOUND	MOL. FORMULA	B. PT.	INDEX No.
Amyl butanoate	$C_9H_{18}O_2$	-	**227**

CELL TEMP.	APPROX AMOUNT	CELL PATH	REMARKS
~125°C	2μl	Length 9·2cm Volume 25ml	Commercial. Probably Isopentyl

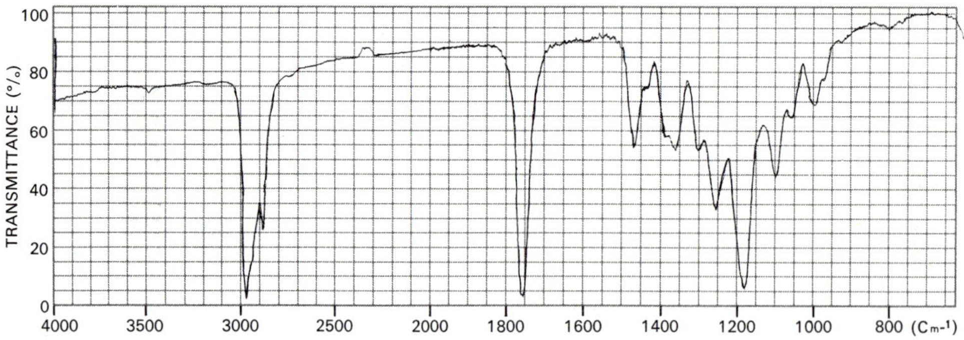

COMPOUND	MOL. FORMULA	B. PT.	INDEX No.
Allyl butanoate	$C_7H_{12}O_2$	143°C	**228**

CELL TEMP.	APPROX AMOUNT	CELL PATH	REMARKS
~ 135°C	2μl	Length 9·2cm Volume 25ml	

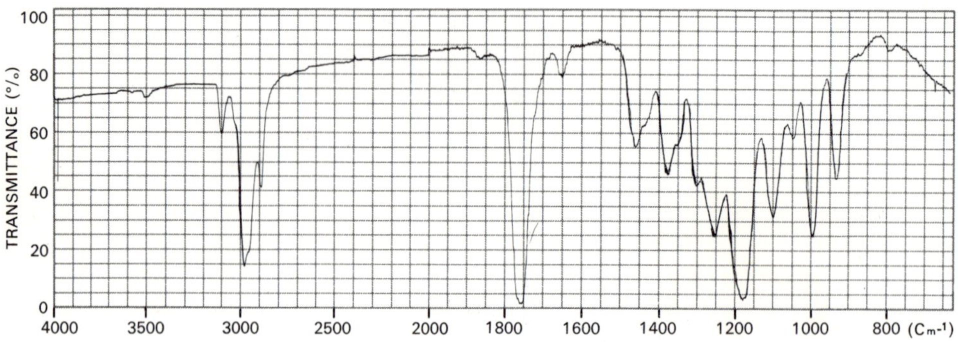

COMPOUND	MOL. FORMULA	B. PT.	INDEX No.
Methyl pentanoate	$C_6H_{12}O_2$	127°C	**229**

CELL TEMP.	APPROX AMOUNT	CELL PATH	REMARKS
~ 145°C	2μl	Length 9·2cm Volume 25ml	

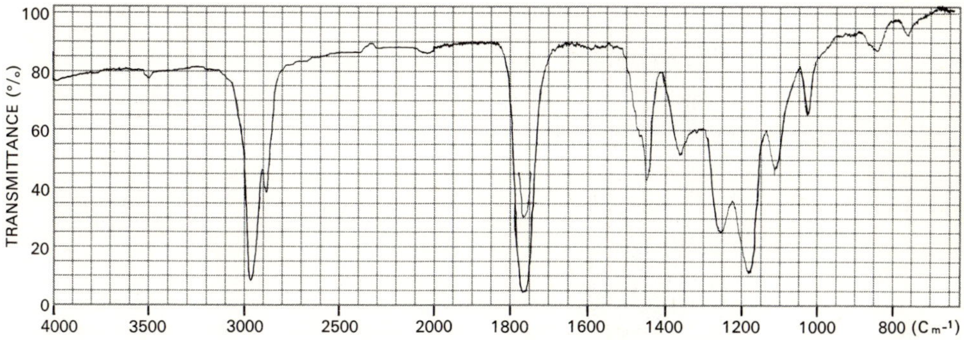

COMPOUND	MOL. FORMULA	B. PT.	INDEX No.
Propyl pentanoate	$C_8H_{16}O_2$	168°C	**230**

CELL TEMP.	APPROX AMOUNT	CELL PATH	REMARKS
~ 145°C	2μl	Length 9·2cm Volume 25ml	

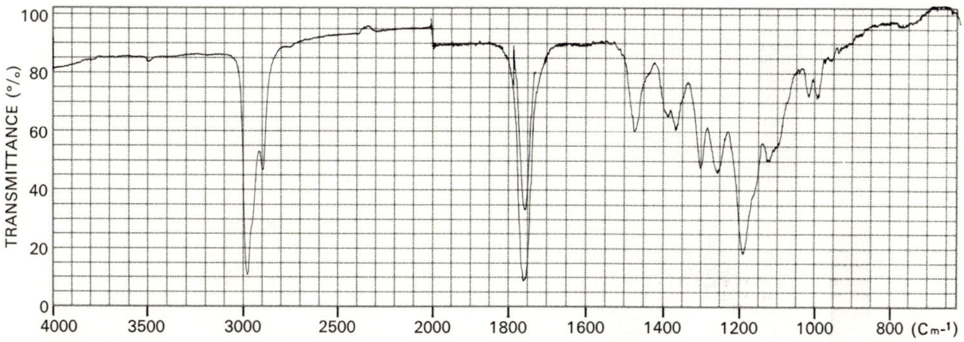

COMPOUND	MOL. FORMULA	B. PT.	INDEX No.
Butyl pentanoate	$C_9H_{18}O_2$	186°C	**231**

CELL TEMP.	APPROX AMOUNT	CELL PATH	REMARKS
~ 145°C	2μl	Length 9·2cm Volume 25ml	

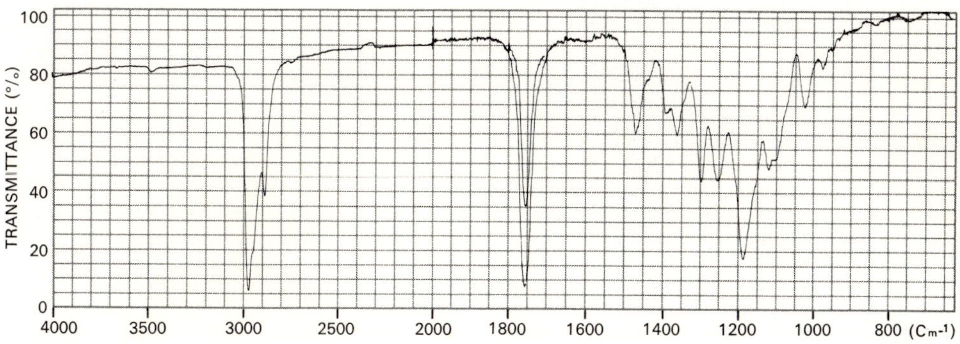

COMPOUND	MOL. FORMULA	B. PT.	INDEX No.
Amyl pentanoate	$C_{10}H_{20}O_2$	-	**232**

CELL TEMP.	APPROX AMOUNT	CELL PATH	REMARKS
~ 135°C	2μl	Length 9·2cm Volume 25ml	Commercial. Probably N-Pentyl

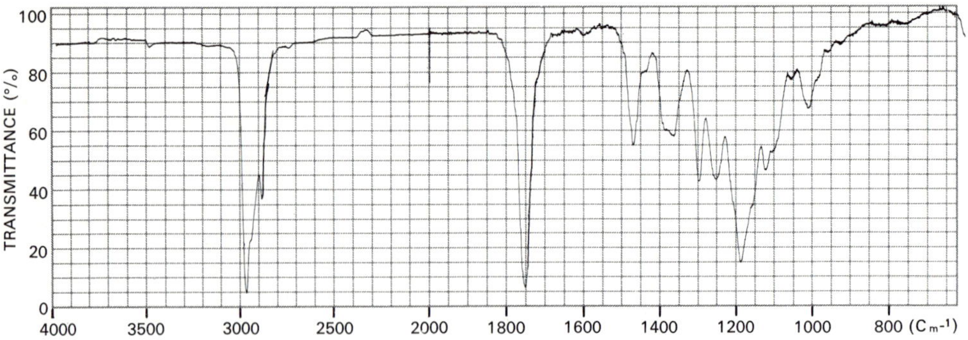

COMPOUND	MOL. FORMULA	B. PT.	INDEX No.
Ethyl-3-methylbutanoate	$C_7H_{14}O_2$	134° C	**233**

CELL TEMP.	APPROX AMOUNT	CELL PATH	REMARKS
~ 115° C	2μl	Length 9·2cm Volume 25ml	

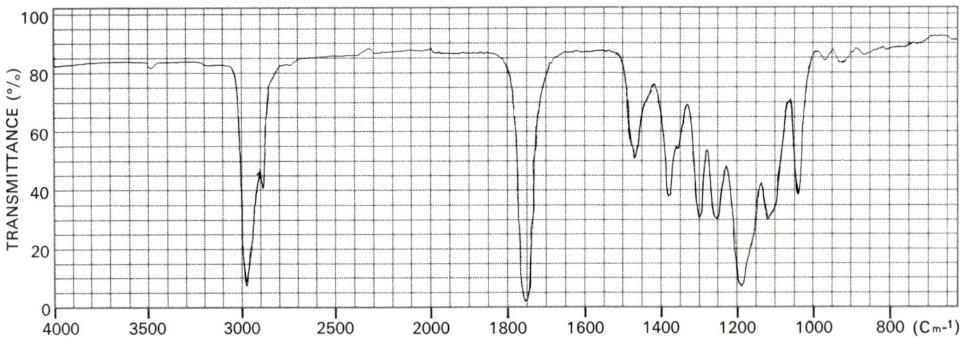

COMPOUND	MOL. FORMULA	B. PT.	INDEX No.
Methyl-2-methylbutanoate	$C_6H_{12}O_2$	-	**234**

CELL TEMP.	APPROX AMOUNT	CELL PATH	REMARKS
~ 145° C	2μl	Length 9·2cm Volume 25ml	

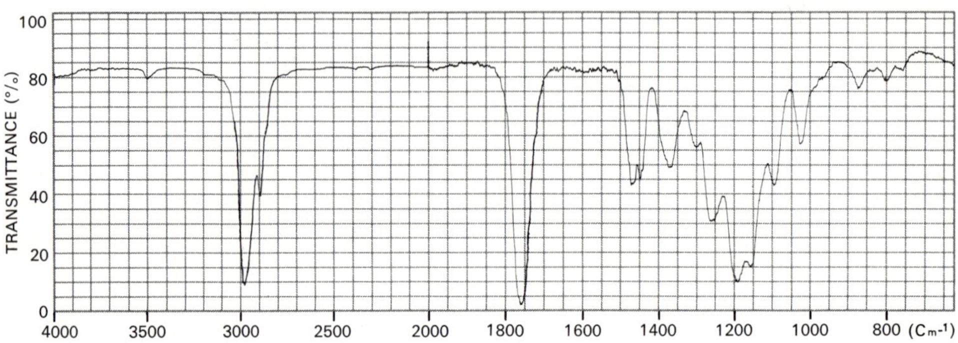

COMPOUND	MOL. FORMULA	B. PT.	INDEX No.
Ethyl 2-methylbutanoate	$C_7H_{14}O_2$	130°C	**235**

CELL TEMP.	APPROX AMOUNT	CELL PATH	REMARKS
~140°C	2μl	Length 9·2cm Volume 25ml	

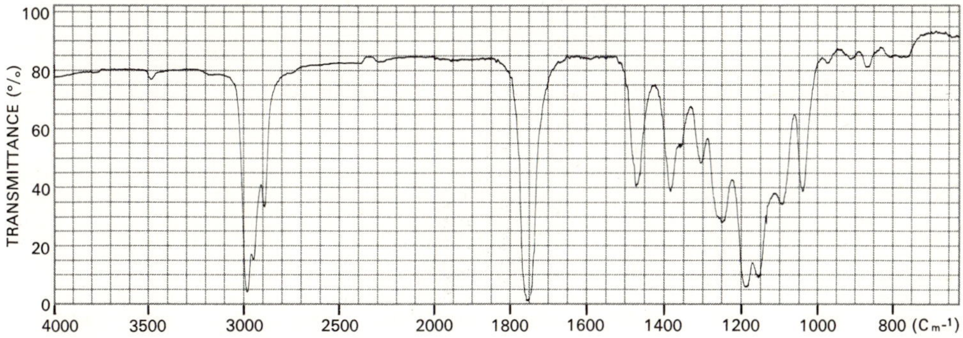

COMPOUND	MOL. FORMULA	B. PT.	INDEX No.
Methyl hexanoate	$C_7H_{14}O_2$	151°C	**236**

CELL TEMP.	APPROX AMOUNT	CELL PATH	REMARKS
~135°C	2μl	Length 9·2cm Volume 25ml	

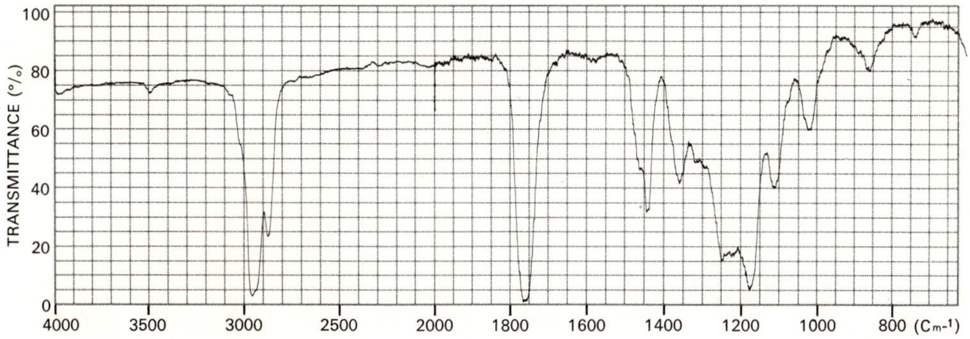

COMPOUND	MOL. FORMULA	B. PT.	INDEX No.
Ethyl hexanoate	$C_8H_{16}O_2$	168°C	**237**

CELL TEMP.	APPROX AMOUNT	CELL PATH	REMARKS
~130°C	2μl	Length 9·2cm Volume 25ml	

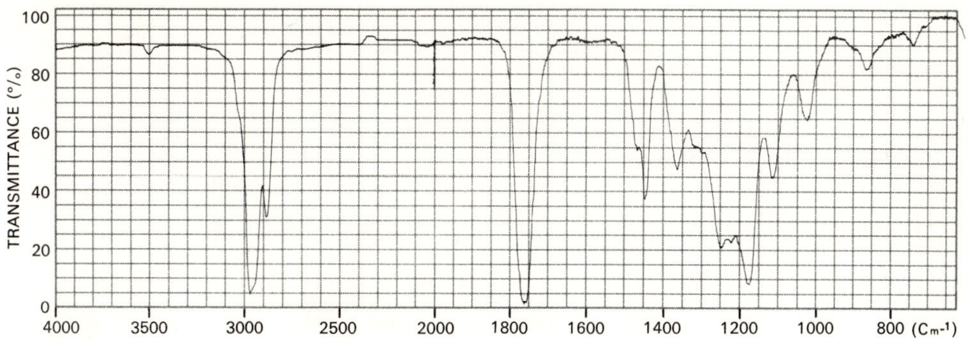

COMPOUND	MOL. FORMULA	B. PT.	INDEX No.
Amyl hexanoate	$C_{11}H_{22}O_2$	-	**238**

CELL TEMP.	APPROX AMOUNT	CELL PATH	REMARKS
~125°C	2µl	Length 9·2cm Volume 25ml	Commercial. Probably N-Pentyl

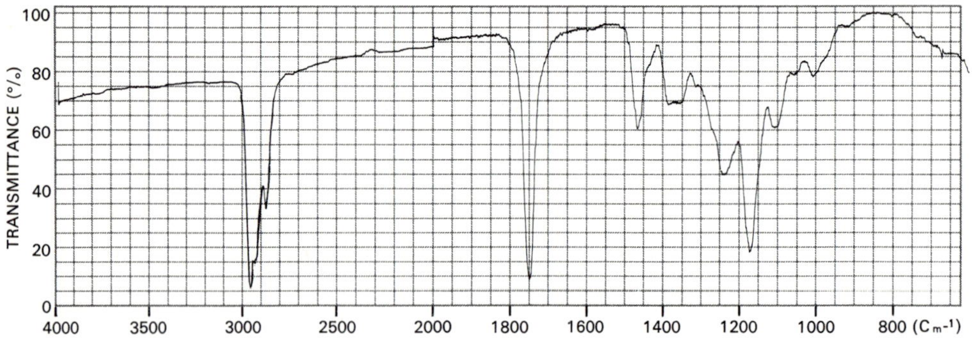

COMPOUND	MOL. FORMULA	B. PT.	INDEX No.
Allyl hexanoate	$C_9H_{16}O_2$	185-187°C	**239**

CELL TEMP.	APPROX AMOUNT	CELL PATH	REMARKS
~135°C	2µl	Length 9·2cm Volume 25ml	

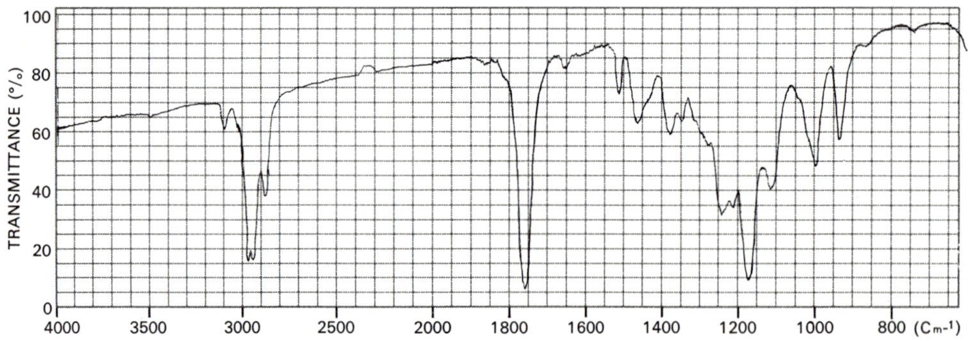

COMPOUND	MOL. FORMULA	B. PT.	INDEX No.
Dimethyl malonate	$C_5H_8O_4$	183°C	**240**

CELL TEMP.	APPROX AMOUNT	CELL PATH	REMARKS
~140°C	0·5µl	Length 9·2cm Volume 25ml	

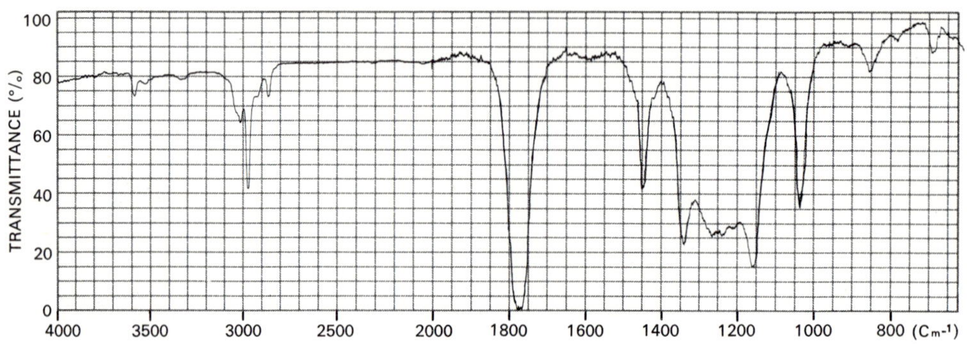

COMPOUND			MOL. FORMULA	B. PT.	INDEX No.
Diethyl malonate			$C_7H_{12}O_4$	199°C	**241**
CELL TEMP.	APPROX AMOUNT	CELL PATH	REMARKS		
~140°C	0·5μl	Length 9·2cm Volume 25ml			

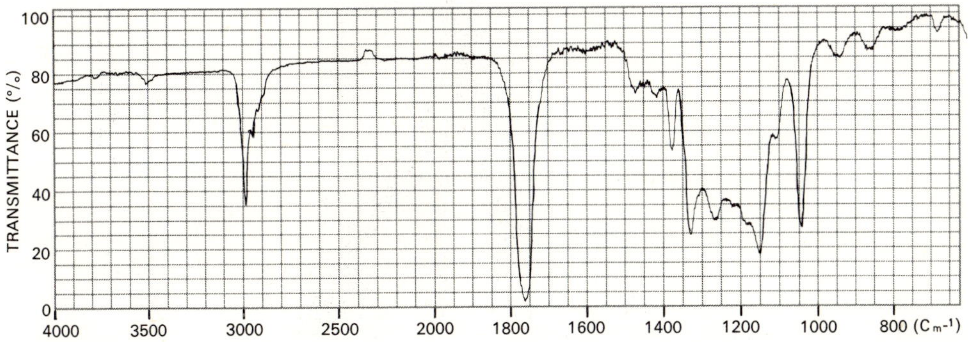

COMPOUND			MOL. FORMULA	B. PT.	INDEX No.
Dimethyl succinate			$C_6H_{10}O_4$	196°C	**242**
CELL TEMP.	APPROX AMOUNT	CELL PATH	REMARKS		
140°C	0·5μl	Length 9·2cm Volume 25ml			

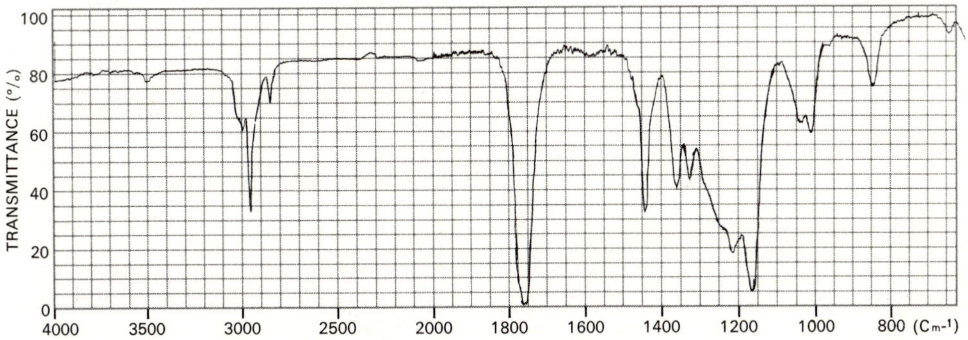

COMPOUND			MOL. FORMULA	B. PT.	INDEX No.
3-Oxobutanoate (Ethyl acetoacetate)			$C_6H_{10}O_3$	180°C	**243**
CELL TEMP.	APPROX AMOUNT	CELL PATH	REMARKS		
~130°C	2μl	Length 9·2cm Volume 25ml			

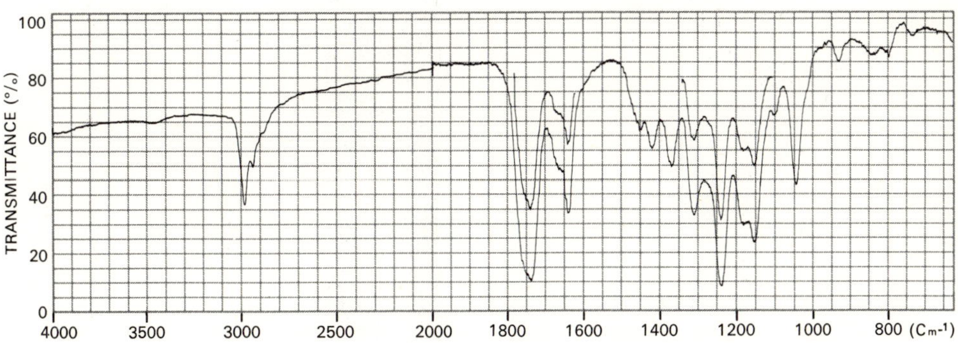

COMPOUND	MOL. FORMULA	B. PT.	INDEX No.
2-Hydroxypropanate (Methyl lactate)	$C_4H_8O_3$	145°C	**244**

CELL TEMP.	APPROX AMOUNT	CELL PATH	REMARKS
~150°C	1μl	Length 9·2cm Volume 25ml	

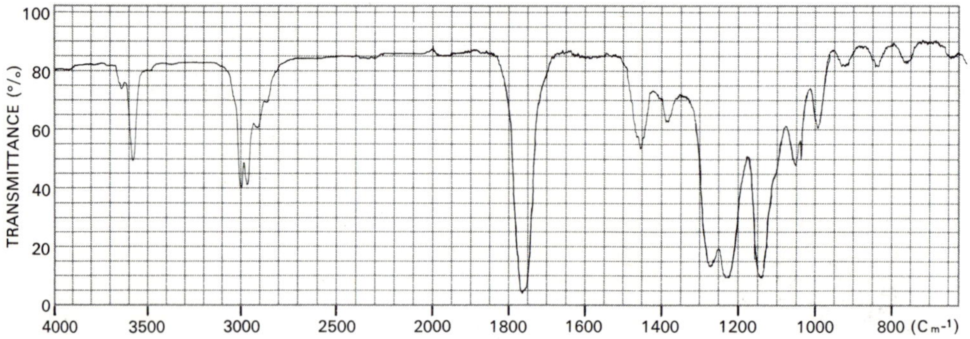

COMPOUND	MOL. FORMULA	B. PT.	INDEX No.
2-Hydroxypropanate (Ethyl lactate)	$C_5H_{10}O_3$	155°C	**245**

CELL TEMP.	APPROX AMOUNT	CELL PATH	REMARKS
~125°C	2μl	Length 9·2cm Volume 25ml	

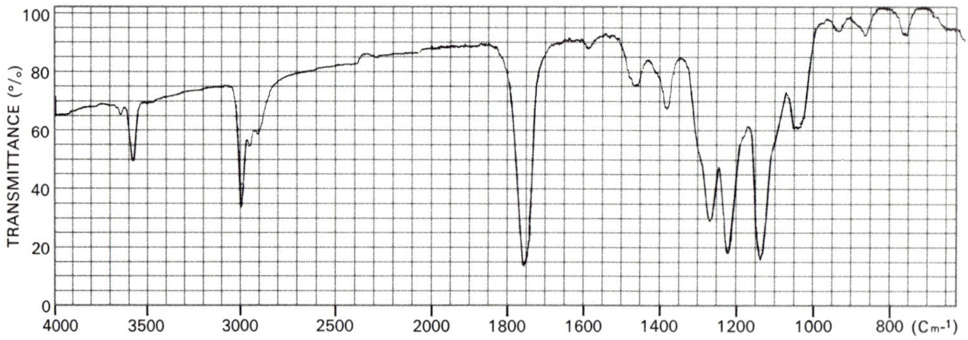

COMPOUND	MOL. FORMULA	B. PT.	INDEX No.
2-Hydroxypropanate (Butyl lactate)	$C_7H_{14}O_3$	83°C/13mm	**246**

CELL TEMP.	APPROX AMOUNT	CELL PATH	REMARKS
~105°C	2μl	Length 9·2cm Volume 25ml	

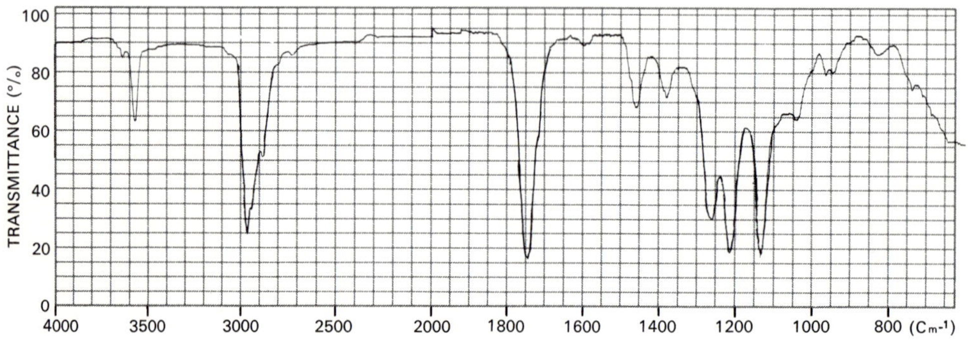

COMPOUND	MOL. FORMULA	B. PT.	INDEX No.
Methyl 2-hydroxy-2-methylpropanoate	$C_5H_{10}O_3$	137°C	**247**

CELL TEMP.	APPROX AMOUNT	CELL PATH	REMARKS
~120°C	1·5μl	Length 9·2cm Volume 25ml	

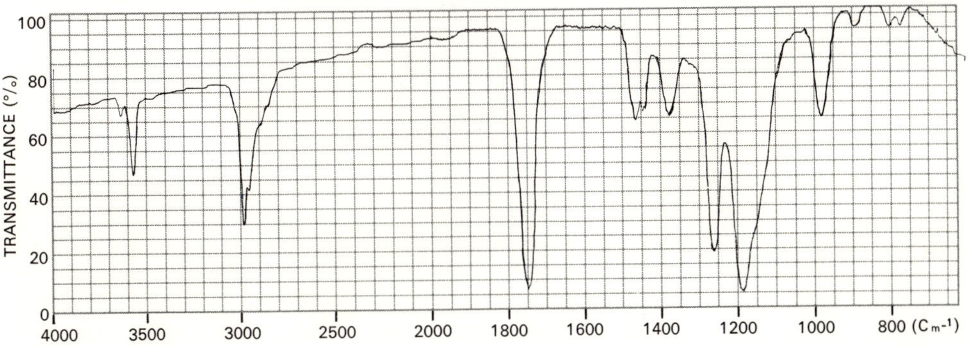

COMPOUND	MOL. FORMULA	B. PT.	INDEX No.
Ethyl 2-hydroxy-2-methylpropanoate	$C_6H_{12}O_3$	150°C	**248**

CELL TEMP.	APPROX AMOUNT	CELL PATH	REMARKS
~120°C	2μl	Length 9·2cm Volume 25ml	

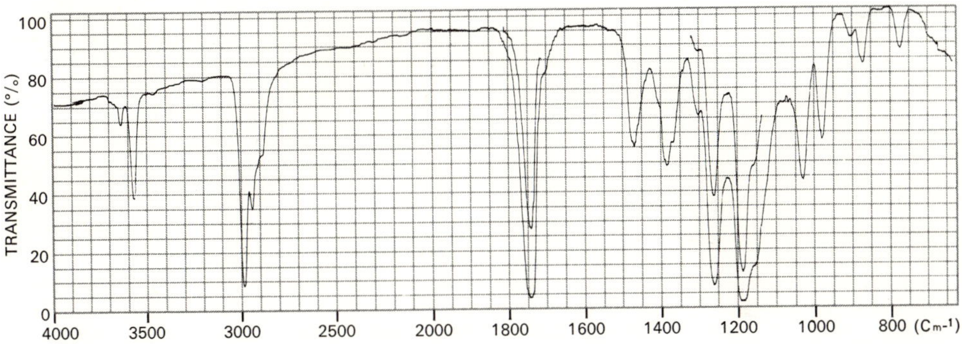

COMPOUND	MOL. FORMULA	B. PT.	INDEX No.
Ethyl 3-hydroxybutanoate	$C_6H_{12}O_3$	185°C	**249**

CELL TEMP.	APPROX AMOUNT	CELL PATH	REMARKS
~120°C	2μl	Length 9·2cm Volume 25ml	

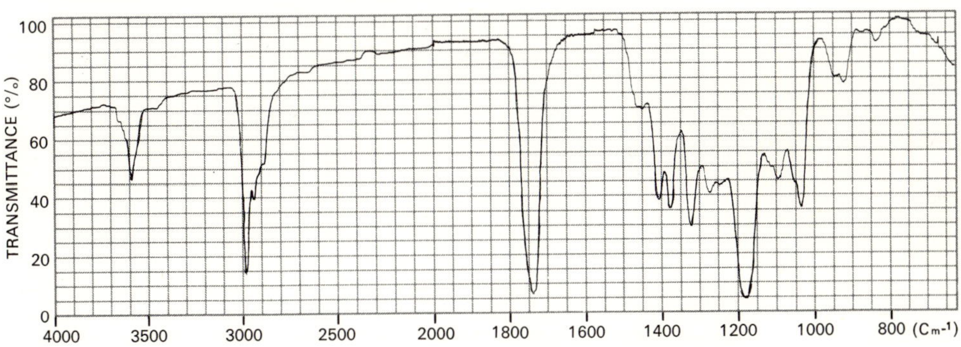

COMPOUND	MOL. FORMULA	B. PT.	INDEX No.
Ethyl 3-aminobutanoate	$C_6H_{13}NO_2$	-	**250**

CELL TEMP.	APPROX AMOUNT	CELL PATH	REMARKS
~140°C	2μl	Length 9·2cm Volume 25ml	

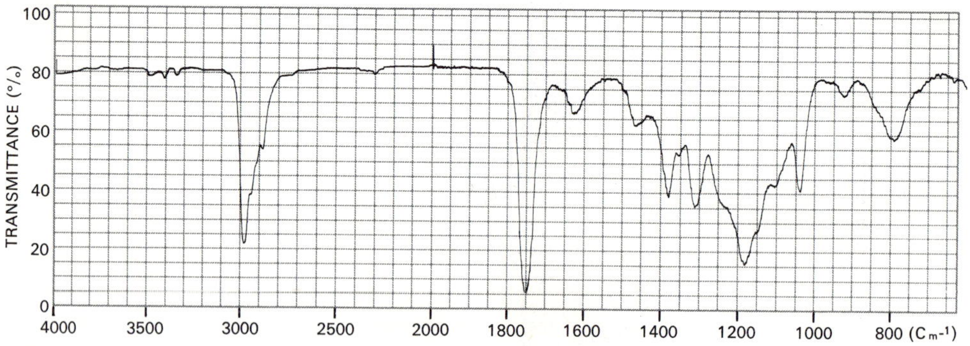

COMPOUND	MOL. FORMULA	B. PT.	INDEX No.
Ethyl but-2-enoate (Ethyl crotonate)	$C_6H_{10}O_2$	143-147°C	**251**

CELL TEMP.	APPROX AMOUNT	CELL PATH	REMARKS
~130°C	1·5μl	Length 9·2cm Volume 25ml	

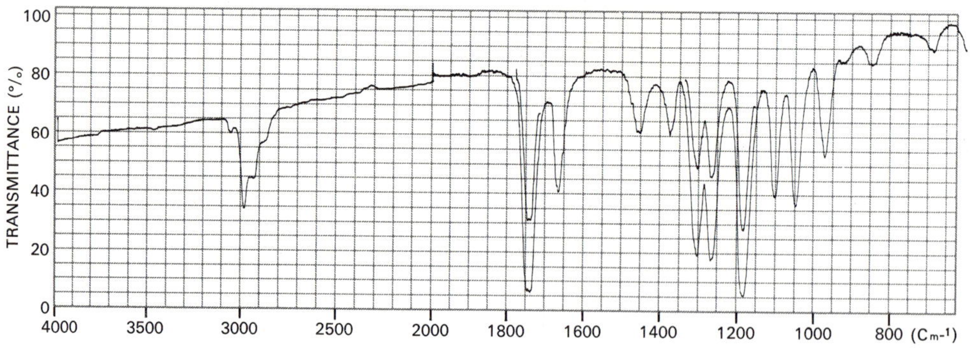

COMPOUND	MOL. FORMULA	B. PT.	INDEX No.
Methyl oct-2-ynoate	$C_9H_{14}O_2$	-	**252**

CELL TEMP.	APPROX AMOUNT	CELL PATH	REMARKS
~135°C	2μl	Length 9·2cm Volume 25ml	

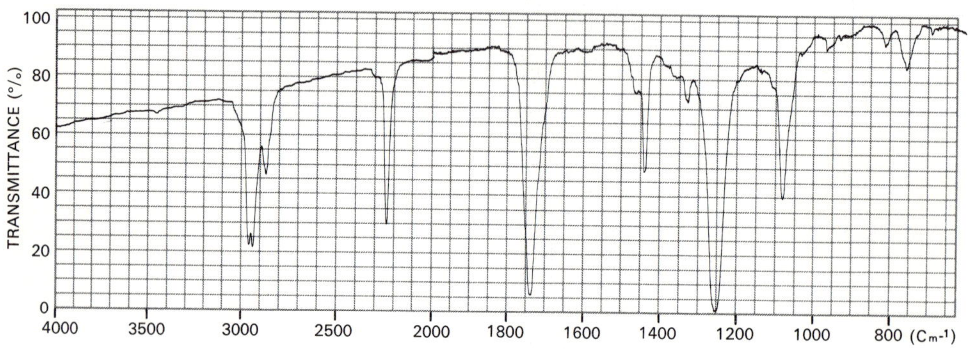

COMPOUND	MOL. FORMULA	B. PT.	INDEX No.
Methyl benzoate	$C_8H_8O_2$	196°C	**253**

CELL TEMP.	APPROX AMOUNT	CELL PATH	REMARKS
~130°C	2μl	Length 9·2cm Volume 25ml	

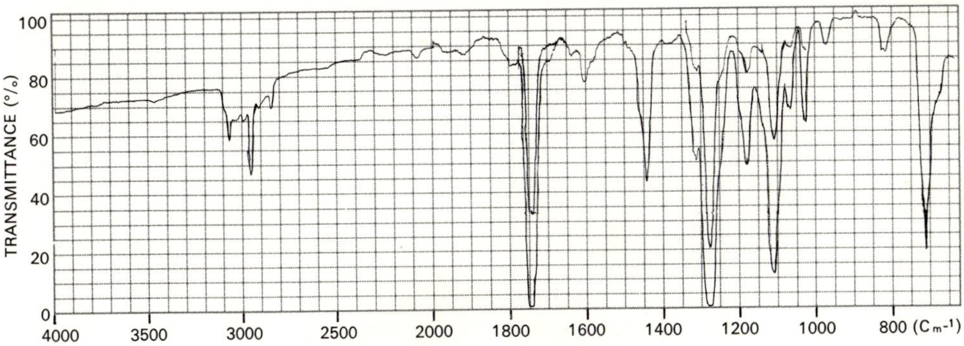

COMPOUND	MOL. FORMULA	B. PT.	INDEX No.
Ethyl benzoate	$C_9H_{10}O_2$	213°C	**254**

CELL TEMP.	APPROX AMOUNT	CELL PATH	REMARKS
~130°C	2μl	Length 9·2cm Volume 25ml	

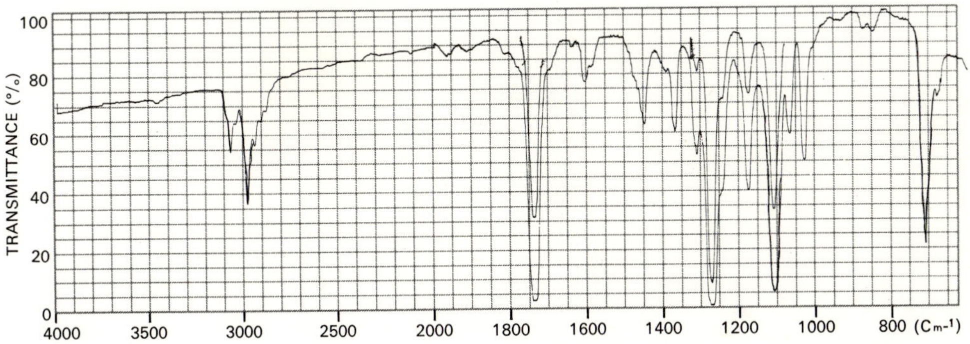

COMPOUND	MOL. FORMULA	B. PT.	INDEX No.
Methyl furoate	$C_6H_6O_3$	181°C	**255**

CELL TEMP.	APPROX AMOUNT	CELL PATH	REMARKS
~130°C	1μl	Length 9·2cm Volume 25ml	

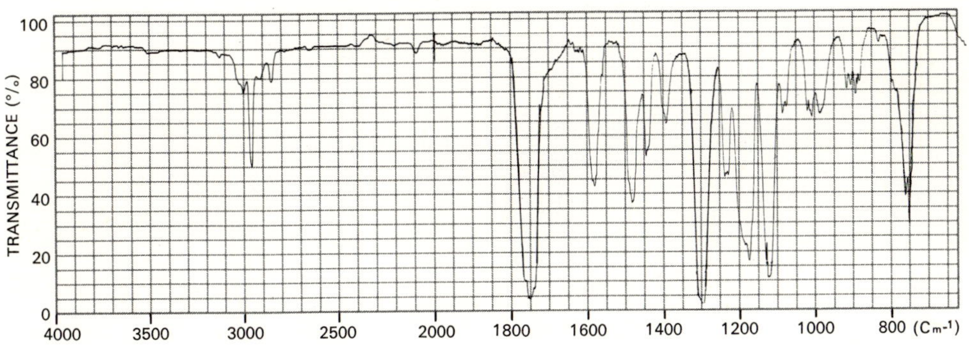

COMPOUND	MOL. FORMULA	B. PT.	INDEX No.
Ethyl furoate	$C_7H_8O_3$	197°C	**256**

CELL TEMP.	APPROX AMOUNT	CELL PATH	REMARKS
~130°C	A few mg	Length 9·2cm Volume 25ml	

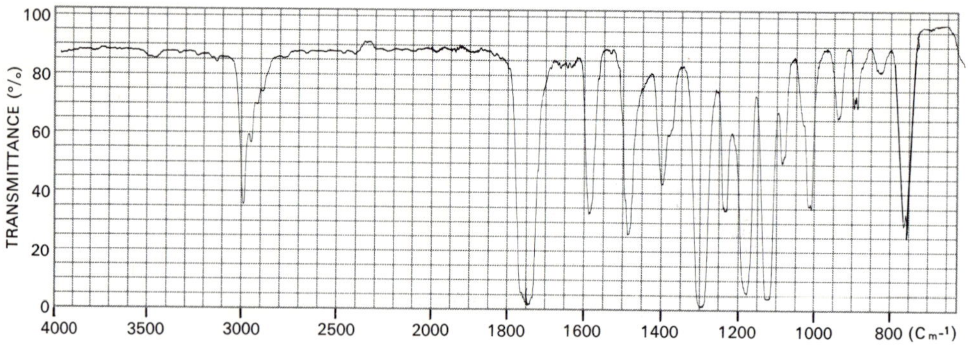

COMPOUND	MOL. FORMULA	B. PT.	INDEX No.
Vinyl acetate	$C_4H_6O_2$	73°C	**257**

CELL TEMP.	APPROX AMOUNT	CELL PATH	REMARKS
~130°C	2μl	Length 9·2cm Volume 25ml	

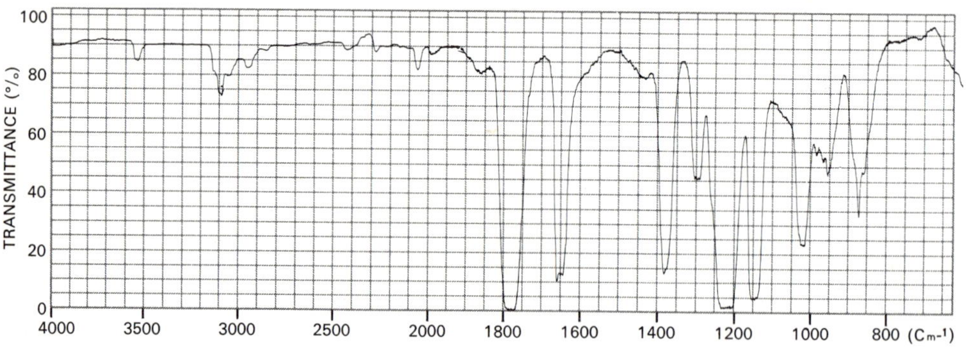

COMPOUND	MOL. FORMULA	B. PT.	INDEX No.
Isopropenyl acetate	$C_5H_8O_2$ 1	~94°C	**258**

CELL TEMP.	APPROX AMOUNT	CELL PATH	REMARKS
~130°C	1μl	Length 9·2cm Volume 25ml	

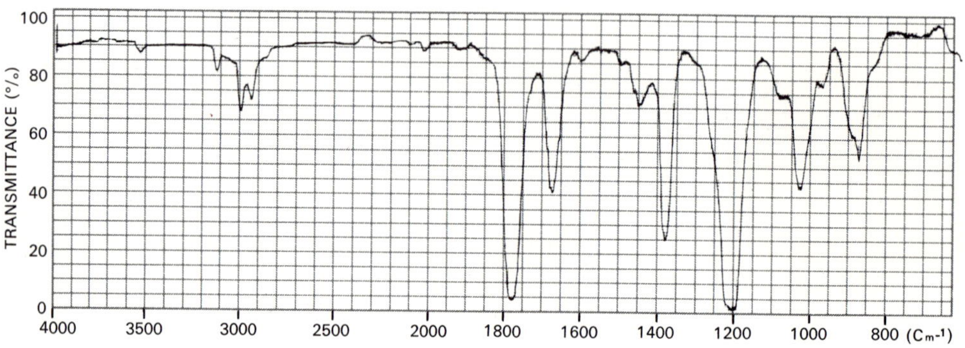

COMPOUND	MOL. FORMULA	B. PT.	INDEX No.
Linalyl acetate	$C_{12}H_{20}O_2$	220°C	**259**

CELL TEMP.	APPROX AMOUNT	CELL PATH	REMARKS
~140°C	2μl	Length 9·2cm Volume 25ml	

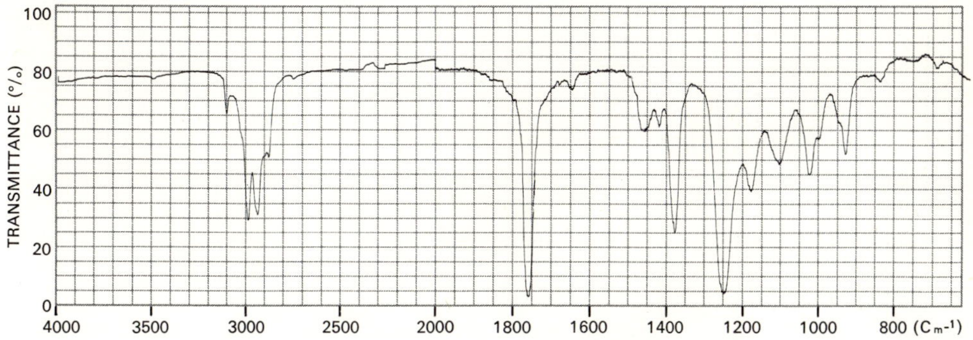

COMPOUND	MOL. FORMULA	B. PT.	INDEX No.
Phenyl acetate	$C_8H_8O_2$	196°C	**260**

CELL TEMP.	APPROX AMOUNT	CELL PATH	REMARKS
~130°C	1μl	Length 9·2cm Volume 25ml	

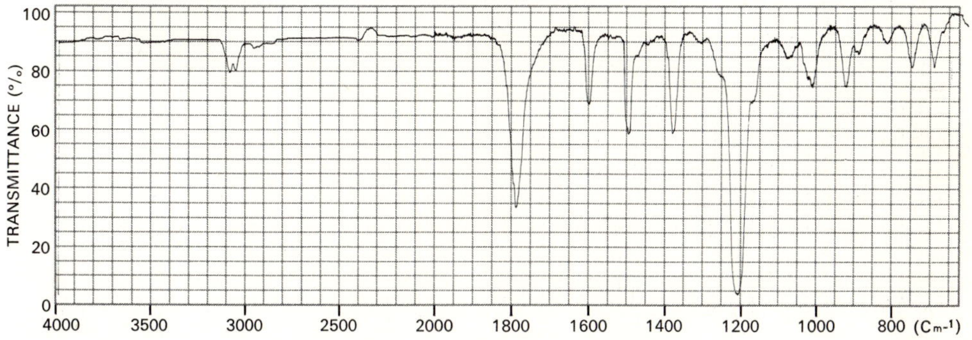

COMPOUND	MOL. FORMULA	B. PT.	INDEX No.
p-Cresyl isobutanoate	$C_{11}H_{14}O_2$	-	**261**

CELL TEMP.	APPROX AMOUNT	CELL PATH	REMARKS
~135°C	2μl	Length 9·2cm Volume 25ml	

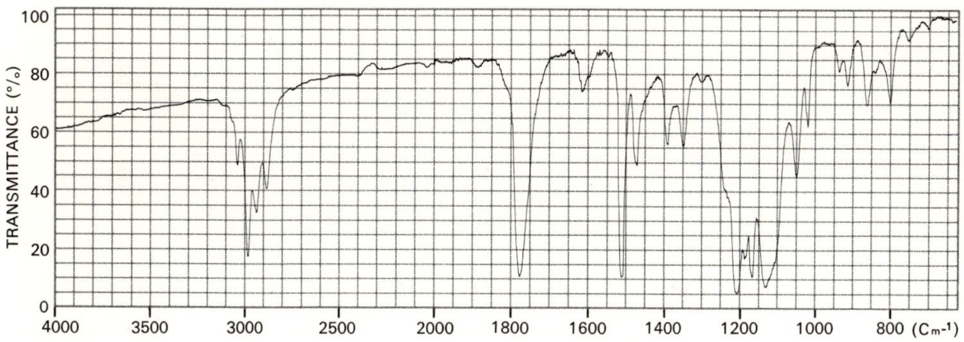

COMPOUND	MOL. FORMULA	B. PT.	INDEX No.
p-Methoxyphenylformate (Anisyl formate)	$C_8H_8O_3$	-	**262**

CELL TEMP.	APPROX AMOUNT	CELL PATH	REMARKS
~130°C	2μl	Length 9·2cm Volume 25ml	

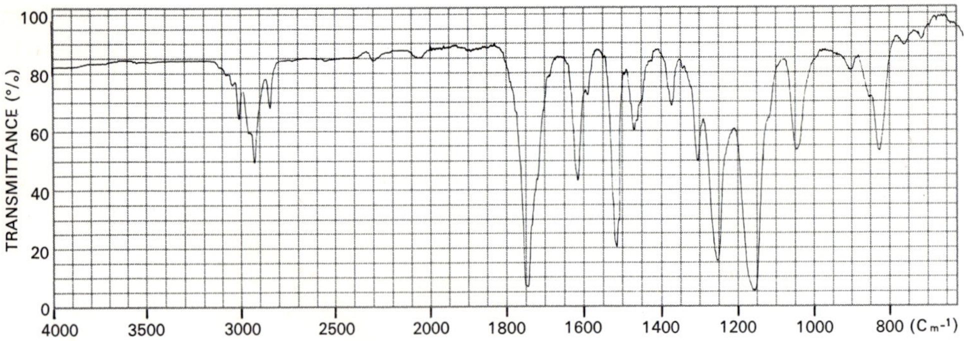

COMPOUND	MOL. FORMULA	B. PT.	INDEX No.
p-Methoxyphenylacetate (Anisyl acetate)	$C_9H_{10}O_3$	-	**263**

CELL TEMP.	APPROX AMOUNT	CELL PATH	REMARKS
~130°C	2μl	Length 9·2cm Volume 25ml	

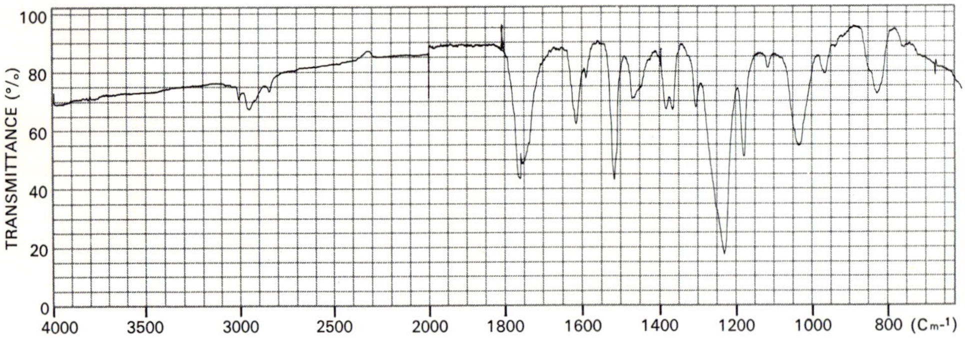

COMPOUND	MOL. FORMULA	B. PT.	INDEX No.
Methyl salicylate	$C_8H_8O_3$	223°C	**264**

CELL TEMP.	APPROX AMOUNT	CELL PATH	REMARKS
~125°C	2μl	Length 9·2cm Volume 25ml	

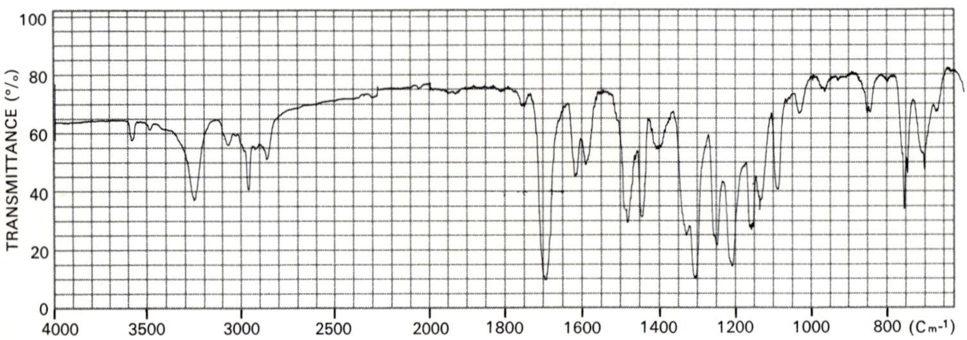

COMPOUND	MOL. FORMULA	B. PT.	INDEX No.
Methyl anthranilate	$C_8H_9NO_2$	256°C	**265**

CELL TEMP.	APPROX AMOUNT	CELL PATH	REMARKS
~150°C	2μl	Length 9·2cm Volume 25ml	

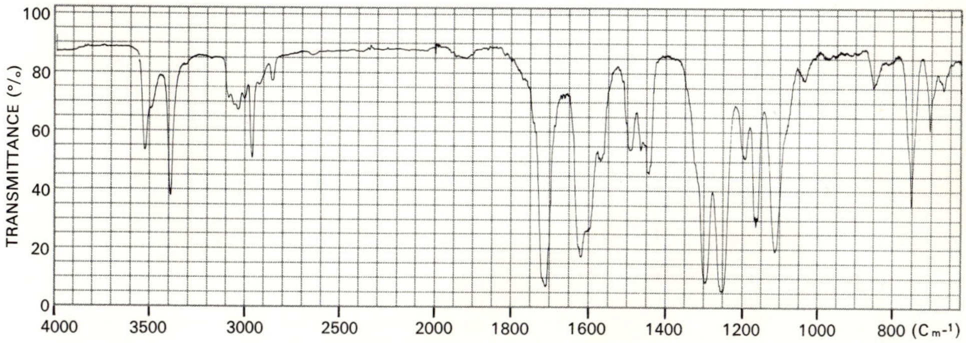

COMPOUND	MOL. FORMULA	B. PT.	INDEX No.
Ethyl anthranilate	$C_9H_{11}NO_2$	268°C	**266**

CELL TEMP.	APPROX AMOUNT	CELL PATH	REMARKS
~150°C	1·5μl	Length 9·2cm Volume 25ml	

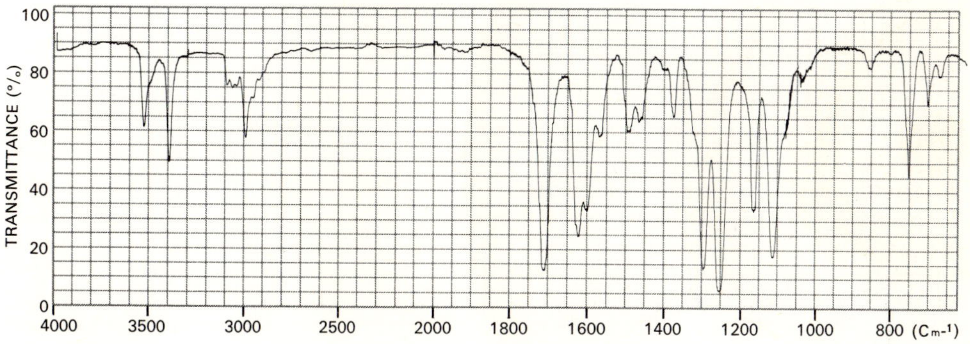

COMPOUND	MOL. FORMULA	B. PT.	INDEX No.
Methyl N-methyl anthranilate	$C_9H_{11}NO_2$	-	**267**

CELL TEMP.	APPROX AMOUNT	CELL PATH	REMARKS
~130°C	2·5μl	Length 9·2cm Volume 25ml	

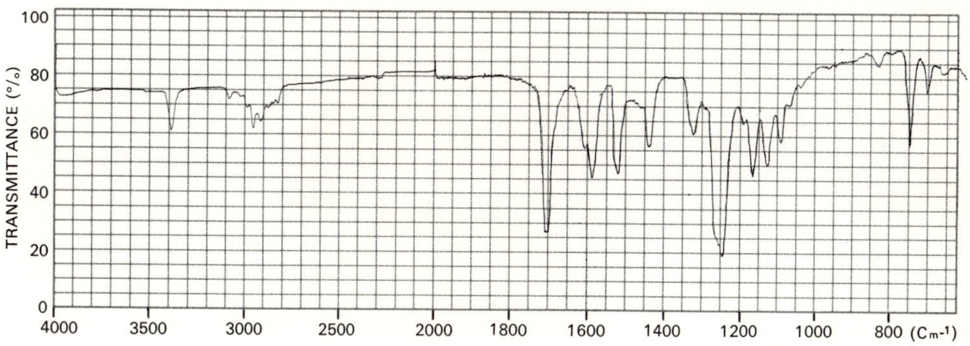

COMPOUND	MOL. FORMULA	B. PT.	INDEX No.
γ-Butyrolactone	$C_4H_6O_2$	206°C	**268**

CELL TEMP.	APPROX AMOUNT	CELL PATH	REMARKS
~150°C	0·5μl	Length 9·2cm Volume 25ml	

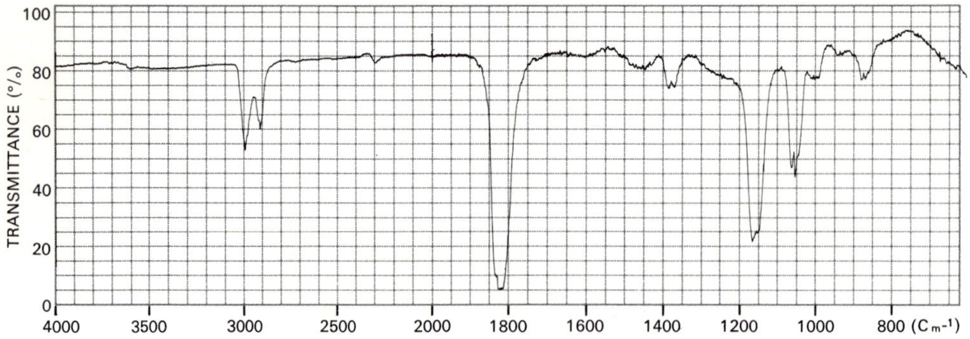

COMPOUND	MOL. FORMULA	B. PT.	INDEX No.
γ-Pentalactone	$C_5H_8O_2$	206°C	**269**

CELL TEMP.	APPROX AMOUNT	CELL PATH	REMARKS
~150°C	0·5μl	Length 9·2cm Volume 25ml	

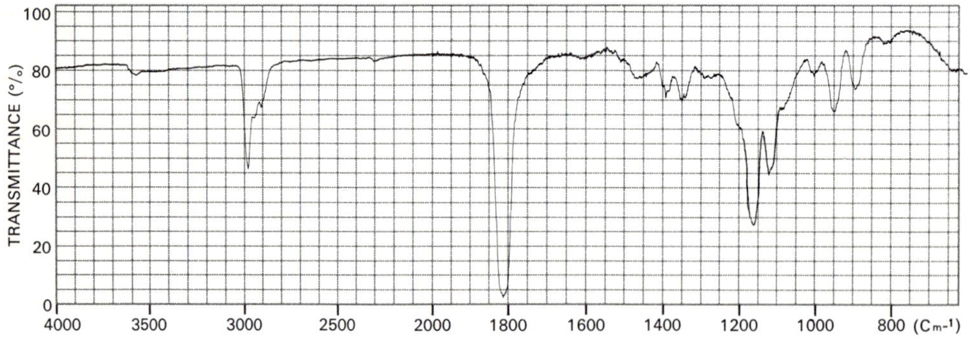

COMPOUND	MOL. FORMULA	B. PT.	INDEX No.
γ-Hexalactone	$C_6H_{10}O_2$	218°C	**270**

CELL TEMP.	APPROX AMOUNT	CELL PATH	REMARKS
~150°C	0·5μl	Length 9·2cm Volume 25ml	

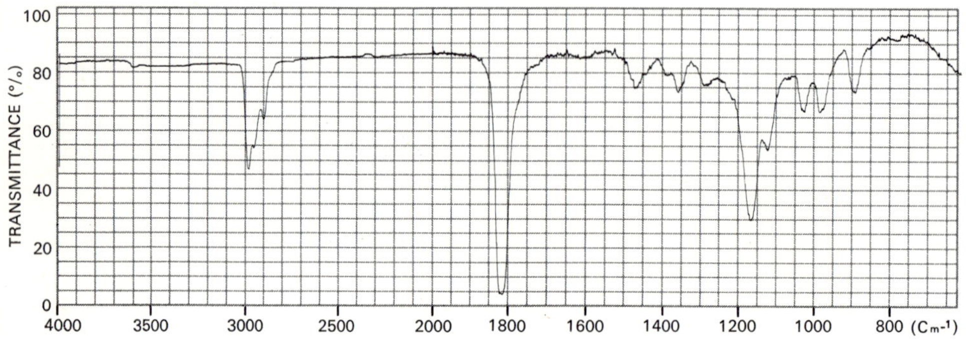

COMPOUND	MOL. FORMULA	B. PT.	INDEX No.
γ-Heptalactone	$C_7H_{12}O_2$	-	**271**

CELL TEMP.	APPROX AMOUNT	CELL PATH	REMARKS
~130°C	2μl	Length 9·2cm Volume 25ml	

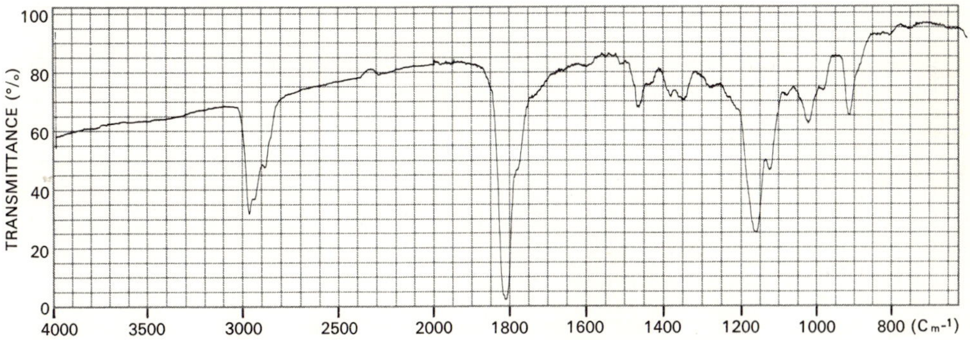

COMPOUND	MOL. FORMULA	B. PT.	INDEX No.
γ-Octalactone	$C_8H_{14}O_2$	133°C/20mm	**272**

CELL TEMP.	APPROX AMOUNT	CELL PATH	REMARKS
~125°C	3μl	Length 9·2cm Volume 25ml	

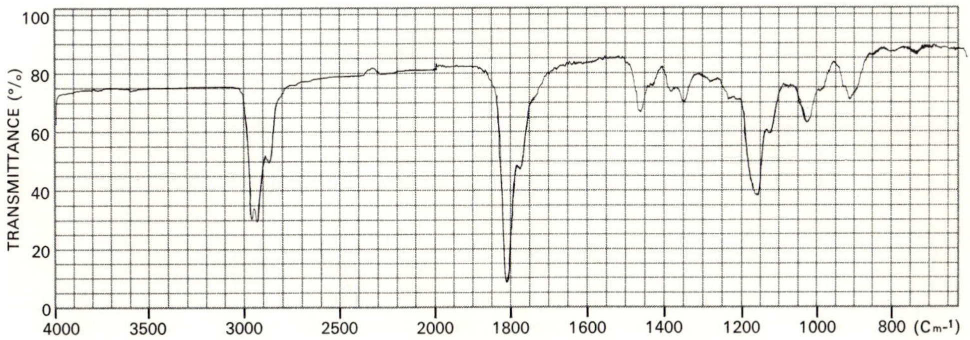

COMPOUND	MOL. FORMULA	B. PT.	INDEX No.
γ-Nonalactone	$C_9H_{16}O_2$	-	**273**

CELL TEMP.	APPROX AMOUNT	CELL PATH	REMARKS
~135°C	1·5μl	Length 9·2cm Volume 25ml	

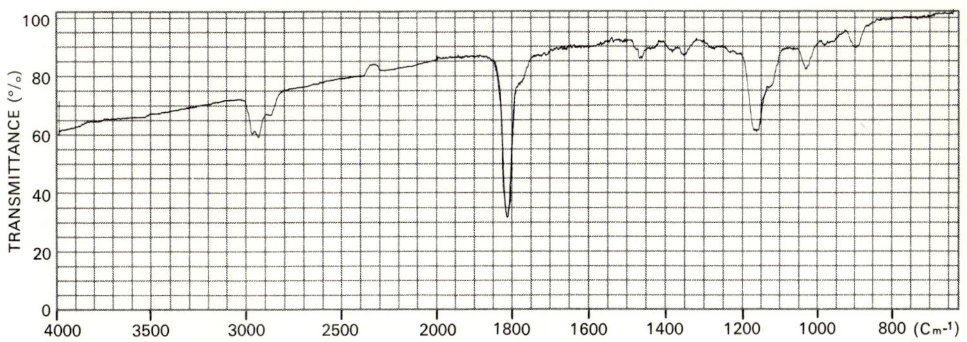

COMPOUND	MOL. FORMULA	B. PT.	INDEX No.
δ-Heptalactone	$C_7H_{12}O_2$	-	**274**

CELL TEMP.	APPROX AMOUNT	CELL PATH	REMARKS
~ 150°C	A few drops	Length 9·2cm Volume 25ml	

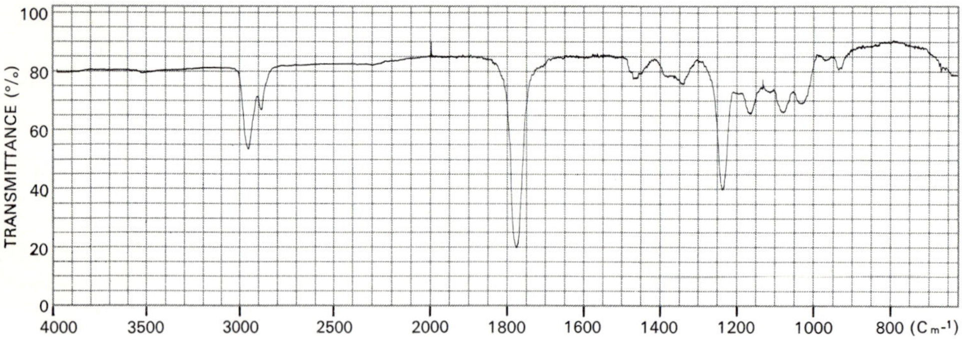

COMPOUND	MOL. FORMULA	B. PT.	INDEX No.
δ-Octalactone	$C_8H_{14}O_2$	-	**275**

CELL TEMP.	APPROX AMOUNT	CELL PATH	REMARKS
~ 140°C	A few drops	Length 9·2cm Volume 25ml	

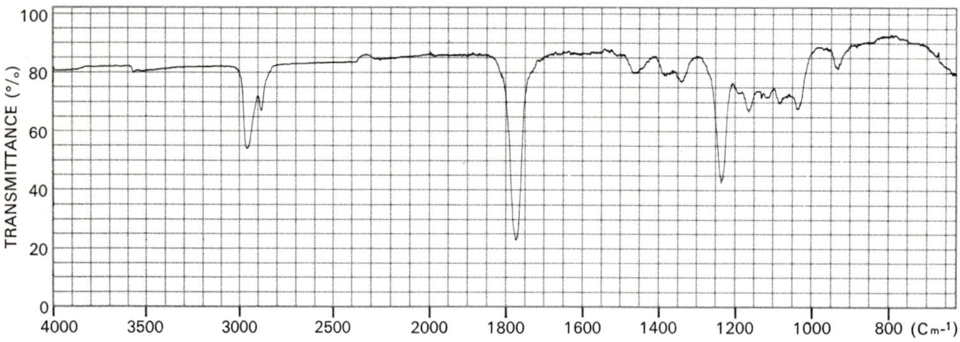

COMPOUND	MOL. FORMULA	B. PT.	INDEX No.
δ-Isooctalactone	$C_8H_{14}O_2$	-	**276**

CELL TEMP.	APPROX AMOUNT	CELL PATH	REMARKS
~ 155°C	A few drops	Length 9·2cm Volume 25ml	

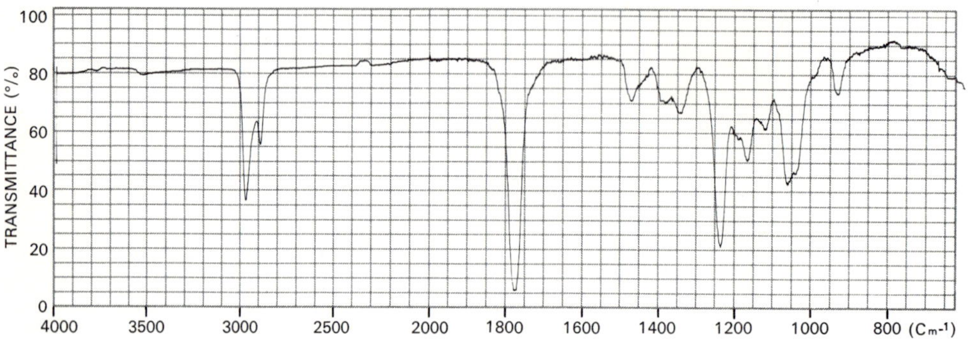

COMPOUND	MOL. FORMULA	B. PT.	INDEX No.
δ-Nonalactone	$C_9H_{16}O_2$	-	**277**

CELL TEMP.	APPROX AMOUNT	CELL PATH	REMARKS
~ 155°C	A few drops	Length 9·2cm Volume 25ml	

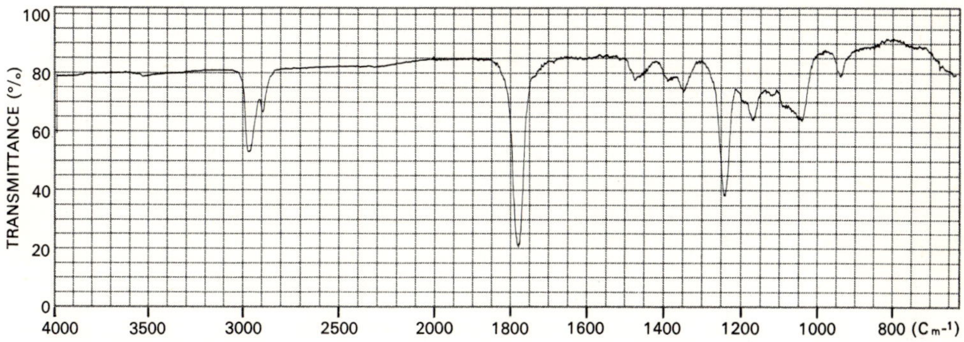

COMPOUND	MOL. FORMULA	B. PT.	INDEX No.
Diethyl ether	$C_4H_{10}O$	35°C	**278**

CELL TEMP.	APPROX AMOUNT	CELL PATH	REMARKS
~ 120°C	1μl	Length 9·2cm Volume 25ml	

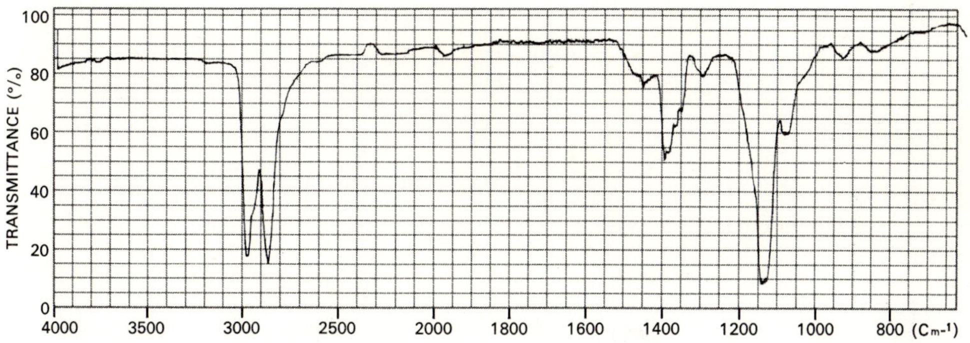

COMPOUND	MOL. FORMULA	B. PT.	INDEX No.
Dibutyl ether	$C_8H_{18}O$	142°C	**279**

CELL TEMP.	APPROX AMOUNT	CELL PATH	REMARKS
~ 130°C	2μl	Length 9·2cm Volume 25ml	

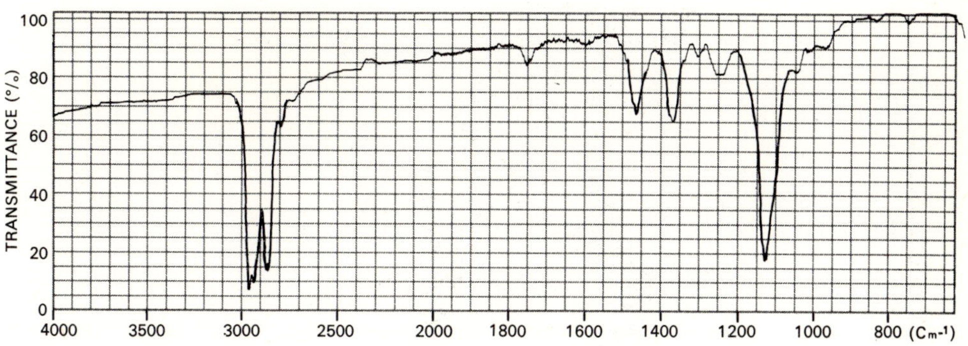

COMPOUND	MOL. FORMULA	B. PT.	INDEX No.
Ethyl vinyl ether	C_4H_8O	36°C	**280**

CELL TEMP.	APPROX AMOUNT	CELL PATH	REMARKS
~ 130°C	2μl	Length 9·2cm Volume 25ml	

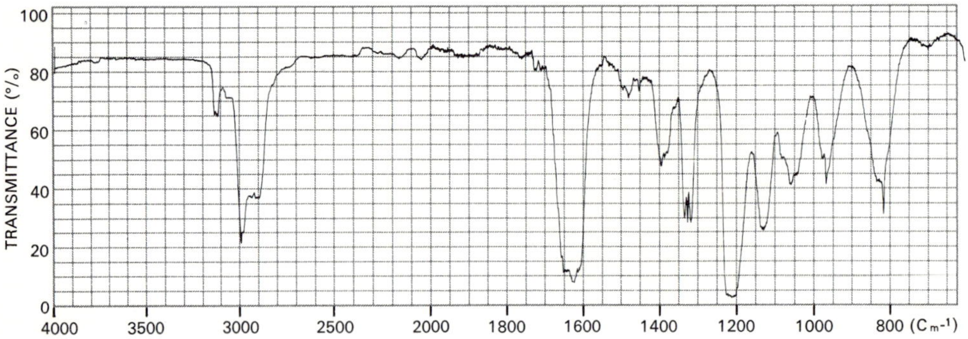

COMPOUND	MOL. FORMULA	B. PT.	INDEX No.
1, 1-Diethoxyethane	$C_6H_{14}O_2$	102°C	**281**

CELL TEMP.	APPROX AMOUNT	CELL PATH	REMARKS
~ 100°C	2μl	Length 9·2cm Volume 25ml	

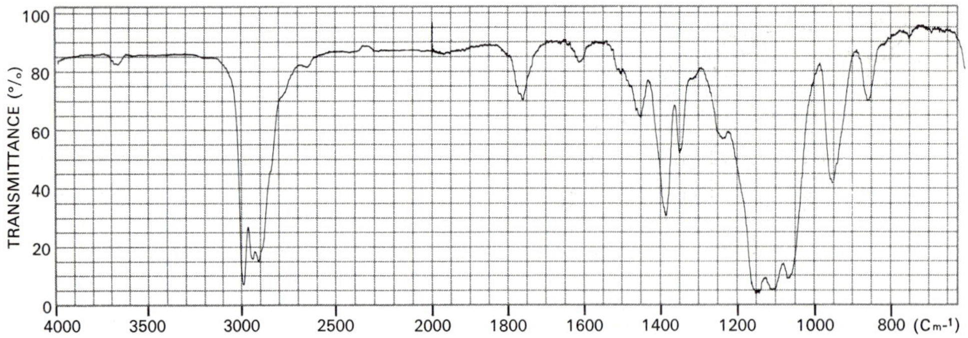

COMPOUND	MOL. FORMULA	B. PT.	INDEX No.
Dioxane	$C_4H_8O_2$	101°C	**282**

CELL TEMP.	APPROX AMOUNT	CELL PATH	REMARKS
~ 130°C	2μl	Length 9·2cm Volume 25ml	

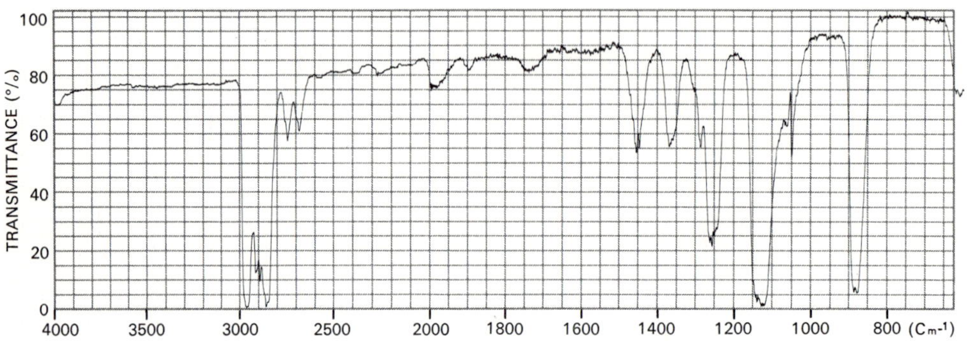

COMPOUND	MOL. FORMULA	B. PT.	INDEX No.
2,3-Dihydropyran	C_5H_8O	87°C	**283**

CELL TEMP.	APPROX AMOUNT	CELL PATH	REMARKS
~125°C	2μl	Length 9·2cm Volume 25ml	

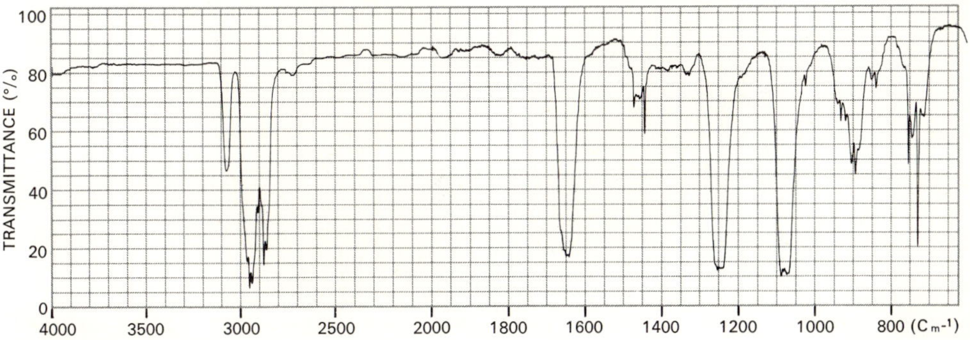

COMPOUND	MOL. FORMULA	B. PT.	INDEX No.
1,8-Cineole	$C_{10}H_{18}O$	176°C	**284**

CELL TEMP.	APPROX AMOUNT	CELL PATH	REMARKS
~140°C	2μl	Length 9·2cm Volume 25ml	

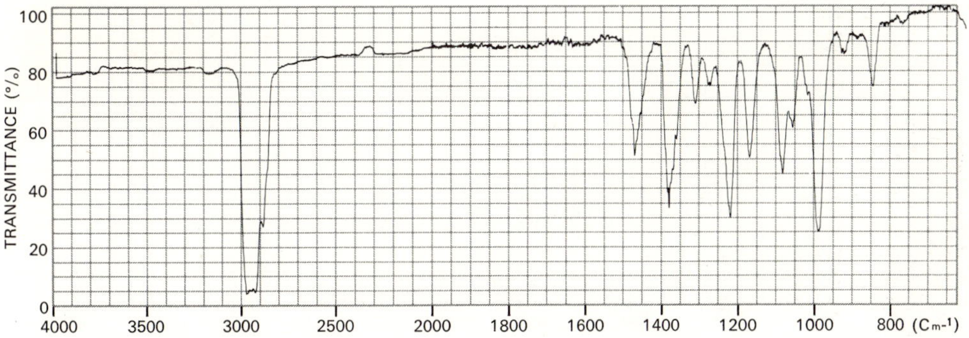

COMPOUND	MOL. FORMULA	B. PT.	INDEX No.
Methoxybenzene (Anisole)	C_7H_8O	154°C	**285**

CELL TEMP.	APPROX AMOUNT	CELL PATH	REMARKS
~135°C	2μl	Length 9·2cm Volume 25ml	

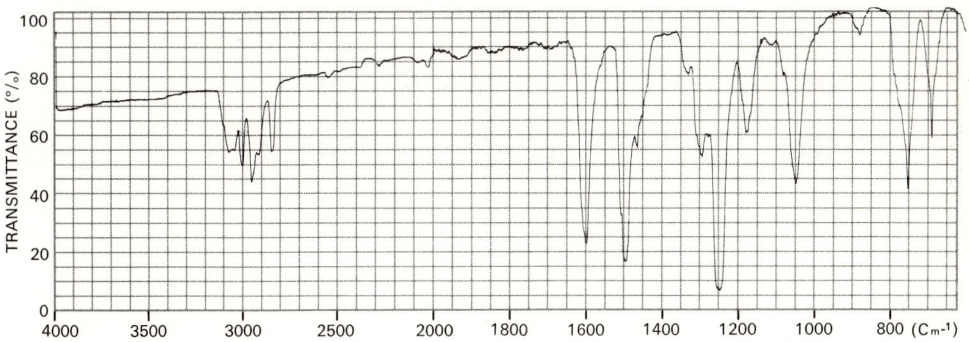

COMPOUND	MOL. FORMULA	B. PT.	INDEX No.
1, 2-Dimethoxybenzene (Veratrole)	$C_8H_{10}O_2$	206°C	**286**

CELL TEMP.	APPROX AMOUNT	CELL PATH	REMARKS
~135°C	2μl	Length 9·2cm Volume 25ml	

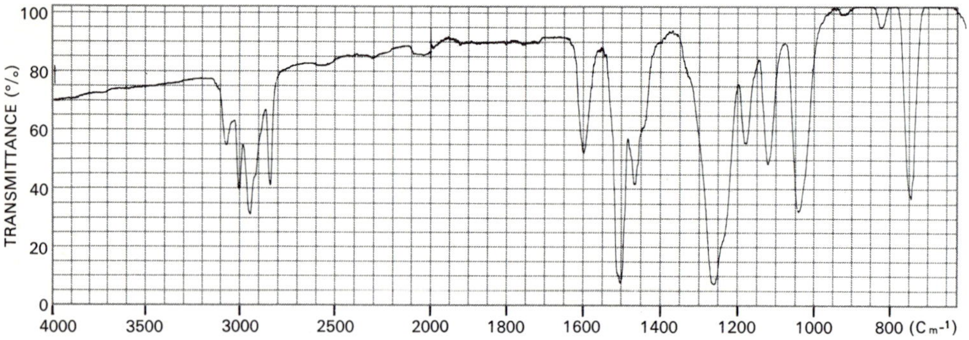

COMPOUND	MOL. FORMULA	B. PT.	INDEX No.
2-Methylfuran	C_5H_6O	66°C	**287**

CELL TEMP.	APPROX AMOUNT	CELL PATH	REMARKS
~145°C	2μl	Length 9·2cm Volume 25ml	

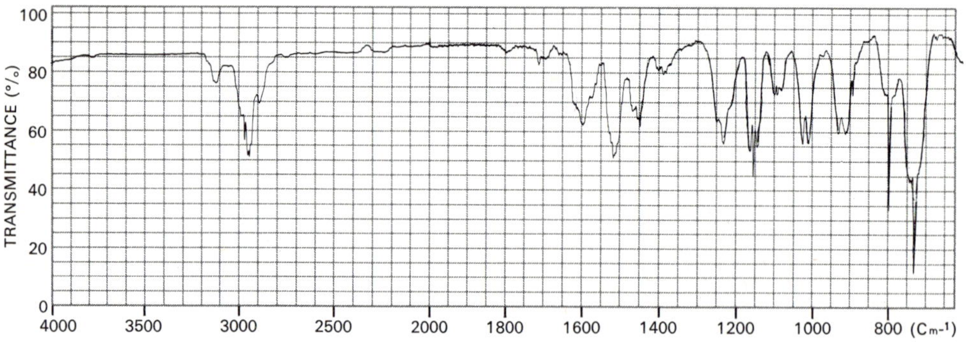

COMPOUND	MOL. FORMULA	B. PT.	INDEX No.
n-Heptane	C_7H_{16}	98·4°C	**288**

CELL TEMP.	APPROX AMOUNT	CELL PATH	REMARKS
~130°C	3μl	Length 9·2cm Volume 25ml	

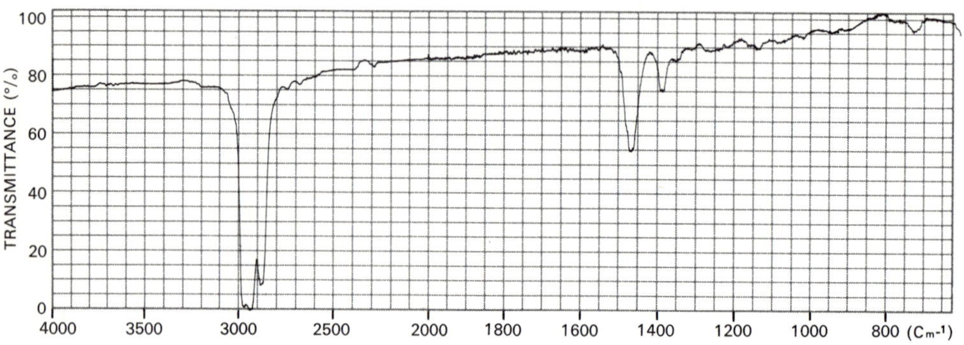

COMPOUND	MOL. FORMULA	B. PT.	INDEX No.
n-Nonane	C_9H_{20}	150·8°C	**289**

CELL TEMP.	APPROX AMOUNT	CELL PATH	REMARKS
~130°C	6μl	Length 9·2cm Volume 25ml	

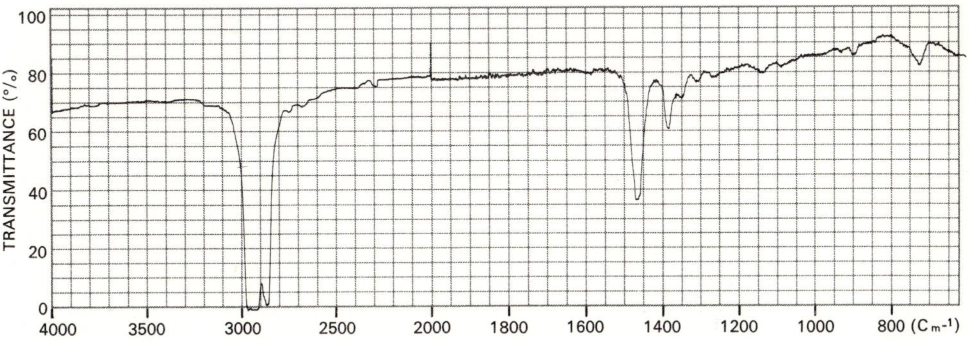

COMPOUND	MOL. FORMULA	B. PT.	INDEX No.
n-Decane	$C_{10}H_{22}$	174°C	**290**

CELL TEMP.	APPROX AMOUNT	CELL PATH	REMARKS
~130°C	4μl	Length 9·2cm Volume 25ml	

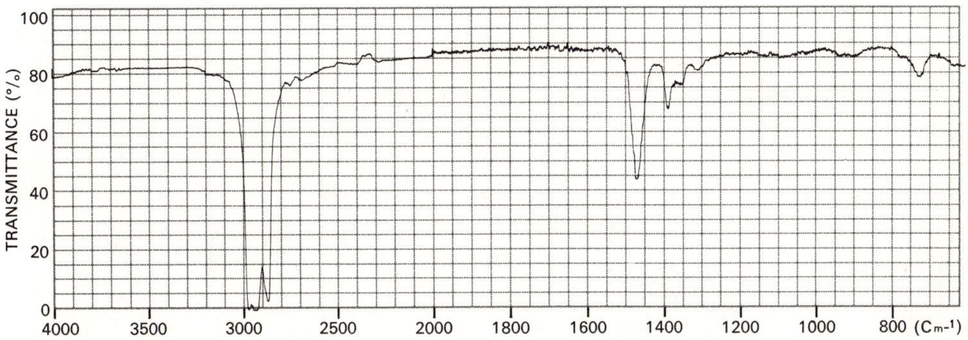

COMPOUND	MOL. FORMULA	B. PT.	INDEX No.
n-Undecane	$C_{11}H_{24}$	195°C	**291**

CELL TEMP.	APPROX AMOUNT	CELL PATH	REMARKS
~130°C	6μl	Length 9·2cm Volume 25ml	

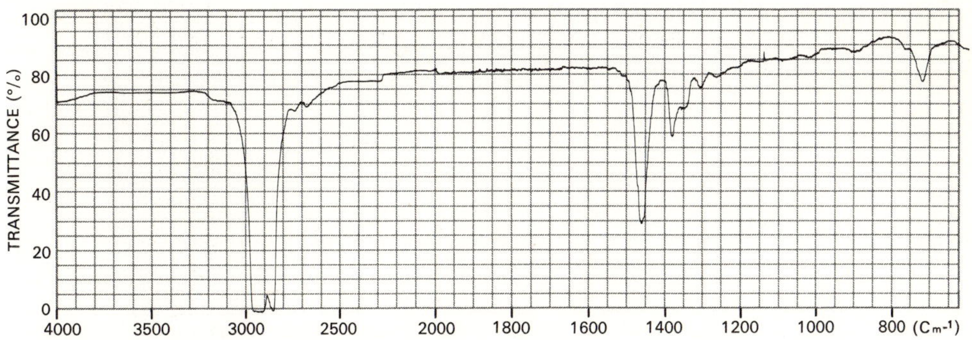

COMPOUND	MOL. FORMULA	B. PT.	INDEX No.
n-Dodecane	$C_{12}H_{26}$	216·3°C	**292**

CELL TEMP.	APPROX AMOUNT	CELL PATH	REMARKS
~125°C	4μl	Length 9·2cm Volume 25ml	

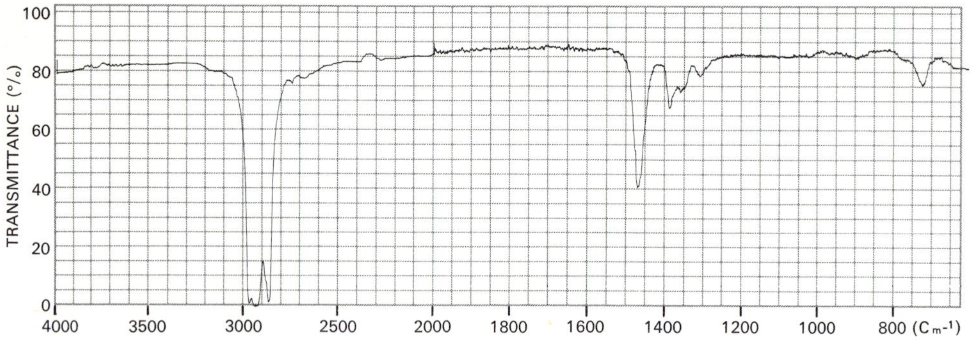

COMPOUND	MOL. FORMULA	B. PT.	INDEX No.
1-Decene	$C_{10}H_{20}$	· 170°C	**293**

CELL TEMP.	APPROX AMOUNT	CELL PATH	REMARKS
~130°C	6μl	Length 9·2cm Volume 25ml	

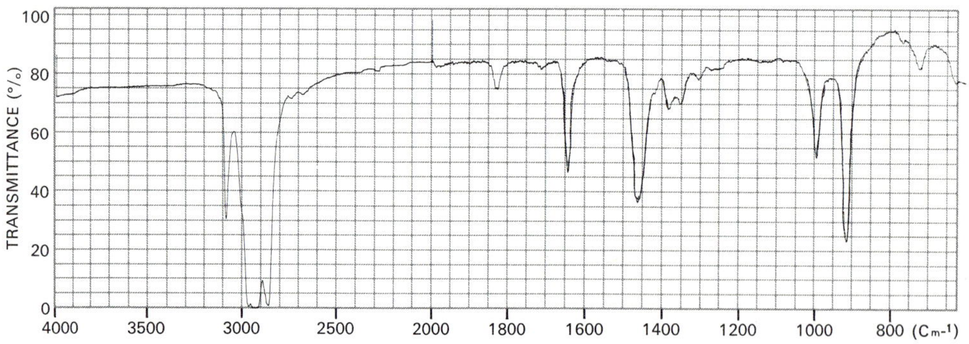

COMPOUND	MOL. FORMULA	B. PT.	INDEX No.
1-Dodecene	$C_{12}H_{24}$	213·4°C	**294**

CELL TEMP.	APPROX AMOUNT	CELL PATH	REMARKS
~125°C	4μl	Length 9·2cm Volume 25ml	

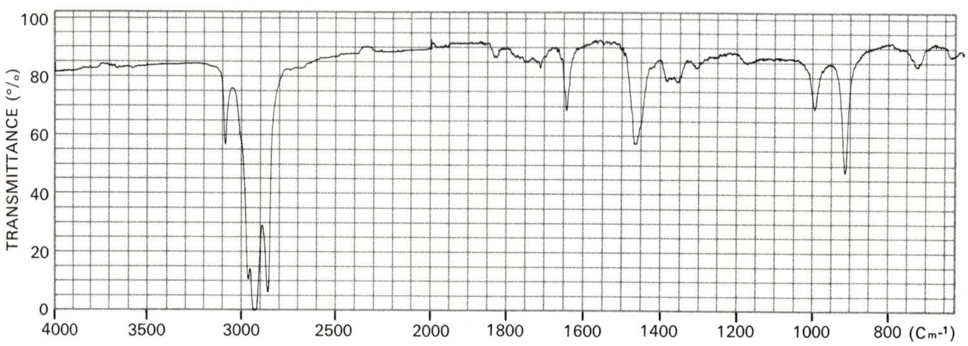

COMPOUND	MOL. FORMULA	B. PT.	INDEX No.
cis-Ocimene (α & β)	$C_{10}H_{16}$	-	**295**

CELL TEMP.	APPROX AMOUNT	CELL PATH	REMARKS
~140°C	3μl	Length 9·2cm Volume 25ml	

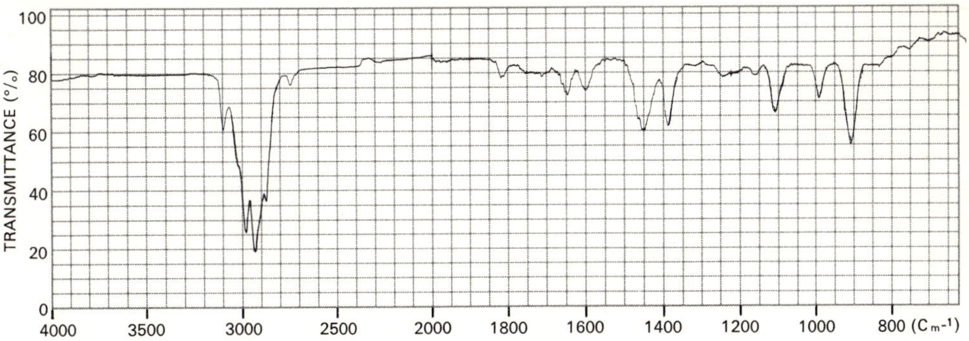

COMPOUND	MOL. FORMULA	B. PT.	INDEX No.
d-Limonene	$C_{10}H_{16}$	178°C	**296**

CELL TEMP.	APPROX AMOUNT	CELL PATH	REMARKS
~125°C	3μl	Length 9·2cm Volume 25ml	

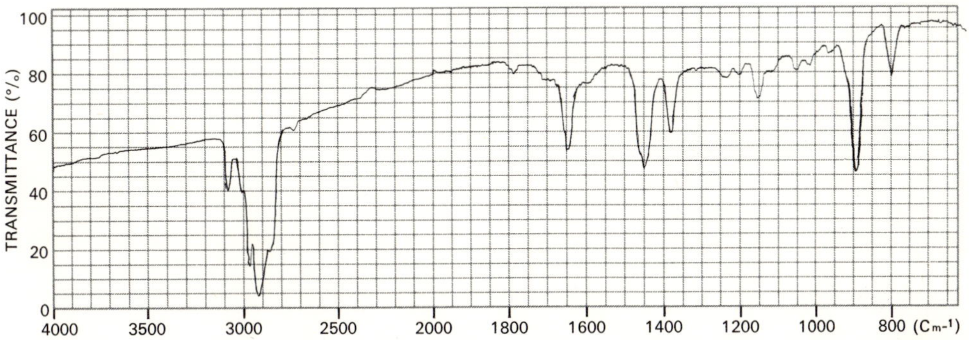

COMPOUND	MOL. FORMULA	B. PT.	INDEX No.
β-Pinene	$C_{10}H_{16}$	164°C	**297**

CELL TEMP.	APPROX AMOUNT	CELL PATH	REMARKS
~135°C	3μl	Length 9·2cm Volume 25ml	

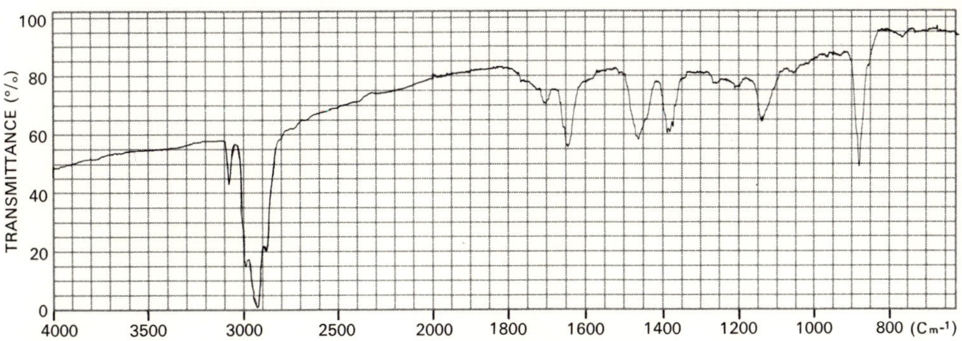

COMPOUND	MOL. FORMULA	B. PT.	INDEX No.
dl-Camphene	$C_{10}H_{16}$	160-162°C	**298**

CELL TEMP.	APPROX AMOUNT	CELL PATH	REMARKS
~135°C	A few mg	Length 9·2cm Volume 25ml	

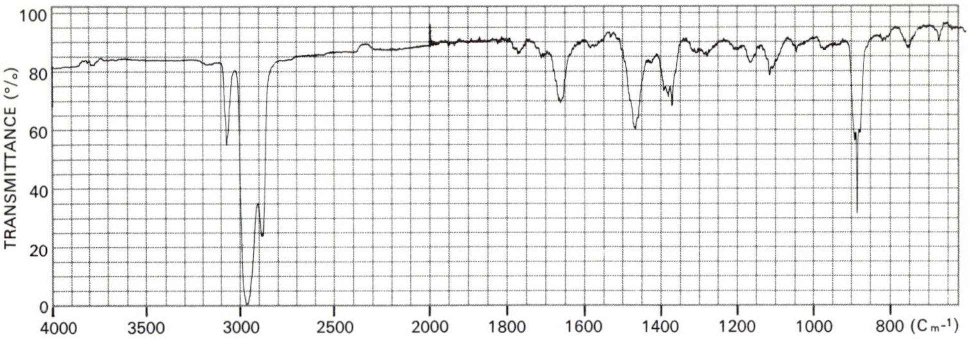

COMPOUND	MOL. FORMULA	B. PT.	INDEX No.
Caryophyllene	$C_{15}H_{24}$	155-157°C/36mm	**299**

CELL TEMP.	APPROX AMOUNT	CELL PATH	REMARKS
~135°C	2μl	Length 9·2cm Volume 25ml	

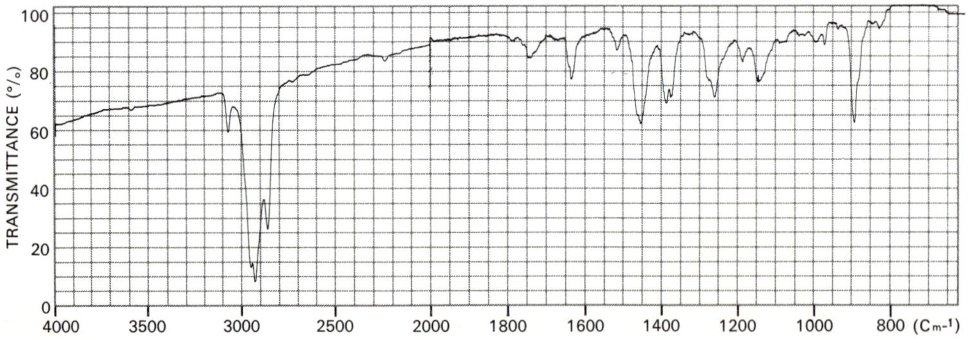

COMPOUND	MOL. FORMULA	B. PT.	INDEX No.
1-Pentyne	C_5H_8	39·3°C	**300**

CELL TEMP.	APPROX AMOUNT	CELL PATH	REMARKS
~135°C	2μl	Length 9·2cm Volume 25ml	* Impurity

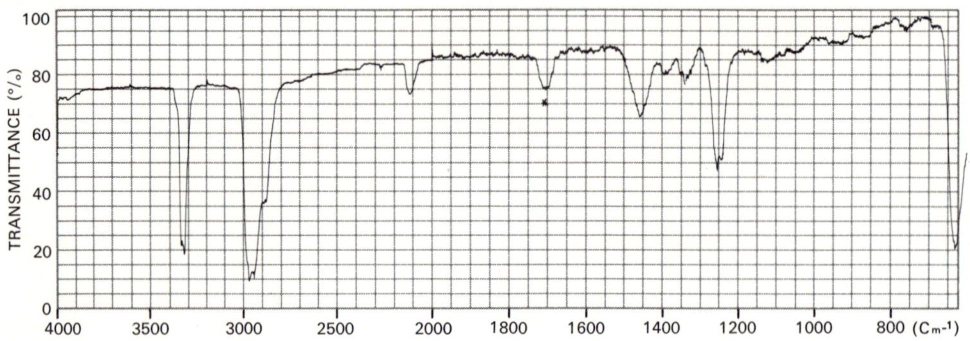

COMPOUND	MOL. FORMULA	B. PT.	INDEX No.
Phenylacetylene	C_8H_6	142-144°C	**301**

CELL TEMP.	APPROX AMOUNT	CELL PATH	REMARKS
~135°C	2μl	Length 9·2cm Volume 25ml	

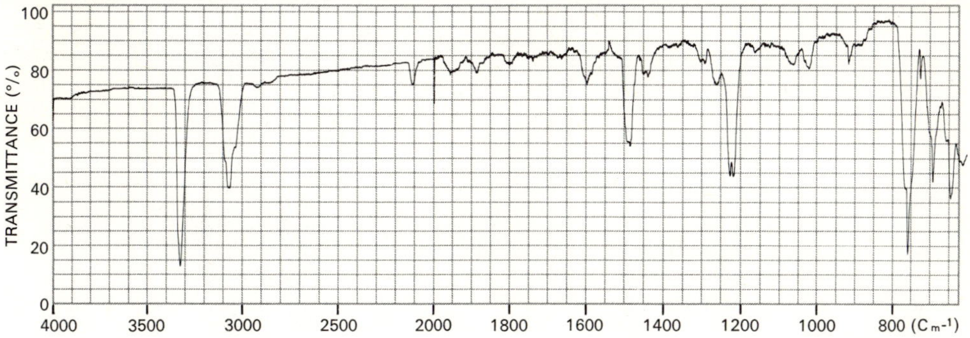

COMPOUND	MOL. FORMULA	B. PT.	INDEX No.
Benzene	C_6H_6	80·1°C	**302**

CELL TEMP.	APPROX AMOUNT	CELL PATH	REMARKS
~145°C	2μl	Length 9·2cm Volume 25ml	

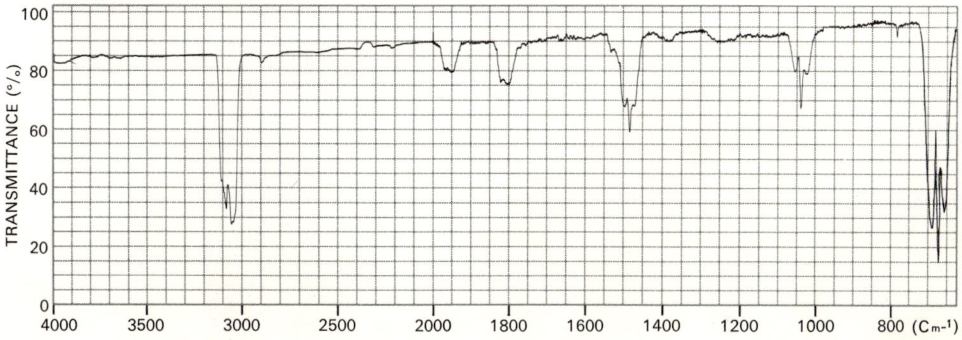

COMPOUND	MOL. FORMULA	B. PT.	INDEX No.
Ethylbenzene	C_8H_{10}	136°C	**303**

CELL TEMP.	APPROX AMOUNT	CELL PATH	REMARKS
~130°C	2μl	Length 9·2cm Volume 25ml	

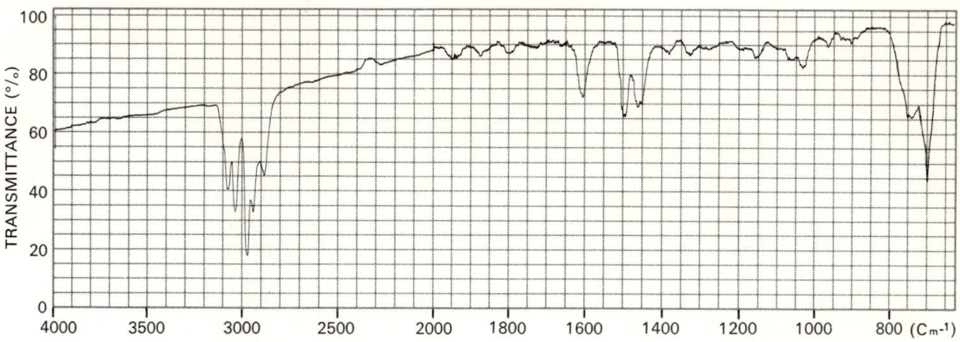

COMPOUND	MOL. FORMULA	B. PT.	INDEX No.
1-Phenyloctane	$C_{14}H_{22}$	264-265°C	**304**

CELL TEMP.	APPROX AMOUNT	CELL PATH	REMARKS
~140°C	2μl	Length 9·2cm Volume 25ml	

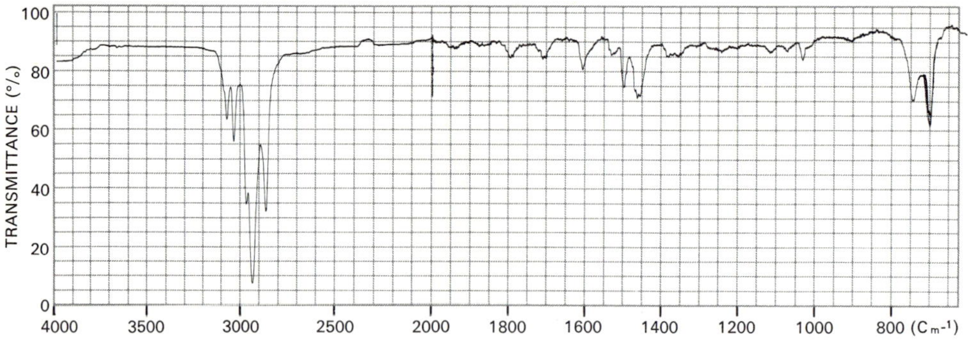

COMPOUND	MOL. FORMULA	B. PT.	INDEX No.
1-Isopropyl-4-methylbenzene (p-Cymene)	$C_{10}H_{14}$	177°C	**305**

CELL TEMP.	APPROX AMOUNT	CELL PATH	REMARKS
~130°C	2μl	Length 9·2cm Volume 25ml	

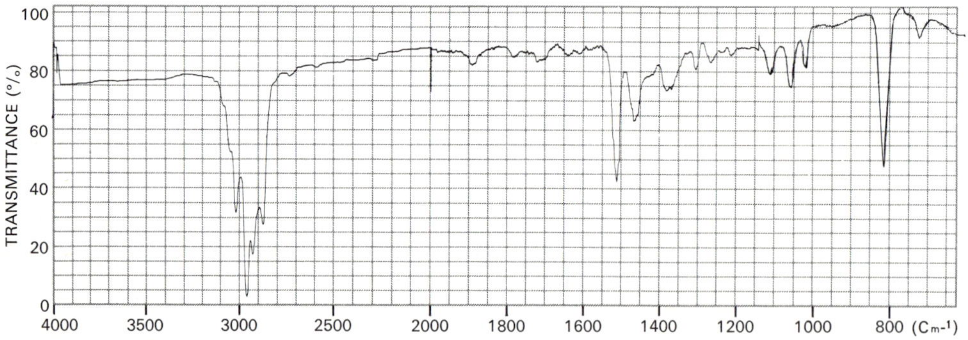

COMPOUND	MOL. FORMULA	B. PT.	INDEX No.
1,3,5-Trimethyl benzene (Mesitylene)	C_9H_{12}	164·7°C	**306**

CELL TEMP.	APPROX AMOUNT	CELL PATH	REMARKS
~120°C	2μl	Length 9·2cm Volume 25ml	

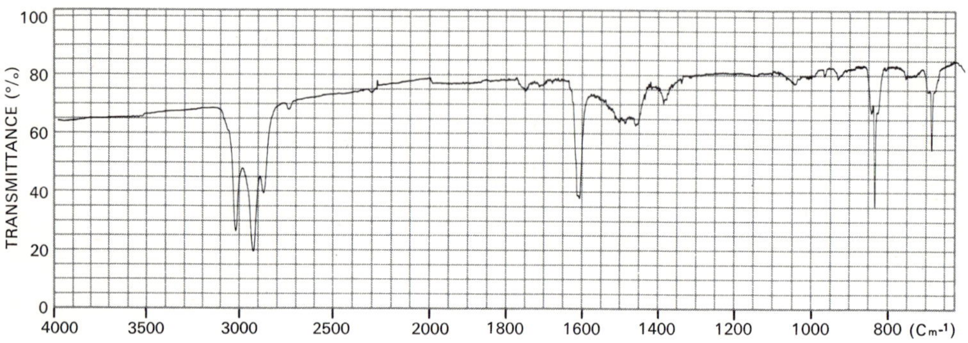

6

Index to vapour spectra found in the commercial reference spectra and the Wyandotte/ASTM punched cards

This index includes all the vapour spectra found in the Sadtler Standard Spectra[121] collection up to Spectrum No. 29000 in the prism spectra and Nos. 5000–8000; 9000–10000; 11000–15000 in the grating spectra;* the vapour spectra in the DMS[39] collection up to Spectrum No. 17711 and No. A 1029 (inorganic); the vapour spectra in the Coblentz[122] collection up to Spectrum No. 6000. The ASTM abstracted literature references[123] referring to vapour spectra in the Wyandotte/ASTM[26] punched card system are also included up to Spectrum No. 9800. These were recovered from the cards themselves by a 'group selection' sort on a card sorter. The vapour state hole is punched both when the compound is a natural vapour and when the spectrum has been run as a vapour. An attempt has been made to remove all the spectra of gases which were obviously examined in a different state, but it is possible that one or two erroneous spectra are included in the index.

The spectra were put into a molecular formula order by an IBM 360/50 computer. Compounds with the same molecular formula are arranged according to the alphabetical order of the first letter of the compound name. The elements are ordered sequentially from 1 to 99 to 0, i.e. all formulae containing no hydrogen appear after those containing hydrogen, and so on.

In accordance with the Wyandotte/ASTM system the recognised collections of spectra are distinguished by the letter appearing in the penultimate column. The letter A in the last column refers to prism infrared spectra in the range 1–15 microns. The following combinations of letters appear:

AA refers to the API collection[124]
BA refers to the user's own collection
CA refers to the Sadtler collection[121]
DA refers to the NRC/NBS collection[125]
EA refers to the ASTM literature spectra[123]
FA refers to the DMS collection[39]
GA refers to the Coblentz collection[122]
HA refers to the Manufacturing Chemists Association collection[126]
JA refers to the Infrared Data Committee of Japan collection[127].

In addition, the reference letters for the Sadtler grating collection are CK, in line with the use of K in column 80 of the ASTM cards to refer to spectra obtained from grating spectrometers.

The DMS inorganic spectra have been given a prefix letter A, i.e. A 123FA.

The spectra included in this book are referred to as BA without a prefix letter.

The Dow collection of vapour spectra[128] have been included in the BA category with prefixes P and G for the prism and grating spectra respectively, i.e. P 234BA and G 356BA.

* At the end of 1969, the Sadtler spectra prism collection consists of Spectrum Nos. 1–37000, and the grating collection consists of Spectrum Nos. 1–16000. The author changed his subscription from prism to grating spectra after prism Spectrum No. 29000. The gaps in the grating spectrum numbers quoted above occur when the grating spectra duplicated some of the original 29000 prism spectra.

C	H	BR	CL	F	I	B	N	O	P	S	SI	M	NAME		SPECTRUM NO.	LOCATION
1	1	1	2										DICHLOROBROMOMETHANE		12	EA
1	1	1		2									BROMODIFLUOROMETHANE		1499	FA
1	1	1		3									BROMODIFLUOROMETHANE	G	68	BA
1	1	2	1										CHLORODIBROMOMETHANE		13	EA
1	1	3											BROMOFORM		14	EA
1	1		1	2									CHLORODIFLUOROMETHANE		3889	FA
1	1		1	2									CHLORODIFLUOROMETHANE		3701	CA
1	1		1	2									DIFLUOROCHLOROMETHANE		6452	EA
1	1		1	2									DIFLUOROCHLOROMETHANE		9	EA
1	1		2	1									DICHLOROFLUOROMETHANE		3890	FA
1	1		2	1									DICHLOROFLUOROMETHANE		10	EA
1	1		2	1									DICHLORO FLUOROMETHANE		1129	CA
1	1		3										CHLOROFORM	G	73	BA
1	1		3										CHLOROFORM	P	79	BA
1	1		3										CHLOROFORM		11	EA
1	1			1				1					DEUTEROFORMYL FLUORIDE		2918	FA
1	1			1				1					FORMYL FLUORIDE,DEUTERO		3490	EA
1	1			1				1					FORMYL FLUORIDE		2917	FA
1	1			1				1					FORMYL FLUORIDE		3489	EA
1	1			3									DEUTEROTRIFLUOROMETHANE		2082	FA
1	1			3									DEUTERO TRIFLUORO METHANE		1993	EA
1	1			3									FLUOROFORM		8	EA
1	1			3									FLUOROFORM		3722	CA
1	1			3						1			TRIFLUOROMETHANETHIOL		25	FA
1	1						1						HYDROGEN CYANIDE		3940	EA
1	1						1						HYDROGEN CYANIDE		1642	FA
1	2	1	1										CHLOROBROMOMETHANE	P	78	BA
1	2	2											DIBROMOMETHANE	G	78	BA
1	2	2											DIBROMOMETHANE	P	95	BA
1	2		2										DICHLOROMETHANE	G	85	BA
1	2		2										DICHLOROMETHANE		6620	CA
1	2		2										DICHLOROMETHANE	P	96	BA
1	2		2										DICHLOROMETHANE		17	EA
1	2		2										DICHLOROMETHANE		1851	EA
1	2			2									DIFLUOROMETHANE		16	EA
1	2			2									DIFLUOROMETHANE		3552	EA
1	2			3								AS	TRIFLUOROMETHYL ARSINE		59	FA
1	2			3					1				TRIFLUOROMETHYL PHOSPHINE		262	FA
1	2				2								DIIODOMETHANE		18	EA
1	2						2						DIAZOMETHANE		1628	FA
1	2						2						DIAZOMETHANE		3920	EA
1	2							1					FORMALDEHYDE		1640	FA
1	2							1					FORMALDEHYDE		3935	EA
1	2							1					FORMALDEHYDE		2538	CA
1	2							2					FORMIC ACID	G	28	BA
1	2							2					FORMIC ACID		1641	FA
1	2							2					FORMIC ACID (-OD)		5125	EA
1	2							2					FORMIC ACID-D		5124	EA
1	2							2					FORMIC ACID		5127	EA
1	2							2					FORMIC ACID DIMER,HYDROXY DEUTERO		6056	EA
1	2							2					FORMIC ACID		3936	EA
1	2							2					FORMIC ACID DIMER,DIDEUTERO		6055	EA
1	2							2					FORMIC ACID DIMER		6054	EA
1	2							2					FORMIC ACID		12931	FA
1	2							2					FORMIC ACID (130 DEG C)		74	BA
1	2							2					FORMIC ACID (40 DEG C)		73	BA
1	2							2					HYDROFORMIC ACID DIMER,DEUTEROXY		6057	EA
1	3	1											BROMOMETHANE		1893	FA
1	3	1											BROMOTRIDEUTEROMETHANE		1894	FA
1	3	1											BROMOMETHANE	G	93	BA
1	3	1											BROMOMETHANE	P	92	BA
1	3	1											BROMOMETHANE		1098	CA
1	3		1										CHLOROMETHANE		3947	EA
1	3		1										CHLOROMETHANE	G	94	BA
1	3		1										CHLOROMETHANE		842	CA
1	3		1										CHLOROMETHANE	P	93	BA
1	3		1										CHLOROMETHANE		1647	FA
1	3		1										CHLOROMETHANE		3759	EA
1	3		1					2		1			METHANESULPHONYL CHLORIDE		91	FA
1	3		2			1		1					DICHLOROMETHOXY BORANE		6484	EA
1	3		2			1		1					METHYL DICHLOROBORINATE		6372	FA
1	3		2					1	1				METHYL PHOSPHONIC DICHLORIDE		13534	FA
1	3		2						1				DICHLORO METHYL PHOSPHINE		13533	FA
1	3		2						1	1			METHYL PHOSPHONOTHIONIC DICHLORIDE		13535	FA
1	3		3									GE	METHYL TRICHLORO GERMANE		13158	FA
1	3		3								1		METHYL TRICHLORO SILANE		1841	EA
1	3		3								1		METHYL TRICHLORO SILANE		2064	FA
1	3			1									FLUOROMETHANE-D3		4463	EA
1	3			1									FLUOROMETHANE		3553	EA
1	3			1									FLUOROTRIDEUTEROMETHANE		1998	FA
1	3			1				2		1			METHANESULPHONYL FLUORIDE		89	FA
1	3			1				3		1			METHYL FLUOROSULPHONATE		4445	EA
1	3			2		1							METHYL DIFLUORO BORANE		4969	FA
1	3			2					1				DIFLUOROMETHYL PHOSPHINE		15355	FA
1	3			3							1		METHYL TRIFLUORO SILANE		407	FA
1	3			4					1				METHYL TETRAFLUORO PHOSPHORANE		13540	FA
1	3				1								IODOMETHANE		15267	FA
1	3				1								IODOMETHANE	G	95	BA
1	3				1								IODOMETHANE		645	EA
1	3					1		1					BORINE CARBONYL (10B,D3)	A	54	FA
1	3					1		1					BORINE CARBONYL	A	51	FA
1	3					1		1					BORINE CARBONYL (D3)	A	53	FA
1	3					1		1					BORINE CARBONYL (10B)	A	52	FA
1	3					1		1					BORINE(10)CARBONYL,DEUTERATED		6419	EA
1	3					1		1					BORINE(11)CARBONYL,DEUTERATED		6420	EA
1	3					1		1					BORINE(10)CARBONYL		6415	EA
1	3					1		1					BORINE(11)CARBONYL		6416	EA
1	3						1	1					FORMAMIDE		2307	EA
1	3						1	2					METHYL NITRITE		5193	FA
1	3						1	2					METHYL NITRITE		47	EA

C H BR CL F I B N O P S SI M

C	H	BR	CL	F	I	B	N	O	P	S	SI	M	NAME	SPECTRUM NO.	LOCATION
1	3						1	2					NITRO METHANE	3957	EA
1	3						1	2					NITROMETHANE	1654	FA
1	3						1	2					NITROMETHANE	G 63	BA
1	3						1	2		1			METHYL THIONITRITE	8859	EA
1	3						1	3					METHYL NITRATE	3781	FA
1	3						1			1		GE	GERMANYL ISOTHIOCYANATE	A 895	FA
1	3						1			1		GE	PERDEUTERO GERMANYL ISOTHIOCYANATE	A 896	FA
1	3						1				1		SILYL-D-3-CYANIDE	6719	EA
1	3						1				1		SILYL CYANIDE	6718	EA
1	3						3						METHYL AZIDE	3946	EA
1	3						3						METHYL AZIDE	1646	FA
1	4												METHANE	1644	FA
1	4												METHANE	3944	EA
1	4												METHANE	4063	CA
1	4												METHANE	G 123	BA
1	4												METHANE,TRIDEUTERATED	3344	EA
1	4												METHANE,TETRADEUTERATED	3345	EA
1	4												METHANE,MONODEUTERATED	3342	EA
1	4												METHANE	3341	EA
1	4												METHANE	3758	EA
1	4												METHANE,DIDEUTERATED	3343	EA
1	4		2								1		(DICHLOROMETHYL) SILANE	5569	FA
1	4							1					METHANOL	3945	EA
1	4							1					METHANOL	6238	EA
1	4							1					METHANOL	3422	EA
1	4							1					METHANOL	1645	FA
1	4							1					METHANOL	G 41	BA
1	4							1					METHANOL	1919	CA
1	4							1					METHANOL	P 44	BA
1	4							1					METHANOL	1	BA
1	4									1			METHANETHIOL	3948	EA
1	4									1			METHANETHIOL	1648	FA
1	5	1									1		METHYL BROMO SILANE	8533	FA
1	5		1								1		(CHLOROMETHYL) SILANE	5568	FA
1	5			1							1		METHYL FLUORO SILANE	8532	FA
1	5				1						1		METHYL IODO SILANE	8534	FA
1	5						1						AMINO METHANE	5101	EA
1	5						1						AMINO(-D2)METHANE-D3	5104	EA
1	5						1						1-AMINOMETHANE-(D5)	1383	FA
1	5						1						AMINO(-D2)METHANE	5102	EA
1	5						1						AMINO METHANE-D3	5103	EA
1	5						1						1-AMINO-TRIDEUTEROMETHANE	1382	FA
1	5						1						AMINOMETHANE	3951	EA
1	5						1						1-AMINOMETHANE	1651	FA
1	5						1						1-DIDEUTEROAMINOMETHANE	1381	FA
1	5								1				METHYL PHOSPHINE	6894	EA
1	6											GE	METHYL GERMANE	16318	FA
1	6						2						METHYL HYDRAZINE	3952	EA
1	6						2						METHYLHYDRAZINE	1652	FA
1	6							1			1		METHOXY SILANE	13554	FA
1	6							1			1		METHOXY TRIDEUTEROSILANE	13556	FA
1	6							1			1		PERDEUTERO (METHOXYSILANE)	13558	FA
1	6										1		METHYL SILANE	1928	EA
1	8					2							METHYL PENTADEUTERODIBORANE	8959	FA
1	8					2							METHYL DIBORANE-D-5	8716	EA
1	8					2							METHYL DIBORANE-D-8	8718	EA
1	8					2							METHYL DIBORANE	8715	EA
1	8					2							METHYL DIBORANE	8958	FA
1	8					2							METHYL DIBORANE-D-3	8717	EA
1	8					2							PERDEUTERO (METHYLDIBORANE)	8961	FA
1	8					2							(TRIDEUTEROMETHYL) DIBORANE	8960	FA
1		1	3										BROMO TRICHLOROMETHANE	1845	CA
1		1	3										BROMOTRICHLOROMETHANE	G 70	BA
1		1		1				1					CARBONYL BROMOFLUORIDE	6872	EA
1		1		3									BROMOTRIFLUOROMETHANE	3563	EA
1		1		3									BROMOTRIFLUOROMETHANE	2094	FA
1		1		3									BROMOTRIFLUOROMETHANE	2478	CA
1		1		3									BROMOTRIFLUOROMETHANE	2581	EA
1		1		3									TRIFLUOROBROMOMETHANE	1856	EA
1		1					1						CYANOGEN BROMIDE	3378	EA
1		2		2									DIBROMODIFLUOROMETHANE	3565	EA
1		2		2									DIBROMODIFLUOROMETHANE	2081	FA
1		2		2									DIBROMODIFLUOROMETHANE	1861	EA
1		2		2									DIBROMODIFLUOROMETHANE	6523	CA
1			1	3									CHLOROTRIFLUOROMETHANE	G 75	BA
1			1	3									CHLOROTRIFLUOROMETHANE	4	EA
1			1	3									CHLOROTRIFLUOROMETHANE	3562	EA
1			1	3				1		1			TRIFLUOROMETHANESULPHONYL CHLORIDE	1	FA
1			1	3						1			TRIFLUOROMETHANESULPHENYL CHLORIDE	28	FA
1			1				1						CYANOGEN CHLORIDE	3379	EA
1			2	2									DICHLORODIFLUOROMETHANE	3255	FA
1			2	2									DICHLORODIFLUOROMETHANE	1	EA
1			2	2									DICHLORODIFLUOROMETHANE	G 82	BA
1			2	2									DICHLORO FLUOROMETHANE	6451	EA
1			2	2									DICHLORODIFLUOROMETHANE	P 83	BA
1			2	2									DICHLORODIFLUOROMETHANE	6837	CK
1			2					1					CARBONYL CHLORIDE	4719	EA
1			2					1					PHOSGENE	3962	EA
1			2					1					PHOSGENE	1657	FA
1			2					1					PHOSGENE	2858	GA
1			2					1					PHOSGENE	P 8	BA
1			2					1					PHOSGENE	G 34	BA
1			2							1			THIOCARBONYL CHLORIDE	82	FA
1			2							1			THIOPHOSGENE	6609	CA
1			3	1									FLUOROTRICHLOROMETHANE	P 91	BA
1			3	1									FLUOROTRICHLOROMETHANE	1862	EA
1			3	1									TRICHLOROFLUOROMETHANE	G 104	BA
1			3	1									TRICHLORO FLUOROMETHANE	1113	CA
1			3	1									TRICHLOROFLUOROMETHANE	2	EA
1			3				1	1			1		TRICHLOROSILYL ISOCYANATE	A 0967	FA

C H BR CL F I B N O P S SI M

C	H	BR	CL	F	I	B	N	O	P	S	SI	M	NAME	SPECTRUM NO.	LOCATION
1			3				1	2					TRICHLORO-NITRO METHANE	P 62	BA
1			3				1	2					TRICHLORONITROMETHANE	11589	CA
1			3				1	2					TRICHLORONITROMETHANE	6249	FA
1			3				1			1			THICYANOGEN TRICHLORIDE	6613	EA
1			3				1			1	1		TRICHLOROSILYL ISOTHIOCYANATE	A 0968	FA
1			4										CARBON TETRACHLORIDE	5	EA
1			4										CARBON TETRACHLORIDE	G 71	BA
1			4										CARBON TETRACHLORIDE	P 76	BA
1			4										CARBON TETRACHLORIDE	6521	CA
1				1			1						FLUORINE CYANIDE	A 0941	FA
1				2			1	1		1			FLUOROFORMIDOSULPHUR DIFLUORIDE	A 755	FA
1				2			1	2	1				DIFLUOROPHOSPHINYL ISOCYANATE	A 798	FA
1				2			2						DIAZODIFLUOROMETHANE	15076	FA
1				2				1					DIFLUORO FORMALDEHYDE	3617	EA
1				2						1			THIOCARBONYL FLUORIDE	A 491	FA
1				3	1								TRIFLUOROIODOMETHANE	2479	CA
1				3	1								TRIFLUOROIODOMETHANE	2580	EA
1				3	1								TRIFLUOROIODOMETHANE	1855	EA
1				3	1								TRIFLUOROIODOMETHANE	2095	FA
1				3	1								TRIFLUOROIODOMETHANE	3564	EA
1				3			1	1					TRIFLUORONITROSOMETHANE	70	FA
1				3			1	1					TRIFLUORONITROSOMETHANE	3023	EA
1				3			1	1					TRIFLUORONITROSOMETHANE	3022	EA
1				3			1	2					FLUOROPICRIN	3026	EA
1				3			1	2					FLUOROPICRIN	3024	EA
1				3			1	2					FLUOROPICRIN	3025	EA
1				3			1	2					TRIFLUORONITROMETHANE	71	FA
1				4									CARBON TETRAFLUORIDE	1924	FA
1				4									CARBON TETRAFLUORIDE	3	EA
1				4									CARBON TETRAFLUORIDE	366	EA
1				4									CARBON TETRAFLUORIDE	3537	EA
1				4									TETRAFLUOROMETHANE	4082	CA
1				4									TETRAFLUOROMETHANE	G 100	BA
1				4				1					METHFORANE FLUOROXY	1030	EA
1				4				1					TRIFLUOROMETHYL OXYFLUORIDE	2905	FA
1				4				2		1			TRIFLUOROMETHANESULPHONYL FLUORIDE	88	FA
1				8						1			TRIFLUOROMETHYL PENTAFLUOROSULPHIDE	A 879	FA
1							4	8					TETRANITROMETHANE	2068	FA
1								1					CARBON MONOXIDE	3917	EA
1								1					CARBON MONOXIDE	G 5	BA
1								1					CARBON MONOXIDE	1142	CA
1								1					CARBON MONOXIDE	P 4	BA
1								1		1			CARBONYL SULPHIDE	P 5	BA
1								1		1			CARBONYL SULPHIDE	G 6	BA
1								1		1			CARBONYL SULPHIDE	83	FA
1								2					CARBON DIOXIDE	3916	EA
1								2					CARBON DIOXIDE	P 3	BA
1								2					CARBON DIOXIDE	A 7	FA
1								2					CARBON DIOXIDE	G 3	BA
1								2					CARBON DIOXIDE	1924	CA
1										2			CARBON DISULPHIDE	G 4	BA
1										2			CARBON DISULPHIDE	P 75	BA
2	1	1	1	3									1-BROMO-2-CHLORO-1,1,2-TRIFLUOROETHANE	6303	CK
2	1	1	1	3									2-BROMO-2-CHLORO-1,1,1-TRIFLUOROETHANE	19787	CA
2	1	1	1	3									2-BROMO-2-CHLORO-1,1,1-TRIFLUOROETHANE	3178	FA
2	1	1	1	3									1,1,1-TRIFLUORO-2-CHLORO-2-BROMO ETHANE	5118	EA
2	1	1		2									2-BROMO-1,1-DIFLUOROETHYLENE	3693	FA
2	1	1		2									1,1-DIFLUORO-2-BROMO ETHYLENE	6433	EA
2	1		1	2									2-CHLORO-1,1-DIFLUOROETHYLENE	3554	FA
2	1		1	4									1-CHLORO-1,1,2,2,-TETRAFLUOROETHANE	3797	FA
2	1		1	8						1			1,2,2-TRIFLUORO-2-CHLOROETHYL SULPHUR -PENTAFLUORIDE	7940	FA
2	1		2	1									CIS-1,2-DICHLORO-1-FLUOROETHYLENE	12194	FA
2	1		2	1									TR-1,2-DICHLORO-1-FLUOROETHYLENE	12195	FA
2	1		2	1									TR-1,2-DICHLORO-1-DEUTERO-2-FLUOROETHYLENE	12197	FA
2	1		2	1									CIS-1,2-DICHLORO-1-DEUTERO-2-FLUOROETHYLENE	12196	FA
2	1		2	1									1,1-DICHLORO-2-FLUOROETHYLENE	16409	FA
2	1		2	1									1,1-DICHLORO-2-DEUTERO-2-FLUOROETHYLENE	16410	FA
2	1		2	3									1,2-DICHLORO-1,1,2-TRIFLUOROETHANE	5701	FA
2	1		2	3									2,2-DICHLORO-1,1,1-TRIFLUOROETHANE	3798	FA
2	1		3										TRICHLOROETHYLENE	P 100	BA
2	1		3										TRICHLOROETHYLENE	G 103	BA
2	1		3	2									1,1,2-TRICHLORO-1,2-DIFLUOROETHANE	3290	FA
2	1		3					2					TRICHLOROACETIC ACID,DEUTERATED	6786	EA
2	1		3					2					TRICHLOROACETIC ACID	6780	EA
2	1			1									FLUOROACETYLENE	8567	EA
2	1			1									FLUORO ACETYLENE	4774	EA
2	1			1									MONOFLUOROACETYLENE	5499	FA
2	1			3									TRIFLUORO ETHYLENE	3573	EA
2	1			3									TRIFLUOROETHYLENE	461	FA
2	1			3									TRIFLUOROETHYLENE	6841	CK
2	1			3									TRIFLUOROETHYLENE	2645	EA
2	1			3									TRIFLUOROETHYLENE-D1	2646	EA
2	1			3				1					TRIFLUORO ACETALDEHYDE	3658	EA
2	1			3				2					TRIFLUORO ACETIC ACID	3632	EA
2	1			3				2					TRIFLUOROACETIC ACID	6437	EA
2	1			3				2					TRIFLUOROACETIC ACID	3697	FA
2	1			5									PENTAFLUOROETHANE	940	FA
2	1			6									PENTAFLUOROETHANE	3554	EA
2	1			6								AS	BISTRIFLUOROMETHYL ARSINE	54	FA
2	1			6			1						DI-TRIFLUOROMETHYLAMINE	553	FA
2	1			6					1				BIS-TRIFLUOROMETHYL PHOSPHINE	261	FA
2	2												ACETYLENE	3908	EA
2	2												ACETYLENE	1620	FA
2	2												ACETYLENE	3961	CA
2	2												ACETYLENE	G 110	BA
2	2	1	1	2									1-BROMO-2-CHLORO-1,1-DIFLUOROETHANE	78	FA
2	2	1	1					1					CHLOROACETYL BROMIDE	12962	FA
2	2	1		1									CIS-1-BROMO-2-FLUOROETHYLENE	6818	FA
2	2	1		1									TRANS-1-BROMO-2-FLUOROETHYLENE	6817	FA

C H BR CL F I B N O P S SI M

C	H	BR	CL	F	I	B	N	O	P	S	SI	M	NAME	SPECTRUM NO.	LOCA-TION
2	2	1		1									CIS-1-BROMO-2-FLUORO ETHYLENE	9287	EA
2	2	1		1									TRANS-1-BROMO-2-FLUORO ETHYLENE	9286	EA
2	2	1		1				1					FLUOROACETYL BROMIDE	12967	FA
2	2	1		2	1								1-BROMO-1,1-DIFLUORO-2-IODOETHANE	76	FA
2	2	1		3									2-BROMO-1,1,1-TRIFLUOROETHANE	3179	FA
2	2	1		3									1,1,1-TRIFLUORO-2-BROMO ETHANE	5119	EA
2	2	2											CIS-1,2-BROMOETHYLENE	3125	FA
2	2	2											1,1-DIBROMOETHYLENE	G 107	BA
2	2	2											CIS-1,2-DIBROMODIDEUTEROETHYLENE	3127	FA
2	2	2											CIS-1,2-DIBROMODEUTEROETHYLENE	3368	EA
2	2	2											CIS-1,2-DIBROMOETHYLENE	3366	EA
2	2	2											TR-1,2-DIBROMOETHYLENE	3362	EA
2	2	2											TR-1,2-DIBROMODEUTEROETHYLENE	3364	EA
2	2	2						1					BROMOACETYL BROMIDE	12965	FA
2	2		1	1									TRANS-1-CHLORO-2-FLUOROETHYLENE	6819	FA
2	2		1	1									CIS-1-CHLORO-2-FLUOROETHYLENE	6820	FA
2	2		1	1									CIS-1-CHLORO-2-FLUORO ETHYLENE	9289	EA
2	2		1	1									TRANS-1-CHLORO-2-FLUORO ETHYLENE	9288	EA
2	2		1	1									1-CHLORO-1-FLUOROETHYLENE	1489	FA
2	2		1	1				1					FLUOROACETYL CHLORIDE	12949	FA
2	2		1	3									2-CHLORO-1,1,1-TRIFLUOROETHANE	3794	FA
2	2		1	3									1,1,1-TRIFLUORO-2-CHLOROETHANE	7457	CA
2	2		1	5						1			TRANS-2-CHLOROVINYLSULPHUR PENTAFLUORIDE	7929	FA
2	2		2										CIS-1,2-DICHLOROETHYLENE	G 84	BA
2	2		2										1,1-DICHLOROETHYLENE	G 108	BA
2	2		2										1,1-DICHLOROETHYLENE	P 103	BA
2	2		2										TR-1,2-DICHLOROETHYLENE	P 86	BA
2	2		2										CIS-1,2-DICHLOROETHYLENE	P 85	BA
2	2		2					1					CHLOROACETYL CHLORIDE	12944	FA
2	2		2					2					DICHLOROACETIC ACID,DEUTERATED	6785	EA
2	2		2					2					DICHLOROACETIC ACID	6779	EA
2	2			1	1								CIS-2-FLUORO-1-IODO ETHYLENE	9291	EA
2	2			1	1								TRANS-1-FLUORO-2-IODOETHYLENE	11849	FA
2	2			1	1								CIS-TR-2-FLUORO-1-IODO ETHYLENE	9290	EA
2	2			1	1								CIS-1-FLUORO-2-IODOETHYLENE	6821	FA
2	2			2									2-DEUTERO-1,1-DIFLUOROETHYLENE	2004	FA
2	2			2									1,1-DIDEUTERO-2,2-DIFLUOROETHYLENE	2003	FA
2	2			2									1,1-DIFLUORO ETHYLENE	3572	EA
2	2			2									TR-1,2-DIFLUOROETHYLENE	6822	FA
2	2			2									CIS-1,2-DIFLUOROETHYLENE	6823	FA
2	2			2									1,1-DIFLUOROETHYLENE	G 109	BA
2	2			2									CIS-1,2-DIFLUORO ETHYLENE	9293	EA
2	2			2									TRANS-1,2-DIFLUORO ETHYLENE	9292	EA
2	2			2									1,1-DIFLUOROETHYLENE	3552	FA
2	2			2									1,1-DIFLUORO-2-DEUTERO ETHYLENE	2340	EA
2	2			2									1,1-DIFLUORO-2,2-DIDEUTERO ETHYLENE	2308	EA
2	2			3									DEUTEROTRIFLUOROETHYLENE	462	FA
2	2			3	1								2-IODO-1,1,1-TRIFLUOROETHANE	23721	CA
2	2			4									1,1,2,2-TETRAFLUOROETHANE	8713	EA
2	2			6			1		1				AMINO-BIS-TRIFLUOROMETHYL PHOSPHINE	6609	EA
2	2						2	1					1,2,4-OXADIAZOLE	16244	FA
2	2						2			1			PERDEUTERIO-1,2,5-THIADIAZOLE	15754	FA
2	2						2			1			1,2,5-THIADIAZOLE	15753	FA
2	2							1					DIDEUTEROKETENE	5106	EA
2	2							1					KETENE	3591	FA
2	2							1					KETENE	5107	EA
2	2							1					MONODEUTEROKETENE	5105	EA
2	3	1											BROMOETHYLENE	8458	EA
2	3	1											BROMOETHYLENE	6435	FA
2	3	1											1-BROMOPROP-2-YNE	P 98	BA
2	3	1						1					ACETYL BROMIDE	19762	CA
2	3	1						1					ACETYL BROMIDE	12937	FA
2	3		1										CHLOROETHYLENE	6434	FA
2	3		1										CHLOROETHYLENE	P 102	BA
2	3		1										CHLOROETHYLENE	8709	EA
2	3		1										CHLOROETHYLENE	3755	EA
2	3		1										CHLORO ETHYLENE	3971	EA
2	3		1										CHLOROETHYLENE-D-3	8708	EA
2	3		1										CHLOROETHYLENE,DEUTERATED	6520	EA
2	3		1										CHLOROETHYLENE	1660	FA
2	3		1										CHLOROETHYLENE	8457	FA
2	3		1										CHLOROETHYLENE-D1	2678	EA
2	3		1										VINYL CHLORIDE	1119	CA
2	3		1	2									1-CHLORO-1,1-DIFLUOROETHANE	3792	FA
2	3		1	2									1,1-DIFLUORO-1-CHLOROETHANE	G 86	BA
2	3		1					1					ACETYL CHLORIDE	G 18	BA
2	3		1					1					ACETYL CHLORIDE	12934	FA
2	3		1					1					ACETYL CHLORIDE	6030	CA
2	3		1					1					ACETYL CHLORIDE	8967	FA
2	3		1					1					MONODEUTEROACETYL CHLORIDE	8971	FA
2	3		1					1					PERDEUTEROACETYL CHLORIDE	12935	FA
2	3		1					1					PERDEUTEROACETYL CHLORIDE	8969	FA
2	3		1					2					MONOCHLOROACETIC ACID	6778	FA
2	3		1					2					MONOCHLOROACETIC ACID,DEUTERATED	6784	EA
2	3		2	1									1,1,1-DICHLOROFLUOROETHANE	536	EA
2	3		2	1									1,1-DICHLORO-1-FLUOROETHANE	3196	FA
2	3		2	5						1			2,2-DICHLOROETHYLSULPHUR PENTAFLUORIDE	7926	FA
2	3		2	5						1			1,2-DICHLOROETHYLSULPHUR PENTAFLUORIDE	7927	FA
2	3		2				1	2					1,1-DICHLORO-1-NITRO ETHANE	P 63	BA
2	3		2						1				VINYL DICHLORO PHOSPHINE	5565	FA
2	3		3										1,1,1-TRICHLOROETHANE	1716	EA
2	3		3										1,1,1-TRICHLOROETHANE	533	EA
2	3		3										1,1,1-TRICHLOROETHANE	G 101	BA
2	3		3										1,1,2-TRICHLOROETHANE	G 102	BA
2	3		3										1,1,1-TRICHLOROETHANE	3560	FA
2	3		3										1,1,1-TRICHLOROETHANE	P 94	BA
2	3		3								1		VINYL TRICHLORO SILANE	1430	FA
2	3			1									FLUOROETHYLENE-D-3	6810	EA
2	3			1									FLUOROETHYLENE-D-2-CIS	6812	EA
2	3			1									FLUOROETHYLENE-D-2	6809	EA
C	H	BR	CL	F	I	B	N	O	P	S	SI	M			

C	H	BR	CL	F	I	B	N	O	P	S	SI	M	NAME	SPECTRUM NO.	LOCATION
2	3			1									FLUOROETHYLENE-D	6811	EA
2	3			1									FLUOROETHYLENE	6807	EA
2	3			1									FLUOROETHYLENE-D	6808	EA
2	3			1				1					ACETYL FLUORIDE	11912	CA
2	3			3									1,1,1-TRIFLUOROETHANE	3561	FA
2	3			3				3		1			METHYL TRIFLUOROMETHANESULPHONATE	99	FA
2	3			5						1			VINYLSULPHUR PENTAFLUORIDE	7928	FA
2	3						1						ACETONITRILE	P 59	BA
2	3						1						ACETONITRILE	G 54	BA
2	3						1						ACETONITRILE	14288	FA
2	3						1						METHYL ISOCYANIDE	3351	EA
2	3						1						METHYL ISOCYANIDE	3023	FA
2	3						1	1					METHYL ISOCYANATE	14462	FA
2	4												CIS-DIDEUTERO-ETHYLENE	3449	EA
2	4												ETHYLENE	3933	EA
2	4												ETHYLENE,MONO-DEUTERATED	1695	EA
2	4												ETHYLENE,ASYM-DI-DEUTERATED	1694	EA
2	4												ETHYLENE	1638	FA
2	4												ETHYLENE	G 119	BA
2	4												ETHYLENE	1131	CA
2	4												ETHYLENE	P 6	BA
2	4	2											1,2-DIBROMOETHANE	P 89	BA
2	4	2											1,2-DIBROMOETHANE	G 77	BA
2	4		1	5						1			2-CHLOROETHYLSULPHUR PENTAFLUORIDE	7924	FA
2	4		1				1	1					METHYL-FORMYL-CHLORIDE OXIME	8473	EA
2	4		2										1,2-DICHLOROETHANE	3922	EA
2	4		2										1,1-DICHLOROETHANE	3921	EA
2	4		2										1,1-DICHLOROETHANE	663	FA
2	4		2										1,2-DICHLOROETHANE	G 83	BA
2	4		2										1,2-DICHLOROETHANE	1629	FA
2	4		2										1,1-DICHLOROETHANE	P 84	BA
2	4		2										1,2-DICHLOROETHANE	P 90	BA
2	4		2										1,1-DICHLOROETHANE	6336	CA
2	4			2									1,1-DIFLUOROETHANE	3557	FA
2	4			2									1,1-DIFLUOROETHANE	6838	CK
2	4			2									1,2-DIFLUOROETHANE	8721	EA
2	4			2									1,1-DIFLUORO ETHANE	1443	EA
2	4			3			1						2,2,2-TRIFLUOROETHYLAMINE	6305	CK
2	4							1					ACETALDEHYDE	P 10	BA
2	4							1					ACETALDEHYDE	G 15	BA
2	4							1					ACETALDEHYDE	1301	GA
2	4							1					ACETALDEHYDE	5824	CA
2	4							1					ACETALDEHYDE	12933	FA
2	4							1					ACETALDEHYDE	152	BA
2	4							1					ETHYLENE OXIDE	1639	FA
2	4							1					ETHYLENE OXIDE	3934	EA
2	4							1					ETHYLENE OXIDE	3448	EA
2	4							1					ETHYLENE OXIDE-D4	3447	EA
2	4							1					ETHYLENE OXIDE	1109	CA
2	4							1					ETHYLENE OXIDE	G 49	BA
2	4							1					ETHYLENE OXIDE	P 52	BA
2	4							1					TETRADEUTEROETHYLENE OXIDE	2093	FA
2	4							1		1			THIOACETIC ACID	222	FA
2	4							2					ACETIC ACID	3498	EA
2	4							2					ACETIC ACID	1618	FA
2	4							2					ACETIC ACID	3906	EA
2	4							2					ACETIC ACID,DEUTERATED	6782	EA
2	4							2					ACETIC ACID	6776	EA
2	4							2					ACETIC ACID	3754	EA
2	4							2					ACETIC ACID-D	5469	EA
2	4							2					ACETIC ACID	5468	EA
2	4							2					ACETIC ACID	3753	EA
2	4							2					ACETIC ACID	G 16	BA
2	4							2					ACETIC ACID	P 11	BA
2	4							2					ACETIC ACID (130 DEG C)	76	BA
2	4							2					ACETIC ACID (35 DEG C)	75	BA
2	4							2					METHYL FORMATE	206	BA
2	4							2					METHYL FORMATE	P 30	BA
2	4							2					METHYL FORMATE	G 30	BA
2	4							3					ETHYLENE OZONIDE	3762	EA
2	4							3					ETHYLENE OZONIDE	1925	FA
2	4									2			DITHIOACETIC ACID	1105	FA
2	4										1		SILYL ACETYLENE	12316	FA
2	4										1		(TRIDEUTEROSILYL) ACETYLENE	12317	FA
2	5											62	ETHYL LITHIUM	6896	EA
2	5	1											BROMOETHANE	P 87	BA
2	5	1											BROMOETHANE	4631	CA
2	5	1											BROMOETHANE	7336	EA
2	5	1											BROMOETHANE1,1,2-TRIDEUTERO	7339	EA
2	5	1											BROMOETHANE,1,2-DIDEUTERO	7338	EA
2	5	1											BROMOETHANE,1-DEUTERO	7337	EA
2	5		1										CHLOROETHANE	3929	EA
2	5		1										CHLOROETHANE	3757	EA
2	5		1										CHLOROETHANE	P 88	BA
2	5		1										CHLOROETHANE	G 87	BA
2	5		1										CHLOROETHANE	661	FA
2	5		1										CHLOROETHANE	533	CA
2	5		1					1					CHLOROMETHYL METHYL ETHER	G 42	BA
2	5		1					1					2-CHLORO ETHANOL	P 40	BA
2	5			1									FLUOROETHANE,1,1,2-TRIDEUTERO	7334	EA
2	5			1									FLUOROETHANE,2-DEUTERO	7335	EA
2	5			1									FLUOROETHANE,1,2-DIDEUTERO	7333	EA
2	5			1									FLUOROETHANE,1-DEUTERO	7332	EA
2	5			1									FLUOROETHANE	7331	EA
2	5			1									FLUOROETHANE	1441	EA
2	5				1								IODOETHANE	G 88	BA
2	5						1						ETHYLENE IMINE	P 66	BA
2	5						1						ETHYLENE IMINE	G 60	BA
2	5						1						ETHYLENEIMINE	14001	FA
2	5						1	1					ACETAMIDE	142	BA

C H BR CL F I B N O P S SI M

C	H	BR	CL	F	I	B	N	O	P	S	SI	M	NAME	SPECTRUM NO.	LOCATION
2	5						1	1					N-METHYL FORMAMIDE,DEUTERATED	3409	EA
2	5						1	1					N-METHYL FORMAMIDE	3408	EA
2	5						1	1					N-METHYL FORMAMIDE	8469	EA
2	5						1	1		1			ETHYL THIONITRITE	8955	FA
2	5						1	2					ETHYL NITRITE	1352	EA
2	5						1	2					O-METHYL FORMHYDROXAMIC ACID	8466	EA
2	5						1	2					NITROETHANE	1653	FA
2	5						1	2					NITRO ETHANE	3954	EA
2	5						1	2					NITROETHANE	P 68	BA
2	5						1	2					NITROETHANE	P 67	BA
2	5						1	3					ETHYL NITRATE	G 61	BA
2	5						1	3					ETHYL NITRATE	1637	FA
2	5						1	3					ETHYL NITRATE	3932	EA
2	6												ETHANE	3756	EA
2	6												ETHANE	1633	FA
2	6												ETHANE	3927	EA
2	6												ETHANE	G 118	BA
2	6												ETHANE	7989	CA
2	6											GE	VINYL GERMANE	7820	EA
2	6											SN	VINYL STANNANE	7821	EA
2	6		1			1		2					CHLORODIMETHOXY BORANE	6485	EA
2	6		1			1		2					DIMETHYL CHLOROBORATE	6371	FA
2	6		2									GE	DIMETHYL DICHLORO GERMANE	12345	FA
2	6		2					2			1		DICHLORO DIMETHYL SILANE	4869	FA
2	6		2								1		DIMETHYL DICHLORO SILANE	1842	EA
2	6		2								1		DIMETHYL DICHLORO SILANE	2065	FA
2	6			1		1							DIMETHYL FLUORO BORANE	4970	FA
2	6			2							1		DIMETHYL DIFLUORO SILANE	5192	FA
2	6			6			2		2				222444-HEXAFLUORO-1,3-DIMETHYL-1,3,2,4--DIAZADIPHOSPHETIDINE	A 673	FA
2	6						1						ETHANOL	3427	EA
2	6						2						AZOMETHANE	3912	EA
2	6						2						AZOMETHANE	1623	FA
2	6						2	2					N,N-DIMETHYL NITRAMIDE	5312	EA
2	6							1					DIMETHYL ETHER	3924	EA
2	6							1					DIMETHYL ETHER	4444	EA
2	6							1					DIMETHYL ETHER	1630	FA
2	6							1					DIMETHYL ETHER	1128	CA
2	6							1					ETHANOL	P 39	BA
2	6							1					ETHANOL	6407	CA
2	6							1					ETHANOL	G 39	BA
2	6							1					ETHANOL	1026	EA
2	6							1					ETHANOL-(D1)	4954	FA
2	6							1					ETHANOL	3928	EA
2	6							1					ETHANOL,DEUTERATED	4314	EA
2	6							1					ETHANOL	1634	FA
2	6							1					ETHANOL	6239	EA
2	6							1					ETHANOL	2917	EA
2	6							1					ETHANOL-D	1028	EA
2	6							1					ETHANOL	2	BA
2	6							2					ETHANE-1,2-DIOL	47	BA
2	6							2		1			DIMETHYL SULPHONE	3143	FA
2	6									1			ETHANETHIOL	1636	FA
2	6									1			ETHANETHIOL	3931	EA
2	6									1			ETHANETHIOL	6306	CA
2	6									1			METHYL SULPHIDE	6300	CA
2	6										1		VINYL SILANE	13889	FA
2	6										1		VINYL SILANE	7819	EA
2	6										1		VINYL (TRIDEUTEROSILANE)	13890	FA
2	6										1		VINYL SILANE	1938	EA
2	6										2		DISILYL ACETYLENE	6775	EA
2	7					1	4						2,5-DIMETHYLTETRAZABOROLE	A 777	FA
2	7					1		2					DIMETHOXY BORANE	829	GA
2	7					1		2					DIMETHOXY BORANE	6482	EA
2	7					1		2					DIMETHYL BORONATE	6378	FA
2	7					1		2					DIMETHOXY BORANE-D	6481	EA
2	7					1		2					DIMETHYL DEUTEROBORONATE	6379	FA
2	7						1						1-AMINOETHANE	1650	FA
2	7						1						1-AMINOETHANE	3950	EA
2	7						1						1-AMINO ETHANE	108	BA
2	7						1						DIMETHYLAMINE	3923	EA
2	7						1						DIMETHYLAMINE	3292	EA
2	7						1						DIMETHYLAMINE	558	FA
2	7						1						DIMETHYLAMINE	1140	CA
2	7						1	1					ETHANOLAMINE	P 38	BA
2	7								1				DIMETHYL PHOSPHINE	3698	FA
2	7								1				DIMETHYL PHOSPHINE	6432	EA
2	8											GE	DIMETHYL GERMANE	16319	FA
2	8											GE	ETHYL TRIDEUTEROGERMANE	16305	FA
2	8											GE	ETHYL GERMANE	16304	FA
2	8		1				1				1		DIMETHYLAMINO CHLOROSILANE	17192	FA
2	8						2						1,2-DIAMINOETHANE	P 65	BA
2	8						2						1,1-DIMETHYL HYDRAZINE	2299	EA
2	8						2						1,1-DIMETHYLHYDRAZINE	1631	FA
2	8						2						1,2-DIMETHYL HYDRAZINE	3926	EA
2	8						2						1,1-DIMETHYL HYDRAZINE	3925	EA
2	8						2						1,2-DIMETHYLHYDRAZINE	1632	FA
2	8										1		DIMETHYL SILANE	1929	EA
2	8										1		DIMETHYL SILANE	5191	FA
2	8										1		ETHYL SILANE	1932	EA
2	9						1				1		DIMETHYLAMINO SILANE	17191	FA
2	10					2							DIMETHYLDIBORANE-D-10	8731	EA
2	10					2							1,1-DIMETHYLDIBORANE (METHYL-D-6)	8730	EA
2	10					2							1,1-DIMETHYLDIBORANE	8728	EA
2	10					2							1,1-DIMETHYLDIBORANE-D-4	8729	EA
2	10					2							ETHYL DIBORANE	7977	FA
2	10					2							ETHYL PENTADEUTERODIBORANE	7978	FA
2	10					2							ETHYL DIBORANE	831	GA
2	10					2							METHYL BORANE DIMER	17301	FA
2	10					2							(PENTADEUTEROETHYL) DIBORANE	7979	FA

C H BR CL F I B N O P S SI M

C	H	BR	CL	F	I	B	N	O	P	S	SI	M	NAME	SPECTRUM NO.	LOCATION
2	10					2							PERDEUTERO ETHYL DIBORANE	7980	FA
2	11					2	1						DIMETHYLAMINO DIBORANE	4043	FA
2	11						1				2		DIMETHYLAMINODISILANE	A 472	FA
2	12					4							ETHYLENE TETRABORANE	844	GA
2		1	1	2									1-BROMO-1-CHLORO-2,2-DIFLUOROETHYLENE	6615	FA
2		1	1	2									1,1-DIFLUORO-2-CHLORO-2-BROMOETHYLENE	6470	EA
2		1	1	4									1-BROMO-2-CHLOROTETRAFLUOROETHANE	9084	FA
2		1		3									BROMOTRIFLUOROETHYLENE	6842	CK
2		1		3									BROMOTRIFLUOROETHYLENE	376	FA
2		1		3									BROMOTRIFLUOROETHYLENE	2304	EA
2		1		3				1					TRIFLUORO ACETYL BROMIDE	3625	EA
2		1		5									1-BROMOPENTAFLUORO ETHANE	3566	EA
2		1		5									1-BROMO-1,1,2,2,2-PENTAFLUORO ETHANE	7035	EA
2		1		5									BROMOPENTAFLUOROETHANE	2439	CA
2		1		5									BROMOPENTAFLUOROETHANE	3462	FA
2		1		5									PENTAFLUOROBROMOETHANE	6492	EA
2		2	1	3									1-CHLORO-1,2-DIBROMO-1,2,2-TRIFLUOROETHANE	6304	CK
2		2	2	2									1,1-DIFLUORO-1,2-DIBROMO-2,2-DICHLOROETHANE	8860	EA
2		2		2									1,1-DIBROMO-2,2-DIFLUOROETHYLENE	4155	FA
2		2		2									1,1-DIFLUORO-2,2-DIBROMO ETHYLENE	6427	EA
2		2		4									1,2-DIBROMOTETRAFLUORO ETHANE	3567	EA
2			1	3									CHLOROTRIFLUOROETHYLENE	1839	EA
2			1	3									1-CHLOROTRIFLUORO ETHYLENE	3576	EA
2			1	3									CHLOROTRIFLUOROETHYLENE	2376	FA
2			1	3									CHLOROTRIFLUOROETHYLENE	6839	CK
2			1	3				1					TRIFLUOROACETYL CHLORIDE	12347	FA
2			1	3				1					TRIFLUORO ACETYL CHLORIDE	3621	EA
2			1	5									1-CHLORO-1,1,2,2,2-PENTAFLUORO ETHANE	7036	EA
2			1	5									CHLOROPENTAFLUOROETHANE	6696	FA
2			1	5									CHLOROPENTAFLUOROETHANE	2440	CA
2			1	9						1			2-CHLOROPERFLUOROETHYLSULPHUR PENTAFLUORIDE	7934	FA
2			2	1			1						DICHLOROFLUORO ACETONITRILE	G 59	BA
2			2	2									TR-1,2-DICHLORO-1,2-DIFLUOROETHYLENE	4319	FA
2			2	2									1,1-DICHLORO-2,2-DIFLUOROETHYLENE	3556	FA
2			2	2									1,1-DICHLORO-2,2-DIFLUOROETHYLENE	G 81	BA
2			2	2									TRANS-DICHLORODIFLUORO ETHYLENE	3825	EA
2			2	2									1,1-DICHLORO-2,2-DIFLUORO ETHYLENE	3575	EA
2			2	2									CIS-1,2-DICHLORO-1, 2-DIFLUOROETHYLENE	17549	FA
2			2	3		1							PERFLUOROVINYLBORON DICHLORIDE	7982	FA
2			2	4									1,1-DICHLORO-1,2,2,2-TETRAFLUOROETHANE	2074	FA
2			2	4									1,2-DICHLORO-1,1,2,2-TETRAFLUOROETHANE	3563	FA
2			2	8						1			1,2,2-TRIFLUORO-1,2-DICHLOROETHYL SULPHUR -PENTAFLUORIDE	7939	FA
2			2	8						1			1,1,2-TRIFLUORO-2,2-DICHLOROETHYL SULPHUR -PENTAFLUORIDE	7938	FA
2			2					2					OXALYL CHLORIDE	43	EA
2			3	1									FLUOROTRICHLOROETHYLENE	999	FA
2			3	1									FLUORO TRICHLOROETHYLENE	G 91	BA
2			3	3									1,1,2-TRICHLORO-1,2,2-TRIFLUOROETHANE	G 105	BA
2			3	3									1,1,1-TRICHLORO-2,2,2-TRIFLUOROETHANE	2076	FA
2			3	3									1,1,2-TRICHLORO-1,2,2-TRIFLUOROETHANE	23717	CA
2			3	3									1,2,2-TRICHLORO-1,1,2-TRIFLUORO ETHANE	8855	EA
2			3				1						TRICHLORO ACETONITRILE	6438	EA
2			3				1						TRICHLORO ACETONITRILE	G 66	BA
2			4										TETRACHLOROETHYLENE	P 97	BA
2			4										TETRACHLOROETHYLENE	G 97	BA
2			4	2									TETRACHLORO-1,2-DIFLUOROETHANE	3291	FA
2				1			1	2					FLUOROCARBONYL ISOCYANATE	A 797	FA
2				2				4					BISFLUOROFORMYL PEROXIDE	9444	FA
2				3			1						PERFLUOROACETONITRILE	3375	EA
2				3			1						TRIFLUORO ACETONITRILE	3608	EA
2				3			1						TRIFLUOROACETONITRILE	1490	FA
2				3			1						TRIFLUOROACETONITRILE	2503	CA
2				4									PERFLUOROETHYLENE	3555	FA
2				4									TETRAFLUOROETHYLENE	G 99	BA
2				4									TETRAFLUORO ETHYLENE	3574	EA
2				4				1					PERFLUOROACETYL FLUORIDE	13282	FA
2				4				1					TETRAFLUORO ACETALDEHYDE	3618	EA
2				4						2			TRIFLUOROMETHYL FLUORODITHIOFORMATE	86	FA
2				5	1								1-IODO PENTAFLUOROETHANE	2434	CA
2				5	1								1-IODO-1,1,2,2,2-PENTAFLUORO ETHANE	7034	EA
2				5	1								PENTAFLUOROIODO ETHANE	[illegible]	FA
2				5		1							PERFLUOROVINYLBORON DIFLUORIDE	7981	FA
2				6									HEXAFLUOROETHANE	3538	EA
2				6									HEXAFLUOROETHANE	6836	CK
2				6			2	1					HEXAFLUOROAZOXYMETHANE	72	FA
2				6			2	2					N-NITROSO-ON-BISTRIFLUOROMETHYLHYDROXYLAMINE	69	FA
2				6				1					BIS-TRIFLUOROMETHYL ETHER	3587	EA
2				6				2					BISTRIFLUOROMETHYL PEROXIDE	9445	FA
2				6				2					PERFLUORO DIMETHYL PEROXIDE	4644	EA
2				6				2		2			TRIFLUOROMETHYL TRIFLUOROMETHANETHIOL -SULPHONATE	4	FA
2				6						3			BISTRIFLUOROMETHYL TRISULPHIDE	23	FA
2				7								AS	FLUOROBISTRIFLUOROMETHYL ARSINE	53	FA
2				8						1			PERFLUOROVINYLSULPHUR PENTAFLUORIDE	7941	FA
2				10						1			PERFLUOROETHYL SULPHURPENTAFLUORIDE	13178	CA
2							2						CYANOGEN	1627	FA
2							2						CYANOGEN	2849	GA
2							2						CYANOGEN	3918	EA
3	1		1	4									3-CHLORO 1,1,3,3-TETRAFLUOROPROPENE	1935	GA
3	1		1	6									1-CHLORO-1,2,2,3,3,3-HEXAFLUOROPROPANE	250	FA
3	1			3									TRIFLUOROMETHYL ACETYLENE	146	EA
3	1			4			1						2,3,3,3-TETRAFLUORO PROPIONITRILE	3610	EA
3	1			5									PENTA(1,2,3,3,3)FLUOROPROP-1-ENE	255	FA
3	1			5									PENTA(1,1,3,3,3)FLUOROPROPENE	77	FA
3	1			5				1					PENTAFLUORO PROPANAL	3659	EA
3	1			5				2					PERFLUOROPROPIONIC ACID	62	FA
3	1			7									2H-HEPTAFLUOROPROPANE	4000	FA
3	1			7									HEPTAFLUOROPROPANE	3555	EA
3	1			7									2-HYDROHEPTAFLUOROPROPANE	3556	EA

C H BR CL F I B N O P S SI M

C	H	BR	CL	F	I	B	N	O	P	S	SI	M	NAME	SPECTRUM NO.	LOCA-TION
3	1			7									1-HYDRO HEPTAFLUOROPROPANE	5937	CA
3	1						1						CYANO-ACETYLENE	6430	EA
3	2	1		5									1-BROMO-1,1,3,3,3-PENTAFLUOROPROPANE	81	FA
3	2		1	3									1-CHLORO-3,3,3-TRIFLUOROPROP-1-ENE	31	FA
3	2		2										1,3-DICHLOROPROPYNE	13277	FA
3	2		3	3									1,1,1-TRIFLUORO-2,3,3-TRICHLOROPROP-1-ENE	G 106	BA
3	2			2									1,1-DIFLUORO-ALLENE	4638	EA
3	2			4									1,3,3,3-TETRAFLUOROPROP-1-ENE	41	FA
3	2			5	1								3-IODO-1,1,1,2,2-PENTAFLUOROPROPANE	6288	CK
3	2			5	1								PENTA(1,1,3,3,3)FLUORO-3-IODOPROPANE	75	FA
3	2			6									2,2-DIHYDRO HEXAFLUOROPROPANE	650	CA
3	2			6				3		1			2,2,2-TRIFLUOROETHYL TRIFLUOROMETHANE-SULPHONATE	1555	FA
3	2							1					3-DEUTEROPROPYNAL	7851	FA
3	2							1					1-DEUTEROPROPYNAL	7850	FA
3	2							1					PROPYNAL	7849	FA
3	3	1											BROMOALLENE	12334	FA
3	3	1		2									3-BROMO-3,3DIFLUORO PROPENE	4641	EA
3	3		1										CHLOROALLENE	12333	FA
3	3		3										3,3,3-TRICHLOROPROP-IENE	6434	EA
3	3			1									1-FLUOROPROP-2-YNE	G 96	BA
3	3			3				1					1,1,1-TRIFLUOROPROPAN-2-ONE	G 36	BA
3	3			3				1					1,1,1-TRIFLUORO ACETONE	1188	EA
3	3			5						1			PROP-1-YNYLSULPHUR PENTAFLUORIDE	7933	FA
3	3			6								AS	METHYL BIS-TRIFLUOROMETHYL ARSINE	56	FA
3	3				1								IODOALLENE	12335	FA
3	3						1						PROP-2-ENE NITRILE	P 60	BA
3	3						1	1					ISOXAZOLE	6879	EA
3	3						1	1					ISOXAZOLE	6183	FA
3	3						1	1					OXAZOLE	15349	FA
3	3						1	1					VINYL ISOCYANATE	10085	FA
3	3						1			1			ISOTHIAZOLE	13159	FA
3	3						1			1			THIAZOLE	16118	FA
3	3						3						PERDEUTERO-1,3,5-TRIAZINE	7853	FA
3	3						3						1,3,5-TRIAZINE	7852	FA
3	3						3						S-TRIAZINE-D-3	9483	EA
3	3						3						S-TRIAZINE	9482	EA
3	4												ALLENE	1622	FA
3	4												ALLENE-D4	1034	EA
3	4												ALLENE-1,1-D2	5450	EA
3	4												ALLENE	3910	EA
3	4												ALLENE	1033	EA
3	4												PROPYNE	G 130	BA
3	4												PROPYNE	3965	EA
3	4		1	3									3-CHLORO-1,1,1-TRIFLUOROPROPANE	35	FA
3	4		1	5						1			2-CHLOROPROP-1-ENYLSULPHUR PENTAFLUORIDE	7930	FA
3	4		2										1,1-DICHLOROCYCLOPROPANE	G 80	BA
3	4		2					1					1,3-DICHLOROACETONE	2133	FA
3	4		2					1					1,1-DICHLOROACETONE	4794	FA
3	4			4									1,1,1,3-TETRAFLUOROPROPANE	40	FA
3	4			6			1		1				METHYLAMINO-BIS-TRIFLUOROMETHYL PHOSPHINE	6610	EA
3	4						2	1					5-METHYL-1,2,4-OXADIAZOLE	16246	FA
3	4							1					PROP-2-ENAL	1621	FA
3	4							1					PROPENAL	3909	EA
3	4							1					PROP-2-ENAL	P 13	BA
3	4							1					PROP-2-ENAL	G 19	BA
3	4							1					PROP-2-ENAL	6645	CA
3	4							1					PROP-2-YN-1-OL	17	BA
3	4							2					PROP-2-ENOIC ACID	16210	FA
3	4							3					2-OXO PROPANOIC ACID (PYRUVIC)	106	BA
3	5	1											CIS-2-BROMO-1-DEUTERIOPROPENE	16269	FA
3	5	1											TRANS-2-BROMO-1-DEUTERIOPROPENE	16270	FA
3	5	1											TRANS-2-BROMO-1,3,3,3-TETRADEUTERIOPROPENE	16273	FA
3	5	1											CIS-2-BROMO-1,3,3,3-TETRADEUTERIOPROPENE	16274	FA
3	5	1											2-BROMOPROPENE	16268	FA
3	5	1											2-BROMO-1,1,-DIDEUTERIOPROPENE	16271	FA
3	5	1											2-BROMO-3,3,3-TRIDEUTERIOPROPENE	16275	FA
3	5	1											BROMOCYCLOPROPANE	G 76	BA
3	5	1											3-BROMOPROP-2-ENE	G 69	BA
3	5	1											PERDEUTERIO-2-BROMOPROPENE	16272	FA
3	5	1						1					EPIBROMOHYDRIN	P 50	BA
3	5	1						1					PROPANYL BROMIDE	12941	FA
3	5		1										CIS-2-CHLORO-1H-PERDEUTEROPROPENE	13274	FA
3	5		1										1-CHLORO PROP-2-ENE	P 72	BA
3	5		1										TRANS-2-CHLORO-1-DEUTEROPROPENE	13270	FA
3	5		1										3-CHLORO PROP-1-ENE	7848	CA
3	5		1										2-CHLORO-1,1-DIDEUTEROPROPENE	13271	FA
3	5		1										2-CHLORO-3,3,3-TRIDEUTEROPROPENE	13275	FA
3	5		1										3-CHLOROPRO-2-ENE	G 74	BA
3	5		1										TRANS-2-CHLORO-1H-PERDEUTEROPROPENE	13273	FA
3	5		1										CIS-2-CHLORO-1-DEUTEROPROPENE	13269	FA
3	5		1										2-CHLOROPERDEUTEROPROPENE	13272	FA
3	5		1										2-CHLOROPROPENE	13268	FA
3	5		1					1					CHLOROACETONE	1847	EA
3	5		1					1					CHLOROACETONE	4791	FA
3	5		1					1					EPICHLOROHYDRIN	P 51	BA
3	5		1					1					PROPANYL CHLORIDE	12939	FA
3	5		1					1					PROPIONYL CHLORIDE	5965	CA
3	5		3										1,2,3-TRICHLOROPROPANE	P 101	BA
3	5			1									3-FLUOROPROP-2-ENE	G 90	BA
3	5			1				2					METHYL FLUOROACETATE	12969	FA
3	5			3									1,1,1-TRIFLUOROPROPANE	2501	CA
3	5			3									1,1,1-TRIFLUOROPROPANE	34	FA
3	5			3				1					DL1,1,1-TRIFLUORO-2-PROPANOL	6309	CK
3	5			3				3		1			ETHYL TRIFLUOROMETHANESULPHONATE	201	FA
3	5			5						1			ALLYLSULPHUR PENTAFLUORIDE	7931	FA
3	5			5						1			PROP-1-ENYLSULPHUR PENTAFLUORIDE	7932	FA
3	5						1						PROPIONITRILE	1667	FA
3	6												CYCLOPROPANE	1338	FA
3	6												CYCLOPROPANE	3919	EA
3	6												CYCLOPROPANE	123	GA

C H BR CL F I B N O P S SI M

C	H	BR	CL	F	I	B	N	O	P	S	SI	M	NAME	SPECTRUM NO.	LOCA-TION
3	6												HEXADEUTEROCYCLOPROPANE	1339	FA
3	6												PROPENE	1659	FA
3	6												PROPYLENE	3964	EA
3	6												PROPYLENE	G 129	BA
3	6												PROPENE	6403	CA
3	6		1	5						1			2-CHLOROPROPYLSULPHUR PENTAFLUORIDE	7925	FA
3	6		2										1,2-DICHLOROPROPANE	P 99	BA
3	6							1					ACETONE	P 12	BA
3	6							1					ACETONE	G 17	BA
3	6							1					1,2-EPOXY PROPANE	9067	FA
3	6							1					METHYL VINYL ETHER	1682	FA
3	6							1					METHYL VINYL ETHER	4426	EA
3	6							1					METHYL VINYL ETHER	G 53	BA
3	6							1					PERDEUTERIOACETONE	15491	FA
3	6							1					PROPAN-2-ONE	6242	EA
3	6							1					PROPAN-2-ONE	1619	FA
3	6							1					PROPAN-2-ONE	3907	EA
3	6							1					1,2-PROPYLENE OXIDE	P 56	BA
3	6							1					PROPANAL	8348	CA
3	6							1					1,2-PROPYLENE OXIDE	G 51	BA
3	6							1					PROPANAL	G 35	BA
3	6							1					PROPAN-2-ONE	6309	CA
3	6							1					PROPYLENE OXIDE	2212	CA
3	6							1					PROP-2-EN-1-OL	13	BA
3	6							1					PROPAN-2-ONE	180	BA
3	6							1					PROPANAL	153	BA
3	6							1					1,1,1-TRIDEUTERIOACETONE	15492	FA
3	6							1					TRIMETHYLENE OXIDE	499	FA
3	6							2					DIOXALANE	9684	CA
3	6							2					ETHYL FORMATE	P 22	BA
3	6							2					ETHYL FORMATE	G 26	BA
3	6							2					ETHYL FORMATE	207	BA
3	6							2					METHYL ACETATE	211	BA
3	6							2					METHYL ACETATE	12968	FA
3	6							2					PROPANOIC ACID	6187	EA
3	6							2					PROPANOIC ACID	77	BA
3	6							3					2-HYDROXY PROPANOIC ACID (LACTIC)	107	BA
3	6							3					PEROXY PROPANOIC ACID	6186	EA
3	6							3					PROPYLENE OZONIDE	3761	EA
3	6							3					PROPENE OZONIDE	1926	FA
3	6							3					1,3,5-TRIOXANE	4005	FA
3	6							3					SYM-TRIOXANE	6436	EA
3	6									1			THIACYCLOBUTANE	1770	FA
3	7		1										2-CHLOROPROPANE	4002	FA
3	7		1										1-CHLOROPROPANE	121	GA
3	7		1										2-CHLORO PROPANE	7849	CA
3	7		1										2-CHLOROPROPANE		
3	7			1									2-FLUOROPROPANE	3806	EA
3	7			1									1-FLUOROPROPANE	3807	EA
3	7				1								2-IODOPROPANE	G 92	BA
3	7						1						AMINOCYCLOPROPANE	G 58	BA
3	7						1						1-AMINOPROP-2-ENE	G 55	BA
3	7						1						1-AMINOPROP-2-ENE	5145	FA
3	7						1						1-AMINO PROP-2-ENE	118	BA
3	7						1						2-METHYLETHYLENEIMINE	649	FA
3	7						1	1					N,N-DIMETHYL FORMAMIDE	146	BA
3	7						1	1					DIMETHYL FORMAMIDE	G 23	BA
3	7						1	1					DIMETHYL FORMAMIDE	P 19	BA
3	7						1	1					N-METHYL ACETAMIDE	6875	EA
3	7						1	1					N-METHYL ACETAMIDE DEUTERATED	3404	EA
3	7						1	1					N-METHYL ACETAMIDE	3403	EA
3	7						1	1					PROPAN-2-ONE OXIME	1317	FA
3	7						1	1					PROPIONAMIDE	143	BA
3	7						1	1		1			ISOPROPYL THIONITRITE	8957	FA
3	7						1	1		1			PROPYL THIONITRITE	8956	FA
3	7						1	2					2-NITROPROPANE	1656	FA
3	7						1	2					1-NITROPROPANE	1655	FA
3	7						1	2					2-NITROPROPANE	G 64	BA
3	7						1	2					1-NITROPROPANE	P 69	BA
3	7						1	2					2-NITROPROPANE	P 70	BA
3	7						1	2					1-NITRO PROPANE	3958	EA
3	7						1	2					2-NITRO PROPANE	3959	EA
3	7						1	3					ISOPROPYL NITRATE	3784	FA
3	7						1	3					PROPYL NITRATE	3783	FA
3	8												2-DEUTEROPROPANE	4017	FA
3	8												1-DEUTEROPROPANE	4016	FA
3	8												2,2-DIDEUTEROPROPANE	4018	FA
3	8												HEXA-1,1,1,3,3,3-DEUTEROPROPANE	14003	FA
3	8												PERDEUTEROPROPANE	14004	FA
3	8												PROPANE,OCTADEUTERO	49	EA
3	8												PROPANE,2,2DIDEUTERO	55	EA
3	8												PROPANE	50	EA
3	8												PROPANE	1658	FA
3	8												PROPANE,1,1,7,3,3,3-HEXADEUTERO	56	EA
3	8												PROPANE	G 128	BA
3	8												PROPANE	3963	EA
3	8												PROPANE	6404	CA
3	8						2						1,1-DIMETHYL 2-METHYLENEHYDRAZINE	1649	FA
3	8						2						METHYLENE DIMETHYL HYDRAZINE	3949	EA
3	8							1					PROPAN-2-OL	6241	EA
3	8							1					PROPAN-1-OL	6240	EA
3	8							1					PROPAN-1-OL	3430	EA
3	8							1					PROPAN-2-OL	3778	FA
3	8							1					PROPAN-1-OL	1969	FA
3	8							1					PROPAN-2-OL	G 40	BA
3	8							1					PROPAN-2-OL	P 43	BA
3	8							1					PROPAN-2-OL	20	BA
3	8							1					PROPAN-1-OL	3	BA
3	8							2					DIMETHOXYMETHANE	G 44	BA
3	8							2					DIMETHOXY METHANE	9685	CA

C H BR CL F I B N O P S SI M

C	H	BR	CL	F	I	B	N	O	P	S	SI	M	NAME	SPECTRUM NO.	LOCA-TION
3	8							2					DIMETHOXY METHANE	P 28	BA
3	8							2					2-METHOXY ETHANOL	45	BA
3	8							2					PROPANE-1,2-DIOL	48	BA
3	8							2					PROPANE-1,3-DIOL	49	BA
3	8									1			1-PROPANETHIOL	5983	CA
3	9		1					3			1		CHLORO TRIMETHOXY SILANE	4870	FA
3	9		1								1		TRIMETHYLCHLORO SILANE	1843	EA
3	9		1								1		TRIMETHYL CHLORO SILANE	2066	FA
3	9					1							TRI (PERDEUTEROMETHYL) BORANE	17293	FA
3	9					1							TRIMETHYL BORON	6903	FA
3	9					1							TRIMETHYL BORANE-D-9	8706	EA
3	9					1							TRIMETHYL BORANE	6445	EA
3	9					1							TRIMETHYL BORANE	832	GA
3	9					1							TRIMETHYL BORANE	1994	FA
3	9					1							TRIMETHYL BORANE	6444	EA
3	9					1		3					TRIMETHOXY BORANE	6483	EA
3	9					1		3					TRIMETHYL BORATE	6377	FA
3	9						1						2-AMINOPROPANE	3813	GA
3	9						1						2-AMINO PROPANE	120	BA
3	9						1						1-AMINO PROPANE	109	BA
3	9						1						METHYL BIS(TRIDEUTEROMETHYL) AMINE	16486	FA
3	9						1						PERDEUTEROTRIMETHYLAMINE	16488	FA
3	9						1						TRIDEUTEROMETHYL DIMETHYLAMINE	16484	FA
3	9						1						TRIMETHYLAMINE	G 67	BA
3	9						1						TRIMETHYLAMINE	5874	EA
3	9						1						TRIMETHYLAMIE	9718	CA
3	9						1						TRIMETHYLAMINE	3969	EA
3	9						1						TRIMETHYLAMINE	3331	EA
3	9						1						TRIMETHYLAMINE	1663	FA
3	9						1						TRIMETHYLAMINE	8474	EA
3	9							3					METHYL BORATE,B-10	6424	EA
3	9							3					METHYL BORATE	6425	EA
3	9							3	1				TRIMETHYL PHOSPHITE	A 692	FA
3	9								1				TRIMETHYL PHOSPHINE	7047	FA
3	10											GE	TRIMETHYL GERMANE	16320	FA
3	10						2						TRIMETHYL HYDRAZINE	2301	EA
3	10						2						TRIMETHYLHYDRAZINE	1664	FA
3	10						2						TRIMETHYL HYDRAZINE	3456	EA
3	10						2						TRIMETHYL HYDRAZINE	3970	EA
3	10							1			1		TRIMETHYL DEUTEROSILANOL	12010	FA
3	10							1			1		TRIMETHYL SILANOL	12003	FA
3	10										1		TRIMETHYL SILANE	1931	EA
3	12					2							TRIMETHYL DIBORANE	8736	EA
3	12					2							TRIMETHYL DIBORANE (BORON-D-3)	8737	EA
3		1		7									2-BROMOHEPTAFLUOROPROPANE	9081	FA
3		1		7									1-BROMOHEPTAFLUOROPROPANE	3463	FA
3		1		7									1-BROMOHEPTAFLUORO PROPANE	3568	EA
3		1		7									1-BROMOHEPTAFLUOROPROPANE	2437	CA
3		2		6									1,3-DIBROMOHEXAFLUOROPROPANE	3098	FA
3		2		6									1,3-DIBROMOHEXAFLUOROPROPANE	2409	CA
3		2		6									1,2-DIBROMOHEXAFLUORO PROPANE	3570	EA
3		2		6									1,3-DIBROMOHEXAFLUOROPROPANE	541	EA
3			1	5				1					PENTAFLUORO PROPIONYL CHLORIDE	3622	EA
3			1	5				1					PENTAFLUORO PROPIONYL CHLORIDE	3095	CA
3			1	5				1					PERFLUOROPROPIONYL CHLORIDE	66	FA
3			1	7									1-CHLOROHEPTAFLUOROPROPANE	2438	CA
3			1	7									1-CHLOROHEPTAFLUOROPROPANE	252	FA
3			1	7									1-CHLOROHEPTAFLUOROPROPANE	3464	FA
3			2	6									1,2-DICHLOROHEXAFLUOROPROPANE	249	FA
3			2	6									1,1-DICHLOROHEXAFLUOROPROPANE	248	FA
3			2	6									1,3-DICHLOROHEXAFLUOROPROPANE	3099	FA
3			2	6									1,3-DICHLOROHEXAFLUOROPROPANE	2410	CA
3			2	6									1,2-DICHLOROHEXAFLUORO PROPANE	3571	EA
3			2	6									1,2-DICHLOROHEXAFLUOROPROPANE	23672	CA
3			2	6									1,3-DICHLOROHEXAFLUOROPROPANE	542	EA
3			2	9								AS	TRIS-TRIFLUOROMETHYL ARSENIC DICHLORIDE	58	FA
3			3	3									1,1,2-TRICHLOROTRIFLUORO PROPENE	3578	EA
3			6					1					HEXACHLOROACETONE	4798	EA
3				3			1						TRIFLUORO ACRYLONITRILE	3611	EA
3				5			1						PENTAFLUORO PROPIONITRILE	3609	EA
3				5			1						PERFLUOROPROPIONITRILE	68	FA
3				6									HEXAFLUOROPROPENE	3577	EA
3				6									HEXAFLUOROPROPENE	7366	CA
3				6									HEXAFLUOROCYCLOPROPANE	15350	FA
3				6									PERFLUOROPROPENE	4023	FA
3				6				1					HEXAFLUORO-2-PROPANONE	6840	CK
3				6				1					HEXAFLUORO PROPANAL	3619	EA
3				6				1					PERFLUOROACETONE	13581	FA
3				6				1					BIS-TRIFLURO METHYL KETONE	3657	EA
3				6						3			BISTRIFLUOROMETHYL TRITHIOCARBONATE	85	FA
3				7	1								HEPTAFLUORO-2-IODOPROPANE	9082	FA
3				7	1								2-IODO-HEPTAFLUOROPROPANE	6290	CK
3				7	1								1-IODOHEPTAFLUORO PROPANE	3569	EA
3				7			1	1					N,N-DI-(TRIFLUOROMETHYL) CARBAMYL FLUORIDE	3793	EA
3				8									OCTAFLUOROPROPANE	3539	EA
3				8									PERFLUOROPROPANE	4001	FA
3				9								AS	TRIS-TRIFLUOROMETHYL ARSINE	52	FA
3								2					CARBON SUBOXIDE	A 902	FA
4	1	1											1-BROMOBUTA-1, 3-DIYNE	16324	FA
4	1		1										1-CHLOROBUTA-1, 3-DIYNE	16323	FA
4	1			5									1H-PENTAFLUOROCYCLOBUT-1-ENE	3297	FA
4	1			5									3H-PENTAFLUOROCYCLOBUTENE	5706	FA
4	1			5									PENTA(3,3,4,4,4)FLUORO BUT-1-YNE	1499	EA
4	1			7				1					BIS-(TRIFLUOROMETHYL) ACETYL FLUORIDE	3797	EA
4	1			9									1-HYDRO-NONAFLUOROBUTANE	3558	EA
4	1			9									TRIS-TRIFLUOROETHYLMETHANE	3557	EA
4	1				1								1-IODOBUTA-1, 3-DIYNE	16325	FA
4	1							4					COBALT CARBONYL HYDRIDE	4471	EA
4	2			3			1						2-(TRIFLUOROMETHYL)ACRYLONITRILE	6306	CK
4	2			4									1,1,4,4-TETRAFLUOROBUTA-1,3-DIENE(TRANS)	13207	FA

C H BR CL F I B N O P S SI M

C	H	BR	CL	F	I	B	N	O	P	S	SI	M	NAME	SPECTRUM NO.	LOCATION
4	2			6									BUT-2-ENE,1,1,1,4,4,4-HEXAFLUORO	292	EA
4	2			6									2H,3H-HEXAFLUORO-CIS-BUT-2-ENE	4026	FA
4	2			6				2					2,2,2-TRIFLUOROETHYL TRIFLUOROACETATE	3292	FA
4	2			8									1,4-DIHYDROOCTAFLUOROBUTANE	3559	EA
4	2									1			DIACETYLENE SULPHIDE	11718	FA
4	3			5				1					PENTA(3,3,4,4,4)FLUORO BUTAN-2-ONE	1189	EA
4	4												BUTATRIENE	13887	FA
4	4		2										1,4-DICHLOROBUTA-1,3-DIENE(CIS,CIS)	12114	FA
4	4		2										1,4-DICHLOROBUTA-1,3-DIENE(TRANS,TRANS)	12118	FA
4	4		2										1,4-DICHLOROBUTA-1,3-DIENE(CIS,TRANS)	12116	FA
4	4		2										PERDEUTERIOPYRIMIDINE	15757	FA
4	4			2									1,4-DIFLUOROBUTA-1,3-DIENE(CIS,CIS)	12111	FA
4	4			2									1,4-DIFLUOROBUTA-1,3-DIENE(TRANS,TRANS)	12113	FA
4	4			2									1,4-DIFLUOROBUTA-1,3-DIENE(CIS,TRANS)	12112	FA
4	4			2				3					FLUOROACETIC ANHYDRIDE	12970	FA
4	4			6				1					2-HYDROHEXAFLUOROPROPYL METHYL ETHER	3593	EA
4	4						2						2,3-DIDEUTEROPYRAZINE	13228	FA
4	4						2						PERDEUTEROPYRAZINE	13229	FA
4	4						2						PYRAZINE	3180	EA
4	4						2						PYRAZINE	5420	FA
4	4						2						PYRIMIDINE	15756	FA
4	4							1					FURAN	G 50	BA
4	4							2					DIKETENE	14605	FA
4	4							2					1,4-DIOXIN	15701	FA
4	4							2					TETRADEUTERODIKETENE	14607	FA
4	4									1			THIOPHENE	G 131	BA
4	5		1					1					CYCLOPROPANE CARBOXYLIC ACID CHLORIDE	G 22	BA
4	5			1									CIS-1-FLUOROBUTA-1,3-DIENE	12109	FA
4	5			1									TRANS-1-FLUOROBUTA-1,3-DIENE	12110	FA
4	5			3									1,1,1-TRIFLUOROBUT-2-ENE	30	FA
4	5			3				1					2,2,2-TRIFLUOROETHYL VINYL ETHER	8955	CA
4	5			3				2					ETHYL TRIFLUOROACETATE	G 27	BA
4	5			6			1						DI-2,2,2-TRIFLUOROETHYLAMINE	555	FA
4	5						1						CYCLOPROPYL CYANIDE	8462	EA
4	5						1						2-METHYL PROP-2-ENE NITRILE	G 62	BA
4	5						1						PYRROLE	5146	FA
4	5						1						PYRROLE	141	BA
4	5						1	1					4-METHYL OXAZOLE	16463	FA
4	6												BUT-2-YNE	4036	FA
4	6												BUTA-1,3-DIENE	3914	EA
4	6												BUTA-1,3-DIENE	1625	FA
4	6												BUTA-1,3-DIENE	893	CA
4	6												BUTA-1,3-DIENE	P 105	BA
4	6												BUTA-1,3-DIENE	G 112	BA
4	6												CYCLOBUTENE	4575	EA
4	6												CYCLOBUTENE-D6	4577	EA
4	6												CYCLOBUTENE	352	EA
4	6												CYCLOBUTENE	2321	FA
4	6												METHYLENE CYCLOPROPANE	2087	EA
4	6												METHYL ALLENE	9202	EA
4	6												METHYLENECYCLOPROPANE	2199	FA
4	6												PERDEUTEROCYCLOBUTENE	2322	FA
4	6		2										3,4-DICHLOROBUT-1-ENE	G 79	BA
4	6			6			1		1				DIMETHYLAMINO-BIS-TRIFLUOROMETHYL PHOSPHINE	6611	EA
4	6						2	1					3,5-DIMETHYL-1,2,4-OXADIAZOLE	16248	FA
4	6							1					BUT-2-ENAL	161	BA
4	6							1					CYCLOBUTANONE	7626	FA
4	6							1					DIVINYL ETHER	16411	FA
4	6							1					DIVINYL ETHER	G 47	BA
4	6							1					3,4-EPOXY-1-BUTENE	G 48	BA
4	6							1					2-METHYL PROP-2-ENAL	8350	CA
4	6							1					METHYL VINYL KETONE	G 32	BA
4	6							1					2,2,4,4-TETRADEUTEROCYCLOBUTANONE	7628	FA
4	6							1					VINYL ACETATE	G 37	BA
4	6							2					BUT-3-ENOIC ACID	101	BA
4	6							2					BUT-2-ENOIC ACID	91	BA
4	6							2					BUTAN-2,3-DIONE	193	BA
4	6							2					GAMMA-BUTYROLACTONE	268	BA
4	6							2					2-METHYL PROP-2-ENOIC ACID	98	BA
4	6							2					METHYL PROP-2-ENOATE	G 29	BA
4	6							2					2-METHYL PROP-2-ENOIC ACID	5129	FA
4	6							2					VINYL ACETATE	257	BA
4	6							3					ACETIC ANHYDRIDE	5173	FA
4	7	1											1-BROMO BUT-1-ENE	13663	CA
4	8												CIS-BUT-2-ENE	3251	FA
4	8												TRANS-BUT-2-ENE	3253	FA
4	8												BUT-1-ENE	3247	FA
4	8												TRANS-BUT-2-ENE	G 115	BA
4	8												C2S-BUT-2-ENE	G 114	BA
4	8												TRANS-BUT-2-ENE	7860	CA
4	8												CIS-BUT-2-ENE	7859	CA
4	8												BUT-1-ENE	7857	CA
4	8												CYCLOBUTANE	2496	CA
4	8												CYCLOBUTANE	9183	EA
4	8												CYCLOBUTANE	2361	FA
4	8												2-METHYLPROPENE	3249	FA
4	8												2-METHYL PROPENE	7858	CA
4	8		2										1,2-DICHLORO-2-METHYL PROPANE	6689	EA
4	8		2					1					DI-2-CHLOROETHYL ETHER	P 48	BA
4	8							1					BUT-1-EN-3-OL	31	BA
4	8							1					N-BUTANAL	1456	CA
4	8							1					N-BUTANAL	154	BA
4	8							1					BUTAN-2-ONE	181	BA
4	8							1					BUTAN-2-ONE	6243	EA
4	8							1					BUTAN-2-ONE	P 15	BA
4	8							1					TR-1,2-DIMETHYL ETHYLENE OXIDE	G 45	BA
4	8							1					ETHYL VINYL ETHER	9152	CA
4	8							1					ETHYL VINYL ETHER	280	BA
4	8							1					2-METHYL PROPANAL	696	CA
4	8							1					OCTADEUTEROFURAN	6813	FA

C H BR CL F I B N O P S SI M

C	H	BR	CL	F	I	B	N	O	P	S	SI	M	NAME	SPECTRUM NO.	LOCA-TION
4	8							1					TETRAHYDROFURAN-D-8	7057	EA
4	8							1					TETRAHYDROFURAN	6815	FA
4	8							1					TETRAHYDROFURAN-2,2,5,5-D-4	7058	EA
4	8							1					2,2,5,5-TETRADEUTEROTETRAHYDROFURAN	6814	FA
4	8							1					TETRAHYDROFURAN	7056	EA
4	8							1					TETRAHYDROFURAN	P 58	BA
4	8							1					TETRAHYDROFURAN-3,3,4,4-D-4	7059	EA
4	8							1					TETRAHYDROFURAN-2,3,4,5-D-4	7060	EA
4	8							1					TETRAHYDROFURAN	G 52	BA
4	8							2					BUTANOIC ACID,DEUTERATED	6783	EA
4	8							2					BUTANOIC ACID	6777	EA
4	8							2					BUTANOIC ACID	6184	EA
4	8							2					N-BUTANOIC ACID	78	BA
4	8							2					DIOXANE	282	BA
4	8							2					DIOXANE	P 49	BA
4	8							2					ETHYL ACETATE	P 20	BA
4	8							2					ETHYL ACETATE	G 24	BA
4	8							2					ETHYL ACETATE	212	BA
4	8							2					3-HYDROXY BUTAN-2-ONE	194	BA
4	8							2					2-METHYL PROPANOIC ACID	84	BA
4	8							2					PROPYL FORMATE	208	BA
4	8							3					ISOBUTENE OZONIDE	1927	FA
4	8							3					ISOBUTYLENE OZONIDE	3760	EA
4	8							3					METHYL 2-HYDROXY PROPANOATE	244	BA
4	8							3					PEROXY BUTANOIC ACID	6185	EA
4	9	1											2-BROMO-2-METHYL PROPANE	4648	CA
4	9		1										1-CHLOROBUTANE	4622	CA
4	9		1										2-CHLORO-2-METHYL PROPANE	4649	CA
4	9		1										2-CHLORO BUTANE	13902	CA
4	9			1									TERT-BUTYL FLUORIDE	8468	EA
4	9						1						2,2-DIMETHYLETHYLENEIMINE	650	FA
4	9						1						TETRAHYDROPYRROLE (PYRROLIDENE)	5147	FA
4	9						1						TETRAHYDROPYRROLE-N-D	8703	EA
4	9						1						TETRAHYDROPYRROLE	8700	EA
4	9						1						TETRAHYDROPYRROLE (PYRROLIDINE)	131	BA
4	9						1	1					BUTYRAMIDE	144	BA
4	9						1	2					N-BUTYL NIRITE	G 57	BA
4	9								1				DIMETHYL VINYL PHOSPHINE	5567	FA
4	10												N-BUTANE	3915	EA
4	10												N-BUTANE	G 113	BA
4	10												N-BUTANE	4095	EA
4	10												(-BUTANE	1626	FA
4	10												N-BUTANE	2286	CA
4	10												2-DEUTERO-2-METHYLPROPANE	4019	FA
4	10												METHYL PROPANE,DEUTERATED	3277	EA
4	10												METHYLPROPANE	3276	EA
4	10												METHYL PROPANE	3943	EA
4	10												2-METHYLPROPANE	1643	FA
4	10												2-METHYLPROPANE	G 122	BA
4	10												2-METHYL PROPANE	51	EA
4	10												METHYL PROPANE	4094	EA
4	10												2-METHYL-2-DEUTEROPROPANE	52	EA
4	10												2-METHYL PROPANE	2285	CA
4	10												TRIMETHYL(-D9)METHANE	4877	EA
4	10												TRIMETHYL(-D9)METHANE-D	4878	EA
4	10							1					TERT-BUTANOL	G 38	BA
4	10							1					BUTAN-1-OL	3433	EA
4	10							1					TERT-BUTANOL	P 36	BA
4	10							1					BUTAN-1-OL	P 35	BA
4	10							1					BUTAN-1-OL	1970	FA
4	10							1					TERT-BUTANOL	2219	CA
4	10							1					TERT-BUTANOL	37	BA
4	10							1					BUTAN-1-OL	4	BA
4	10							1					BUTAN-2-OL	21	BA
4	10							1					DIETHYL ETHER	278	BA
4	10							1					DIETHYL ETHER	2916	EA
4	10							1					DIETHYL ETHER	3930	EA
4	10							1					DIETHYL ETHER	1635	FA
4	10							1					DIETHYL ETHER	6401	CA
4	10							1					DIETHYL ETHER	G 43	BA
4	10							1					DIETHYL ETHER	P 53	BA
4	10							2					2-ETHOXY ETHANOL	P 41	BA
4	10									1			METHYL ISOPROPYL SULPHIDE	1781	FA
4	10									1			METHYL PROPYL SULPHIDE	1775	FA
4	11					1		2					DIETHYL BORONATE	6373	FA
4	11					1		2					DIETHYL DEUTEROBORONATE	6374	FA
4	11					1		2					DIETHOXY BORANE	6488	EA
4	11					1		2					DIETHOXY BORANE-D	6486	EA
4	11					1		2					ETHYL DIMETHOXY BORANE	836	GA
4	11						1						1-AMINOBUTANE	G 56	BA
4	11						1						1-AMINOBUTANE	P 61	BA
4	11						1						2-AMINO BUTANE	121	BA
4	11						1						1-AMINOBUTANE	6647	CA
4	11						1						1-AMINO BUTANE	110	BA
4	11						1						DIETHYLAMINE	125	BA
4	11						1						DIETHYLAMINE	P 64	BA
4	11						1						DIETHYLAMINE	8960	CA
4	11						1						DIETHYLAMINE	557	FA
4	11						1						ETHYLDIMETHYLAMINE	5562	FA
4	11						1	1					N,N-DIMETHYL-2-AMINOETHANOL	5162	FA
4	11						1	1					2-(N,N-DIMETHYL)AMINO ETHANOL	50	BA
4	11						1				1		1-SILYL PYRROLIDINE	17197	FA
4	11								1				ETHYL DIMETHYL PHOSPHINE	5564	FA
4	12											GE	GERMANIUM TETRAMETHYL	1835	EA
4	12											PB	LEAD TETRAMETHYL	1837	EA
4	12											SN	TIN TETRAMETHYL	1838	EA
4	12	1					1				1		DIETHYLAMINO BROMOSILANE	17195	FA
4	12		1				1				1		DIETHYLAMINO CHLOROSILANE	17194	FA
4	12				1		1				1		DIETHYLAMINOIODOSILANE	17196	FA
4	12						2						TETRAMETHYL HYDRAZINE	3967	EA

C H BR CL F I B N O P S SI M

C	H	BR	CL	F	I	B	N	O	P	S	SI	M	NAME	SPECTRUM NO.	LOCATION
4	12						2						TETRAMETHYL HYDRAZINE	6715	EA
4	12						2						TETRAMETHYLHYDRAZINE	1661	FA
4	12						4						TETRAMETHYLTETRAZENE	1662	FA
4	12						4						TETRAMETHYL TETRAZENE	3968	EA
4	12							1			1		METHOXY TRIMETHYL SILANE	4873	FA
4	12							2			1		DIMETHOXY DIMETHYL SILANE	4872	FA
4	12							3			1		METHYL TRIMETHOXY SILANE	4871	FA
4	12							3			1		METHYL TRIMETHOXY SILANE	P 55	BA
4	12							4			1		TETRAMETHOXY SILANE	7852	EA
4	12										1		TETRAMETHYL SILANE	1844	EA
4	12										1		TETRAMETHYL SILANE	1445	EA
4	12										1		TETRAMETHYL SILANE	1930	EA
4	12										1		TETRAMETHYL SILANE	1859	EA
4	12										1		TETRAMETHYL SILANE	2067	FA
4	13		1				2				1		BIS (DIMETHYLAMINO) CHLOROSILANE	17193	FA
4	14					2							UNSYM-DIETHYL DIBORANE	6063	EA
4	14					2							1,1-DIETHYLDIBORANE-D-14	8735	EA
4	14					2							SYM-DIETHYL DIBORANE	6064	EA
4	14					2							SYM-DIETHYL DIBORANE	830	GA
4	14					2							1,1-DIETHYLDIBORANE (ETHYL-D-10)	8734	EA
4	14					2							1,1-DIETHYLDIBORANE	8732	EA
4	14					2							1,1-DIETHYLDIBORANE (BORON-D-4)	8733	EA
4	14					2							ETHYL BORANE DIMER	17302	FA
4	14					2							TETRAMETHYL DIBORANE	8738	EA
4	14					2							TETRAMETHYL DIBORANE (BORON-D-2)	8739	EA
4	16						2				2		1,1-BISDIMETHYLAMINODISILANE	A 473	FA
4			1	7				1					HEPTAFLUORYL BUTANYL CHLORIDE	7863	CA
4			1	13						1			4-CHLOROPERFLUOROBUTYLSULPHUR PENTAFLUORIDE	7935	FA
4			2	4									2,3-DICHLOROTETRAFLUOROBUTA-1,3-DIENE	4087	FA
4			2	6									3,4-DICHLOROHEXAFLUOROBUT-1-ENE	4088	FA
4			2	6									CIS-1,2-DICHLOROHEXAFLUOROCYCLOBUTANE	527	EA
4			2	6									TR-1,2-DICHLOROHEXAFLUOROCYCLOBUTANE	528	EA
4			2	8									1,2-DICHLORO-PERFLUORO METHYL PROPANE	79	FA
4			2	8									2,2-DICHLORO-OCTAFLUOROBUTANE	80	FA
4			2	8									1,2-DICHLORO-PERFLUOROBUTANE	8492	EA
4			2	8									2,3-DICHLORO-2-TRIFLUOROMETHYL PERFLUOROPROPANE	8494	EA
4			4	4									1,3,4,4-TETRACHLORO-1,2,3,4-TETRAFLUOROBUT-1-ENE	G 98	BA
4				4									1,1,4,4-TETRAFLUORO BUTA-1,2,3-TRIENE	6172	EA
4				4			2						TETRAFLUORO SUCCINONITRILE	3612	EA
4				6									HEXAFLUORO-2-BUTYNE	2644	EA
4				6									HEXAFLUORO BUTA-1,3-DIENE	3373	EA
4				6									HEXAFLUOROCYCLOBUTENE	406	FA
4				6									HEXAFLUORO BUTA-1,3-DIENE	3582	EA
4				6									PERFLUOROBUT-2-YNE	4027	FA
4				6									PERFLUOROBUTA-1,3-DIENE	1456	FA
4				6				2					GAMMA-HEXAFLUORO BUTYROLACTONE	2443	CA
4				7			1						HEPTAFLUORO BUTYRONITRILE	3613	EA
4				8									OCTAFLUORO CYCLOBUTANE	3548	EA
4				8									OCTAFLUORO BUT-2-ENE	3580	EA
4				8									OCTAFLUORO BUT-1-ENE	3579	EA
4				8									PERFLUOROCYCLOBUTANE	4028	FA
4				8									PERFLUORO CYCLOBUTANE	16925	CA
4				8									2-TRIFLUOROMETHYLPENTAFLUORO PROPENE	3581	EA
4				8				1					OCTAFLUORO TETRAMETHYLENE OXIDE	3591	EA
4				8				1					OCTAFLUORO BUTANAL	3620	EA
4				8				1					PERFLUORO-OXOLAN	4236	FA
4				10									DECAFLUORO METHYL PROPANE	3541	EA
4				10									DECAFLUORO BUTANE	3540	EA
4				10									PERFLUORO BUTANE	4037	FA
4				10				1					BIS-PENTAFLUOROETHYL ETHER	3588	EA
4				12								AS2	TETRA-TRIFLUOROMETHYLDIARSINE	57	FA
4							2						DICYANOACETYLENE	1440	EA
5	1		1	8									3-CHLORO-1,2,3,4,4,5,5,5-OCTAFLUOROPENT-1-ENE	258	FA
5	1			11									1-HYDRO-UNDECAFLUOROPENTANE	3560	EA
5	4							2					FURFURAL	P 24	BA
5	4							2					FURFURAL	175	BA
5	5			3				2					1,1,1-TRIFLUORO-2,4-PENTANEDIONE	6301	CK
5	5						1						PYRIDINE-2,6-D2	4158	EA
5	5						1						PYRIDINE	5300	EA
5	5						1						PYRIDINE-2,5-D2	4159	EA
5	5						1						PYRIDINE	P 71	BA
5	5						1						PYRIDINE	G 65	BA
5	5						1						PYRIDINE-4-D	4157	EA
5	5						1						PYRIDINE	4160	EA
5	5						1						PYRIDINE	147	BA
5	5						1						PYRIDINE	239	FA
5	6												2-METHYL BUT-1-EN-3-YNE	17302	CA
5	6							1					2-METHYL FURAN	287	BA
5	6							2					FURFURYL ALCOHOL	18	BA
5	6							2					FURFURYL ALCOHOL	P 42	BA
5	7						1	1					2,4-DIMETHYL OXAZOLE	16465	FA
5	8												CYCLOPENTENE	19781	CA
5	8												CYCLOPENTENE	4231	FA
5	8												2-METHYL BUTA-1,3-DIENE	7454	CA
5	8												PENTA-1,4-DIENE	G 126	BA
5	8												CIS-PENTA-1,3-DIENE	4095	FA
5	8												TRANS-PENTA-1,3-DIENE	4096	FA
5	8												PENTA-1,4-DIENE	4097	FA
5	8												PENT-1-YNE	300	BA
5	8												SPIROPENTANE	4232	FA
5	8		1	3				1					2-CHLORO-1,1,2-TRIFLUOROETHYL ISOPROPYL ETHER	3723	FA
5	8		1	3				1					2-CHLORO-1,1,2-TRIFLUOROETHYL PROPYL ETHER	3722	FA
5	8							1					ALLYL VINYL ETHER	8306	CA
5	8							1					CYCLOPENTANONE	196	BA
5	8							1					2,3-DIHYDROPYRAN	283	BA
5	8							1					2,3-DIHYDRO-5-METHYL FURAN	2497	CA
5	8							1					2,3-DIHYDROPYRAN	2498	CA
5	8							2					ETHYL PROP-2-ENOATE	G 25	BA
5	8							2					ETHYL PROP-2-ENOATE	P 21	BA
5	8							2					ISOPROPENYL ACETATE	258	BA

C H BR CL F I B N O P S SI M

C	H	BR	CL	F	I	B	N	O	P	S	SI	M	NAME		SPECTRUM NO.	LOCA-TION
5	8							2					2-METHYL BUT-2-ENOIC ACID		100	BA
5	8							2					3-METHYL BUT-2-ENOIC ACID		99	BA
5	8							2					METHYL 2-METHYL PROP-2-ENOATE	G	31	BA
5	8							2					METHYL 2-METHYL PROP-2-ENOATE	P	31	BA
5	8							2					GAMMA-PENTALACTONE		269	BA
5	8							2					PENTA-2,4-DIONE		195	BA
5	8							4					DIMETHYL MALONATE		240	BA
5	9		1					1					3-METHYL BUTANYL CHLORIDE		4736	CA
5	9		1					1					2-METHYL 4-CHLOROBUTAN-3-ONE		5060	FA
5	9						1						N,N-DIMETHYL PROP-2-YN-YLAMINE		13739	CK
5	10												CYCLOPENTANE		7854	CA
5	10												1,1-DIMETHYLCYCLOPROPANE		11831	CA
5	10												1,1-DIMETHYLCYLOPROPANE		4237	FA
5	10												ETHYL CYCLOPROPANE		4239	FA
5	10												2-METHYLBUT-2-ENE		4093	FA
5	10												2-METHYLBUT-1-ENE		4092	FA
5	10												3-METHYLBUT-1-ENE		4094	FA
5	10												2-METHYL BUT-2-ENE-D		4482	EA
5	10												2-METHYL BUT-2-ENE		12975	CK
5	10												3-METHYL BUT-1-ENE		7856	CA
5	10												2-METHYL BUT-1-ENE		6320	CA
5	10												PENT-2-ENE	P	115	BA
5	10												TRANS-PENT-2-ENE	G	127	BA
5	10												PENT-2-ENE		4616	CA
5	10												CIS-PENT-2-ENE		2780	FA
5	10												PENT-1-ENE		2779	FA
5	10												TRANS-PENT-2-ENE		2782	FA
5	10												PENT-1-ENE		685	CA
5	10							1					CYCLOPENTANOL		29	BA
5	10							1					2-METHYL BUT-3-EN-2-OL		40	BA
5	10							1					2-METHYL BUTAN-3-ONE		5058	FA
5	10							1					2-METHYL TETRAHYDROFURAN		9262	EA
5	10							1					PENTAN-2-ONE	P	32	BA
5	10							1					PENTAN-2-ONE	G	33	BA
5	10							1					PENT-1-EN-4-OL		35	BA
5	10							1					PENT-4-EN-1-OL		9	BA
5	10							1					PENTAN-3-ONE		190	BA
5	10							1					PENTAN-2-ONE		182	BA
5	10							1					N-PENTANAL		155	BA
5	10							1					PROPYL ACETATE	P	33	BA
5	10							2					BUTYL FORMATE		209	BA
5	10							2					ETHYL PROPANOATE		218	BA
5	10							2					METHYL BUTANOATE		223	BA
5	10							2					2-METHYL BUTANOIC ACID		86	BA
5	10							2					3-METHYL BUTANOIC ACID		85	BA
5	10							2					N-PENTANOIC ACID		79	BA
5	10							2					PROPYL ACETATE		213	BA
5	10							2					TETRAHYDROFURFURYL ALCOHOL		19	BA
5	10							3					ETHYL 2-HYDROXY PROPANOATE		245	BA
5	10							3					METHYL 2-HYDROXY 2-METHYL PROPANOATE		247	BA
5	10									1			CYCLOPENTANETHIOL		4234	FA
5	11	1											2-BROMO 2-METHYL BUTANE		13649	CA
5	11		1										2-CHLOROPENTANE		19764	CA
5	11		1										2-CHLORO 2-METHYL BUTANE		6651	CA
5	11						1						PIPERIDINE		117	BA
5	11						1						N-(2-PROPYL)-ETHYLENEIMINE		651	FA
5	11						1	3					ISOPENTYL NITRATE		6486	CA
5	12												2,2-DIMETHYL PROPANE		7986	CA
5	12												3-METHYL BUTANE		4093	EA
5	12												2-METHYLBUTANE	G	124	BA
5	12												2-METHYL BUTANE		691	CA
5	12												N-PENTANE		6406	CA
5	12												PERDEUTEROTETRAMETHYLMETHANE		2079	FA
5	12												TETRAMETHYLMETHANE		2080	FA
5	12												TETRAMETHYL METHANE		1846	EA
5	12												TETRAMETHYL METHANE-D12		1845	EA
5	12		1				1				1		PIPERIDINOCHLOROSILANE		17199	FA
5	12							1					3-METHYL BUTAN-1-OL		7	BA
5	12							1					PENTAN-1-OL		5	BA
5	12							1					PENTAN-2-OL		22	BA
5	12							1					PENTAN-3-OL		25	BA
5	12							1					PENTAN-1-OL		3436	EA
5	12							1					PENTAN-1-OL	P	34	BA
5	12							1					PENTAN-1-OL		1972	FA
5	12									1			TERT-BUTYL METHYL SULPHIDE		4276	FA
5	12									1			ETHYL ISOPROPYL SULPHIDE		1783	FA
5	12									1			PENTANE-3-THIOL		4262	FA
5	13					1		1					DIETHYL METHOXY BORANE		835	GA
5	13						1						1-AMINO PENTANE		111	BA
5	13						1						3-AMINO PENTANE		124	BA
5	13						1						2-AMINO PENTANE		122	BA
5	13						1						N-METHYL BUTYLAMINE		130	BA
5	13						1				1		1-SILYL PIPERIDINE		17198	FA
5	15					1							TRIETHYL BORANE		6446	EA
5	15					1							TRIETHYL BORANE		6447	EA
5		1		11									1-BROMO-UNDECAFLUOROPENTANE		545	EA
5		1		11									1-BROMO-UNDECAFLUOROPENTANE		3460	FA
5		1		11									1-BROMO-UNDECAFLUOROPENTANE		2408	CA
5			1	11									1-CHLORO-UNDECAFLUOROPENTANE		2407	CA
5			1	11									1-CHLORO-UNDECAFLUOROPENTANE		546	EA
5			1	11									1-CHLORO-UNDECAFLUOROPENTANE		3461	FA
5				10									DECAFLUOROPENT-1-ENE		3583	EA
5				10									DECAFLUOROCYCLOPENTANE		3549	EA
5				11	1								1-IODO-UNDECAFLUOROPENTANE		3459	FA
5				11	1								1-IODO-UNDECAFLUOROPENTANE		2405	CA
5				11	1								1-IODO-UNDECAFLUOROPENTANE		543	EA
5				11			1						PERFLUORO-(1-METHYLIMINOBUTANE)		3677	FA
5				11			1						UNDECAFLUORO PIPERIDINE		3607	EA
5				12									DODECAFLUOROPENTANE		3542	EA
5				12									PERFLUOROPENTANE		3564	FA

C H BR CL F I B N O P S SI M

C	H	BR	CL	F	I	B	N	O	P	S	SI	M	NAME	SPECTRUM NO.	LOCA-TION
5				12									PERFLUOROISOPENTANE	3566	FA
5				13			1						DI(PENTAFLUOROETHYL)TRIFLUOROMETHYLAMINE	3599	EA
5								5				FE	IRON PENTACARBONYL	5451	EA
6	1			5									DEUTEROPENTAFLUOROBENZENE	5575	FA
6	1			5									PENTAFLUOROBENZENE	5573	FA
6	1			7									2H-HEPTAFLUOROCYCLOHEXA-1,3-DIENE	4202	FA
6	1			7									1H-HEPTAFLUOROCYCLOHEXA-1,4-DIENE	4205	FA
6	1			7									1H-HEPTAFLUOROCYCLOHEXA-1,3-DIENE	4201	FA
6	1			9									4H-NONAFLUOROCYCLOHEXENE	4069	FA
6	1			9									1H-NONAFLUOROCYCLOHEXENE	4067	FA
6	1			11									UNDECAFLUORO-CYCLOHEXANE	3710	FA
6	2			4									1,2,4,5-TETRAFLUORO BENZENE	1818	EA
6	2			4									1,2,4,5-TETRAFLUOROBENZENE	2716	FA
6	2			6									1H,5H-HEXAFLUOROCYCLOHEXA-1,4-DIENE	4207	FA
6	2			6									1H,4H-HEXAFLUOROCYCLOHEXA-1,4-DIENE	4206	FA
6	2			6									1H,2H-HEXAFLUOROCYCLOHEXA-1,3-DIENE	4203	FA
6	2			8									OCTA(1H,5H)FLUOROCYCLOHEXENE	4073	FA
6	2			8									TRANS-OCTA(4H,5H)FLUOROCYCLOHEXENE	4076	FA
6	2			8									OCTA(1H,4H)FLUOROCYCLOHEXENE	4072	FA
6	2			8									CIS-OCTA(3H,4H)FLUOROCYCLOHEXENE	4074	FA
6	2			8									CIS-OCTA(4H,5H)FLUOROCYCLOHEXENE	4075	FA
6	2			8									OCTA(1H,2H)FLUOROCYCLOHEXENE	4070	FA
6	2			8									OCTA(1H,3H)FLUOROCYCLOHEXENE	4071	FA
6	2			10									TR-1H,4H-DECAFLUORO-CYCLOHEXANE	3716	FA
6	2			10									TR-1H,3H-DECAFLUORO-CYCLOHEXANE	3714	FA
6	2			10									CIS-1H,3H-DECAFLUORO-CYCLOHEXANE	3713	FA
6	2			10									CIS-1H,4H-DECAFLUORO-CYCLOHEXANE	3715	FA
6	2			10									CIS-1H,2H-DECAFLUORO-CYCLOHEXANE	3711	FA
6	2			10									TR-1H,2H-DECAFLUORO-CYCLOHEXANE	3712	FA
6	3											AS	TRIETHYNYL ARSINE	16142	FA
6	3											SB	TRIETHYNYL STIBINE	16143	FA
6	3			3									1,3,5-TRIFLUOROBENZENE	2575	FA
6	3			7									TRANS-HEPTA(1H,4H,5H)FLUOROCYCLOHEXENE	4079	FA
6	3			7									HEPTA(1H,2H,4H)FLUOROCYCLOHEXENE	4077	FA
6	3			7									CIS-HEPTA(1H,4H,5H)FLUOROCYCLOHEXENE	4078	FA
6	3			9									NONA-(1/2,4)-1H,2H,4H-FLUOROCYCLOHEXANE	3719	FA
6	3			9									NONA-(1,4/2)-1H,2H,4H-FLUOROCYCLOHEXANE	3720	FA
6	3			9									NONA-(1,2/4)-1H,2H,4H-FLUOROCYCLOHEXANE	3721	FA
6	3								1				TRIETHYNYL PHOSPHINE	16141	FA
6	4	1	1										P-BROMOCHLOROBENZENE	3350	FA
6	4	1	1										O-BROMOCHLOROBENZENE	3051	FA
6	4	1		1									P-BROMOFLUOROBENZENE	3415	EA
6	4	1		1									P-BROMOFLUOROBENZENE	3052	FA
6	4	1					1	2					P-BROMONITROBENZENE	3376	FA
6	4	2											P-DIBROMOBENZENE	3332	FA
6	4	2											O-DIBROMOBENZENE	3049	FA
6	4	2											M-DIBROMOBENZENE	3330	FA
6	4		1	1									M-CHLOROFLUOROBENZENE	3485	FA
6	4		1	1									O-CHLOROFLUOROBENZENE	1432	FA
6	4		1	1									P-CHLOROFLUOROBENZENE	3413	EA
6	4		1	1									M-CHLOROFLUOROBENZENE	2091	FA
6	4		1	1									O-CHLOROFLUOROBENZENE	3418	EA
6	4		1				1	2					M-CHLORONITROBENZENE	2397	FA
6	4		1				1	2					M-CHLORONITROBENZENE	2397	FA
6	4		1				1	2					P-CHLORONITROBENZENE	2394	FA
6	4		2										1,4-DICHLOROBENZENE	P 82	BA
6	4		2										1,2-DICHLOROBENZENE	P 80	BA
6	4		2										1,3-DICHLOROBENZENE	P 81	BA
6	4		2										P-DICHLOROBENZENE	2853	FA
6	4		2										O-DICHLOROBENZENE	4807	FA
6	4		2										M-DICHLOROBENZENE	2380	FA
6	4			1			1	2					P-FLUORONITROBENZENE	3378	FA
6	4			2									O-DIFLUOROBENZENE	3337	FA
6	4			2									1,4-DIFLUORO BENZENE	1814	EA
6	4			2									P-DIFLUOROBENZENE	2494	FA
6	4			6									TRANS-HEXA(3,3,4,5,6,6)FLUOROCYCLOHEXENE	5207	FA
6	4				2								P-DI-IODOBENZENE	3335	FA
6	4						2	4					1,2-DINITRO BENZENE	2379	FA
6	5	1											BROMOBENZENE	P 74	BA
6	5	1											BROMOBENZENE	3339	FA
6	5	1						1					4-BROMO PHENOL	3358	FA
6	5	1						1					2-BROMO PHENOL	3360	FA
6	5	1						1					4-BROMOPHENOL	67	BA
6	5	1						1					2-BROMOPHENOL	65	BA
6	5		1										CHLOROBENZENE	G 72	BA
6	5		1										CHLOROPENTADEUTEROBENZENE	13538	FA
6	5		1										CHLOROBENZENE	3338	FA
6	5		1										CHLOROBENZENE	P 77	BA
6	5		1					1					4-CHLOROPHENOL	66	BA
6	5		1					1					2-CHLOROPHENOL	64	BA
6	5		1					1					4-CHLORO PHENOL	3354	FA
6	5		1					1					2-CHLORO PHENOL	3356	FA
6	5			1									FLUOROBENZENE	1816	EA
6	5			1									FLUOROBENZENE	2580	FA
6	5			1									FLUOROBENZENE	G 89	BA
6	5			1				1					2-FLUORO PHENOL	3365	FA
6	5			5									TRANS-PENTA(1,4,4,5,6)FLUOROCYCLOHEXENE	5206	FA
6	5				1								IODOBENZENE	3340	FA
6	5				1			1					2-IODO PHENOL	3363	FA
6	5						1	1					NITROSOBENZENE	5360	FA
6	5						1	2					NITROBENZENE	3368	FA
6	5						1	3					3-NITRO PHENOL	3370	FA
6	6												BENZENE	1624	FA
6	6												BENZENE	3913	EA
6	6												BENZENE	P 104	BA
6	6												BENZENE	6402	CA
6	6												BENZENE	G 111	BA
6	6												BENZENE-D6	3451	EA
6	6												BENZENE	302	BA
6	6												HEXADEUTEROBENZENE	1861	FA

C H BR CL F I B N O P S SI M

C	H	BR	CL	F	I	B	N	O	P	S	SI	M	NAME	SPECTRUM NO.	LOCATION
6	6		1				1						3-CHLORO ANILINE	140	BA
6	6		1				1						2-CHLORO ANILINE	139	BA
6	6							1					DIPROP-1-YNYL ETHER	11719	FA
6	6							1					PHENOL	3344	FA
6	6							1					PHENOL	52	BA
6	6							1					PHENOL	9462	EA
6	6							1					PHENOL-O-D	9465	EA
6	6							2					2-ACETYL FURAN	205	BA
6	6							2					1,2-DIHYDROXY BENZENE (CATECHOL)	70	BA
6	6							2					1,4-DIHYDROXY BENZENE (QUINOL)	3341	FA
6	6							2					1,2-DIHYDROXY BENZENE (CATECHOL)	3343	FA
6	6							2					5-METHYL FURFURAL	176	BA
6	6							3					METHYL FUROATE	255	BA
6	7						1						ANILINE	132	BA
6	7						1						ANILINE	3346	FA
6	7						1						N,N-DIDEUTEROANILINE	6876	FA
6	7						1						4-METHYL PYRIDINE	5297	EA
6	7						1						3-METHYL PYRIDINE	5298	EA
6	7						1						2-METHYL PYRIDINE	5299	EA
6	7						1						2-METHYL PYRIDINE	148	BA
6	7						1						4-METHYL PYRIDINE	150	BA
6	7						1						3-METHYL PYRIDINE	149	BA
6	7						1						4-METHYL PYRIDINE	245	FA
6	7						1						3-METHYL PYRIDINE	243	FA
6	7						1						2-METHYL PYRIDINE	241	FA
6	8												CYCLOHEXA-1,4-DIENE	13264	FA
6	8												1,2-DIMETHYLENECYCLOBUTANE	1332	FA
6	8												HEXA-1,3,5-TRIENE	4746	EA
6	8												PERDEUTEROCYCLOHEXA-1,4-DIENE	13266	FA
6	8						2						1,3-DIAMINOBENZENE	4813	FA
6	8						2						1,2-DIAMINOBENZENE	3181	FA
6	8						2						1,4-DIAMINOBENZENE	3182	FA
6	8						2						2,6-DIMETHYL PYRAZINE	10889	FA
6	8							1					HEXA-2,4-DIENAL	167	BA
6	8							2					HEXA-2,4-DIENOIC ACID	97	BA
6	9			5									111,2,2-PENTAFLUOROHEXANE	5703	FA
6	10												CYCLOHEXENE	P 108	BA
6	10												CYCLOHEXENE	G 116	BA
6	10		2					1					2,2-DIMETHYL 4,4-DICHLOROBUTAN-3-ONE	5066	FA
6	10			9									3H-NONAFLUOROCYCLOHEXENE	4068	FA
6	10							1					CYCLOHEXANONE	197	BA
6	10							1					CYCLOHEXANONE	P 17	BA
6	10							1					HEX-2-ENAL	162	BA
6	10							1					4-METHYL PENT-3-EN-2-ONE	P 27	BA
6	10							2					ALLYL PROPANOATE	222	BA
6	10							2					ETHYL BUT-2-ENOATE (CROTONATE)	251	BA
6	10							2					HEX-2-ENOIC ACID	92	BA
6	10							2					HEX-3-ENOIC ACID	102	BA
6	10							2					GAMMA-HEXALACTONE	270	BA
6	10							3					ETHYL 3-OXOBUTANOATE	243	BA
6	10							4					DIMETHYL SUCCINATE	242	BA
6	11		1					1					2,2-DIMETHYL 4-CHLOROBUTAN-3-ONE	5063	FA
6	11			3									1,1,1-TRIFLUOROHEXANE	5702	FA
6	12												CYCLOHEXANE	P 107	BA
6	12												CYCLOHEXANE	6607	CA
6	12												1-ETHYL-1-METHYLCYCLOPROPANE	4718	FA
6	12												HEX-1-ENE	G 121	BA
6	12												1-HEXENE	6323	CA
6	12												CIS-3-HEXENE	36953	CA
6	12												CIS-3-METHYL-2-PENTENE	36954	CA
6	12												TRANS-4-METHYL PENT-2-ENE	6326	CA
6	12												4-METHYL PENT-2-ENE	6325	CA
6	12												METHYL CYCLOPENTANE	7855	CA
6	12												4-METHYL PENT-1-ENE	8302	CA
6	12												MEX-1-ENE	12977	CK
6	12												1,1,2-TRIMETHYLCYCLOPROPANE	4716	FA
6	12						4						HEXAMETHYLENETETRAMINE	1187	FA
6	12							1					TERT-BUTYL VINYL ETHER	8574	FA
6	12							1					CYCLOHEXANOL	P 37	BA
6	12							1					CYCLOHEXANOL	30	BA
6	12							1					HEX-1-EN-3-OL	32	BA
6	12							1					HEX-5-EN-1-OL	8	BA
6	12							1					TRANS-HEX-3-EN-1-OL	11	BA
6	12							1					CIS-HEX-3-EN-1-OL	10	BA
6	12							1					TRANS-HEX-2-EN-1-OL	15	BA
6	12							1					CIS-HEX-2-EN-1-OL	14	BA
6	12							1					HEXAN-2-ONE	183	BA
6	12							1					N-HEXANAL	156	BA
6	12							1					HEXAN-2-ONE	P 25	BA
6	12							1					4-METHYL PENTAN-2-ONE	P 26	BA
6	12							1					3-METHYL PENTAN-2-ONE	184	BA
6	12							2					AMYL FORMATE	210	BA
6	12							2					BUTYL ACETATE	214	BA
6	12							2					BUTYL ACETATE	P 16	BA
6	12							2					N-BUTYL ACETATE	G 20	BA
6	12							2					ETHYL BUTANOATE	224	BA
6	12							2					2-ETHYL BUTANOIC ACID	89	BA
6	12							2					N-HEXANOIC ACID	80	BA
6	12							2					4-HYDROXY 4-METHYL PENTAN-2-ONE	P 18	BA
6	12							2					3-METHYL PENTANOIC ACID	88	BA
6	12							2					4-METHYL PENTANOIC ACID	87	BA
6	12							2					METHYL PENTANOATE	229	BA
6	12							2					METHYL 2-METHYL BUTANOATE	234	BA
6	12							2					PROPYL PROPANOATE	219	BA
6	12							3					2-ETHOXYETHYL ACETATE	P 23	BA
6	12							3					ETHYL 2-HYDROXY 2-METHYL PROPANOATE	248	BA
6	12							3					ETHYL 3-HYDROXY BUTANOATE	249	BA
6	12									1			TR-2,5-DIMETHYL THIOPHANE	3584	EA
6	13						1						N-(2-BUTYL)-ETHYLENEIMINE	652	FA
6	13						1	2					ETHYL 3-AMINO BUTANOATE	250	BA

C H BR CL F I B N O P S SI M

C	H	BR	CL	F	I	B	N	O	P	S	SI	M	NAME	SPECTRUM NO.	LOCA- TION
6	14												HEXANE	6606	CA
6	14												N-HEXANE	P 110	BA
6	14							1					DI-ISOPROPYL ETHER	3787	FA
6	14							1					DI-N-PROPYL ETHER	P 57	BA
6	14							1					DI-N-PROPYLETHER	G 46	BA
6	14							1					DIISOPROPYL ETHER	P 54	BA
6	14							1					DIPROPYL ETHER	5987	CA
6	14							1					HEXAN-1-OL	6	BA
6	14							1					HEXAN-2-OL	23	BA
6	14							1					HEXAN-1-OL	3419	EA
6	14							1					HEXAN-1-OL	1967	FA
6	14							1					2-METHYL PENTAN-2-OL	38	BA
6	14							2					1,1-DIETHOXY ETHANE	281	BA
6	14							3					2-(2-ETHOXY)ETHOXY ETHANOL	46	BA
6	15					1							TRI (PERDEUTEROETHYL) BORANE	17294	FA
6	15					1							TRIETHYL BORANE	833	GA
6	15					1							TRIETHYL BORANE	4438	FA
6	15					1							TRIETHYL BORANE-D-15	8707	EA
6	15					1		2					DIISOPROPOXY BORANE-B-10	6489	EA
6	15					1		2					DIISOPROPYL (10B) BORONATE	6376	FA
6	15					1		3					TRIETHOXY BORANE	6487	EA
6	15					1		3					TRIETHOXY BORANE	837	GA
6	15						1						1-AMINO HEXANE	112	BA
6	15						1						DIPROPYLAMINE	126	BA
6	15						1						TRIETHYLAMINE	5153	FA
6	15						1	1					2-(N,N-DIETHYL)AMINO ETHANOL	51	BA
6	18					2							TRIETHYL DIBORANE	8740	EA
6	18					2							TRIETHYL DIBORANE (BORON-D-3)	8741	EA
6	18							1			2		HEXAMETHYL SILOXANE	1448	EA
6			1	17						1			6-CHLOROPERFLUOROHEXYLSULPHUR PENTAFLUORIDE	7936	FA
6			2	12									3,4-DICHLORODODECAFLUOROHEXANE	251	FA
6				6									HEXAFLUOROBENZENE	7933	EA
6				6									PERFLUOROBENZENE	5577	FA
6				8									PERFLUOROCYCLOHEXA-1,3-DIENE	4200	FA
6				8									PERFLUOROCYCLOHEXA-1,4-DIENE	4204	FA
6				10									DECAFLUOROCYCLOHEXENE	6247	CK
6				10									PERFLUOROCYCLOHEXENE	4066	FA
6				10				3					PERFLUOROPROPIONIC ANHYDRIDE	67	FA
6				11			1						UNDECAFLUORO HEXANONITRILE	3615	EA
6				12									DODECAFLUOROCYCLOHEXANE	3550	EA
6				12									PERFLUOROCYCLOHEXANE	5289	EA
6				12									PERFLUOROCYCLOHEXANE	3708	FA
6				12									PERFLUOROHEX-3-ENE	254	FA
6				13			1						PERFLUORO-2-IMINO HEXANE	5261	EA
6				14									TETRADECAFLUOROHEXANE	3543	EA
6				14									2,3-TRIFLUOROMETHYLPERFLUOROBUTANE	9083	FA
6				14						1			BISPERFLUOROISOPROPYL SULPHIDE	10276	FA
6				14						2			BIS(HEPTAFLUOROPROPYL) DISULPHIDE	87	FA
6				14						2			BIS(HEPTAFLUOROPROPYL) DISULPHIDE	2406	CA
6				14						2			BISPERFLUOROISOPROPYL DISULPHIDE	10277	FA
6				14						2			PERFLUORO,N-PROPYL DISULPHIDE	224	EA
6				14						3			BISPERFLUOROISOPROPYL TRISULPHIDE	10278	FA
6				14						3			PERFLUORO,N-PROPYL TRISULPHIDE	225	EA
6				15			1						TRIS-PENTAFLUOROETHYLAMINE	3600	EA
6							2						1,4-DICYANOBUTA-1,3-DIYNE	16467	FA
6								6				CR	CHROMIUM HEXACARBONYL	2648	EA
6								6				MO	MOLYBDENUM HEXACARBONYL	2649	EA
7	1		1	50									1-HYDROPENTADECAFLUOROHEPTANE	3561	EA
7	1			13									CIS-1H-DECAFLUORO-4-TRIFLUOROMETHYL-CYCLOHEXANE	3717	FA
7	1			13									TRANS-1H-DECAFLUORO-4-TRIFLUOROMETHYL-CYCLOHEXANE	3718	FA
7	5		1					1					BENZOYL CHLORIDE	535	EA
7	5			3									BENZOTRIFLUORIDE	5117	EA
7	5						1						BENZONITRILE	3382	FA
7	6							1					BENZALDEHYDE	3383	FA
7	6							1					BENZALDEHYDE	173	BA
7	6							2					BENZOIC ACID	105	BA
7	6							2					BENZOIC ACID	3183	FA
7	6							2					BENZOIC ACID,DEUTERATED	6787	EA
7	6							2					BENZOIC ACID	6781	EA
7	6							2					2-FORMYLCYCLOPENTADIENYLIDENE METHANOL	12079	FA
7	6							2					SALICYLALDEHYDE	179	BA
7	7		1										M-CHLOROTOLUENE	2388	FA
7	7		1										CHLORO PHENYL METHANE	P 73	BA
7	7		1										O-CHLOROTOLUENE	2392	FA
7	7		1										P-CHLOROTOLUENE	2386	FA
7	7			1									M-FLUOROTOLUENE	3393	FA
7	7						1	1					2-FORMYLCYCLOPENTADIENYLIDENEMETHYLAMINE	12081	FA
7	7						1	2					4-NITROTOLUENE	2406	FA
7	7						1	2					2-NITROTOLUENE	2401	FA
7	7						1	2					3-NITROTOLUENE	2404	FA
7	8												TOLUENE	G 132	BA
7	8												TOLUENE	3380	FA
7	8												TOLUENE-ALPHA-D3	4873	EA
7	8												TOLUENE	P 117	BA
7	8							1					BENZYL ALCOHOL	16	BA
7	8							1					P-CRESOL	3394	FA
7	8							1					P-CRESOL	55	BA
7	8							1					M-CRESOL	54	BA
7	8							1					O-CRESOL	53	BA
7	8							1					O-CRESOL	3397	FA
7	8							1					M-CRESOL	3396	FA
7	8							1					METHOXY BENZENE (ANISOLE)	285	BA
7	8							2					1,2-DIHYDROXY 4-METHYL BENZENE (4-METHYL CATECHOL)	72	BA
7	8							2					1,2-DIHYDROXY 3-METHYL BENZENE (3-METHYL CATECHOL)	71	BA
7	8							2					2-METHOXY PHENOL(GUAICOL)	62	BA
7	8							3					ETHYL FUROATE	256	BA
7	9						1						BENZYLAMINE	119	BA
7	9						1						2,6-DIMETHYL PYRIDINE	246	FA

C H BR CL F I B N O P S SI M

C	H	BR	CL	F	I	B	N	O	P	S	SI	M	NAME	SPECTRUM NO.	LOCA-TION
7	9						1						3-METHYL ANILINE	135	BA
7	9						1						2-METHYL ANILINE	134	BA
7	9						1						N-METHYL ANILINE	133	BA
7	9						1	1					2-METHOXY ANILINE	137	BA
7	10							1					TR,TR-HEPTA-2,4-DIENAL	168	BA
7	11			5									111,2,2-PENTAFLUOROHEPTANE	5705	FA
7	12												HEPTA-1, CIS-5-DIENE	17632	FA
7	12												HEPTA-1, TRANS-5-DIENE	17633	FA
7	12							1					4-METHYL CYCLOHEXANONE	P 29	BA
7	12							2					ALLYL BUTANOATE	228	BA
7	12							2					N-BUTYL PROP-2-ENOATE	G 21	BA
7	12							2					GAMMA-HEPTALACTONE	271	BA
7	12							2					DELTA-HEPTALACTONE	274	BA
7	12							2					HEPT-2-ENOIC ACID	93	BA
7	12							4					DIETHYL MALONATE	241	BA
7	13			3									1,1,1-TRIFLUOROHEPTANE	5704	FA
7	14												CIS-1,2-DIMETHYL CYCLOPENTANE	G 117	BA
7	14												2,4-DIMETHYL PENT-2-ENE	9006	CK
7	14												2,4-DIMETHYL PENT-1-ENE	9005	CK
7	14												METHYL CYCLOHEXANE	G 125	BA
7	14												METHYL CYCLOHEXANE	P 111	BA
7	14												1,1,2,2-TETRAMETHYLCYCLOPROPANE	4717	FA
7	14							1					HEPT-1-EN-4-OL	36	BA
7	14							1					HEPTAN-2-ONE	185	BA
7	14							1					N-HEPTANAL	157	BA
7	14							1					2-METHYL CYCLOHEXANOL	P 45	BA
7	14							1					3-METHYL CYCLOHEXANOL	P 46	BA
7	14							2					AMYL ACETATE	215	BA
7	14							2					AMYL ACETATE	P 14	BA
7	14							2					BUTYL PROPANOATE	220	BA
7	14							2					N-BUTYL GLYCIDYL ETHER	P 47	BA
7	14							2					ETHYL 2-METHYL BUTANOATE	235	BA
7	14							2					ETHYL 3-METHYL BUTANOATE (ISOVALERATE)	233	BA
7	14							2					N-HEPTANOIC ACID	81	BA
7	14							2					METHYL HEXANOATE	236	BA
7	14							2					PROPYL BUTANOATE	225	BA
7	14							3					BUTYL 2-HYDROXY PROPANOATE	246	BA
7	16												2,2-DIMETHYL PENTANE	7976	CK
7	16												N-HEPTANE	288	BA
7	16												N-HEPTANE	G 120	BA
7	16												2-METHYL HEXANE	36947	CA
7	17						1						1-AMINO HEPTANE	113	BA
7				12									PERFLUORO-(1-METHYLCYCLOHEXENE)	5707	FA
7				14									PERFLUOROHEPT-1-ENE	3294	FA
7				16									HEXADECAFLUORO HEPTANE	3544	FA
7				16									PERFLUOROHEPTANE	296	FA
8	4											GE	TETRAETHYNYL GERMANE	16469	FA
8	4											SN	TETRAETHYNYL STANNANE	16470	FA
8	4		4					1					OMEGA,246-TETRACHLOROACETOPHENONE	5537	FA
8	4										1		TETRAETHYNYL SILANE	16468	FA
8	5		1				2						2-CHLOROQUINOXALINE	10874	FA
8	5		3				1						OMEGA,2,4-TRICHLOROACETOPHENONE	5528	FA
8	5		3					1					OMEGA,2,5-TRICHLOROACETOPHENONE	5531	FA
8	5		3					1					OMEGA,3,4-TRICHLOROACETOPHENONE	5534	FA
8	6												1-DEUTERO-2-PHENYLACETYLENE	7318	FA
8	6												PHENYLACETYLENE	4826	FA
8	6												PHENYL ACETYLENE	301	BA
8	6		2					1					OMEGA-DICHLOROACETOPHENONE	5091	FA
8	6		2					1					OMEGA,4-DICHLOROACETOPHENONE	5094	FA
8	6						2						QUINOXALINE	10876	FA
8	6						2	1					1(1H)-PHTHALAZONE	10879	FA
8	6						2	1					2(1H)-QUINOXALONE	10872	FA
8	7		1					1					OMEGA-CHLOROACETOPHENONE	5089	FA
8	8												STYRENE MONOMER	P 116	BA
8	8							1					ACETOPHENONE	202	BA
8	8							2					METHYL BENZOATE	253	BA
8	8							2					4-METHOXY BENZALDEHYDE	178	BA
8	8							2					2-METHOXY BENZALDEHYDE	177	BA
8	8							2					PHENYL ACETATE	260	BA
8	8							2					TROPOLONE METHYL ETHER	4624	FA
8	8							3					METHYL SALICYLATE	264	BA
8	8							3					P-METHOXY PHENYL FORMATE	262	BA
8	9						1	2					METHYL ANTHRANILATE	265	BA
8	10												ETHYL BENZENE	303	BA
8	10												ETHYL BENZENE	P 109	BA
8	10												ALPHA-METHYL STYRENE	P 112	BA
8	10												1,3-XYLENE	3418	FA
8	10												1,3-XYLENE	P 119	BA
8	10												1,2-XYLENE	3416	FA
8	10												M-XYLENE-ALPHA,ALPHA-D6	4876	FA
8	10												1,2-XYLENE	P 118	BA
8	10												1,4-XYLENE	P 120	BA
8	10												1,4-XYLENE	3421	FA
8	10							1					2-PHENYL ETHANOL	12	BA
8	10							1					1-PHENYL ETHANOL	34	BA
8	10							1					2,3-XYLENOL	56	BA
8	10							1					3,5-XYLENOL	61	BA
8	10							1					3,4-XYLENOL	60	BA
8	10							1					2,6-XYLENOL	59	BA
8	10							1					2,5-XYLENOL	58	BA
8	10							1					2,4-XYLENOL	57	BA
8	10							1					2,6-XYLENOL	6067	FA
8	10							1					3,5-XYLENOL	5688	FA
8	10							2					1,3-DIMETHOXY BENZENE	4830	FA
8	10							2					1,2-DIMETHOXY BENZENE (VERATROLE)	4833	FA
8	10							2					1,2-DIMETHOXY BENZENE (VERATROLE)	286	BA
8	10							3					2,6-DIMETHOXY PHENOL	63	BA
8	11						1						2,4-DIMETHYL ANILINE	136	BA
8	11						1	1					4-ETHOXY ANILINE	138	BA
8	14							1					TRANS-OCT-2-ENAL	163	BA

C H BR CL F I B N O P S SI M

C	H	BR	CL	F	I	B	N	O	P	S	SI	M	NAME	SPECTRUM NO.	LOCA-TION
8	14							2					DELTA-ISOOCTALACTONE	276	BA
8	14							2					OCT-2-ENOIC ACID	94	BA
8	14							2					OCT-3-ENOIC ACID	103	BA
8	14							2					DELTA-OCTALACTONE	275	BA
8	14							2					GAMMA-OCTALACTONE	272	BA
8	15						1						1-CYCLOHEXYLETHYLENEIMINE	653	FA
8	16							1					OCT-1-EN-3-OL	33	BA
8	16							1					OCTAN-2-ONE	186	BA
8	16							1					OCTAN-3-ONE	191	BA
8	16							1					N-OCTANAL	158	BA
8	16							1					2,2,4-TRIMETHYL PENTAN-3-ONE	5067	FA
8	16							2					AMYL PROPANOATE	221	BA
8	16							2					BUTYL BUTANOATE	226	BA
8	16							2					ETHYL HEXANOATE	237	BA
8	16							2					2-ETHYL HEXANOIC ACID	90	BA
8	16							2					HEXYL ACETATE	216	BA
8	16							2					N-OCTANOIC ACID	82	BA
8	16							2					PROPYL PENTANOATE	230	BA
8	18												N-OCTANE	P 114	BA
8	18												2,2,4-TRIMETHYL PENTANE	6608	CA
8	18							1					DIBUTYL ETHER	279	BA
8	18							1					OCTAN-4-OL	27	BA
8	18							1					OCTAN-3-OL	26	BA
8	18							1					OCTAN-2-OL	24	BA
8	19						1						1-AMINO OCTANE	114	BA
8	19						1						2-AMINO OCTANE	123	BA
8	19						1						DIBUTYLAMINE	127	BA
8	19						1						1,1,3,3-TETRAMETHYL 1-AMINO BUTANE	115	BA
8	20							1			1		ETHOXYTRIETHYL SILANE	1452	EA
8	20							3			1		TRIETHOXYETHYL SILANE	1453	EA
8	20							4			1		TETRAETHOXY SILANE	1454	EA
8	22					2							TETRAETHYL DIBORANE (BORON-D-2)	8743	EA
8	22					2							TETRAETHYL DIBORANE	8742	EA
8			1	21						1			8-CHLOROPERFLUOROOCTYLSULPHUR PENTAFLUORIDE	7937	FA
8				14									PERFLUORO-(1,4-DIMETHYLCYCLOHEXENE)	5708	FA
9	7						1						QUINOLINE	151	BA
9	10							1					4-METHYLACETOPHENONE	203	BA
9	10							1					PROPIOPHENONE	204	BA
9	10							2					ETHYL BENZOATE	254	BA
9	10							3					P-METHOXY PHENYL ACETATE	263	BA
9	11						1	2					ETHYL ANTHRANILATE	266	BA
9	11						1	2					METHYL N-METHYL ANTHRANILATE	267	BA
9	12												TETRACYCLO(3,3,1,0(2,8),0(4,6))NONANE	15755	FA
9	12												1,3,5-TRIMETHYL BENZENE (MESITYLENE)	306	BA
9	12							1					2,4,6-TRIMETHYL PHENOL	5691	FA
9	14							1					TR,CIS-NONA-2,4-DIENAL	169	BA
9	14							2					METHYL OCT-2-YN OATE	252	BA
9	16							1					TRANS-NON-2-ENAL	164	BA
9	16							2					ALLYL HEXANOATE	239	BA
9	16							2					NON-2-ENOIC ACID	95	BA
9	16							2					DELTA-NONALACTONE	277	BA
9	16							2					GAMMA-NONALACTONE	273	BA
9	18							1					NONAN-2-ONE	187	BA
9	18							1					N-NONANAL	159	BA
9	18							2					AMYL BUTANOATE	227	BA
9	18							2					BUTYL PENTANOATE	231	BA
9	18							2					N-NONANOIC ACID	83	BA
9	20												N-NONANE	289	BA
9	21					1		3					TRIISOPROPYL (10B) BORATE	6375	FA
9	21					1		3					TRIISOPROPOXY BORANE-B-10	6490	EA
9	21					1		3					TRIISOPROPOXY BORANE-B-10	6491	EA
9	21						1						1-AMINO NONANE	116	BA
10	10												NAPHTHALENE	362	EA
10	12											FE	FERROCENE-D-10	6723	EA
10	12											FE	FERROCENE	6721	EA
10	12											RU	RUTHENOCENE	6725	EA
10	12							1					4-ISOPROPYL BENZALDEHYDE(CUMINIC)	174	BA
10	12							2					4-ALLYL 2-METHOXY PHENOL (EUGENOL)	69	BA
10	12							2					4-ISOPROPYLTROPOLONE	4625	FA
10	14												1-ISOPROPYL 4-METHYL BENZENE(P-CYMENE)	305	BA
10	14							1					D-CARVONE	199	BA
10	14							1					2,6-DIEETHYL PHENOL	6068	FA
10	14							1					1,1-DIMETHYL 2-PHENYL ETHANOL	39	BA
10	14							1					2-ISOPROPYL 5-METHYL PHENOL (THYMOL)	68	BA
10	15						1						N,N-DIETHYL ANILINE	145	BA
10	16												DL-CAMPHENE	298	BA
10	16												D-LIMONENE	296	BA
10	16												CIS-OCIMENE	295	BA
10	16												BETA-PINENE	297	BA
10	16							1					TR,TR-DECA-2,4-DIENAL	170	BA
10	16							1					DL-PIPERITONE	198	BA
10	16							1					PULEGONE	200	BA
10	18							1					1,8-CINEOLE	284	BA
10	18							1					TRANS-DEC-2-ENAL	165	BA
10	18							1					GERANIOL	42	BA
10	18							1					LINALOOL	43	BA
10	18							1					A-TERPINEOL	44	BA
10	18							2					DEC-2-ENOIC ACID	96	BA
10	20												DEC-1-ENE	293	BA
10	20							1					CITRONELLOL	41	BA
10	20							1					DECAN-2-ONE	188	BA
10	20							1					N-DECANAL	160	BA
10	20							2					AMYL PENTANOATE	232	BA
10	20							2					OCTYL ACETATE	217	BA
10	22												N-DECANE	290	BA
10	22							1					DECAN-5-OL	28	BA
10	23						1						DIPENTYLAMINE	128	BA
11	14							2					P-CRESYL 2-METHYL PROPANOATE	261	BA
11	16												1-METHYL 4-TERT-BUTYL BENZENE	P 106	BA
11	18							1					TR,TR-UNDECA-2,4-DIENAL	171	BA

C H BR CL F I B N O P S SI M

C	H	BR	CL	F	I	B	N	O	P	S	SI	M	NAME	SPECTRUM NO.	LOCATION
11	20							1					TRANS-UNDEC-2-ENAL	166	BA
11	20							2					UNDEC-10-ENOIC ACID	104	BA
11	22							1					UNDECAN-5-ONE	192	BA
11	22							1					UNDECAN-2-ONE	189	BA
11	22							2					AMYL HEXANOATE	238	BA
11	24												N-UNDECANE	291	BA
12	10							1					2-HYDROXY BIPHENYL	5693	FA
12	12											CR	DIBENZENE CHROMIUM	A 257	FA
12	18							1					2,6-DI-ISOPROPYL PHENOL	6070	FA
12	20							1					TR,TR-DODECA-2,4-DIENAL	172	BA
12	20							2					LINALYL ACETATE	259	BA
12	24												DODEC-1-ENE	294	BA
12	26												N-DODECANE	292	BA
12	27						1						DIHEXYLAMINE	129	BA
13	10												FLUORENE	17634	FA
13	20							1					BETA-IONONE	201	BA
14	22												N-OCTYL BENZENE	304	BA
15	24												CARYOPHYLLENE	299	BA
	1	1											HYDROGEN BROMIDE	3938	EA
	1	1											HYDROGEN BROMIDE	G 7	BA
	1		1										HYDROGEN CHLORIDE	G 8	BA
	1		1										HYDROGEN CHLORIDE	3939	EA
	1		1										HYDROGEN CHLORIDE	10558	CA
	1		1					3		1			CHOROSULPHONIC ACID	A 626	FA
	1		2			1							DICHLOROBORANE	A 476	FA
	1			1				3		1			FLUOROSULPHONIC ACID	A 625	FA
	1						1	2					NITROUS ACID,DEUTERO	365	EA
	1						1	2					NITROUS ACID	364	EA
	1						1	3					NITRIC ACID	3941	FA
	1						1	3					NITRIC ACID	6439	EA
	1						1	3					NITRIC ACID	A 804	FA
	2		2								1		DICHLOROSILANE	1858	EA
	2			2							1		DIDEUTERODIFLUOROSILANE	A 630	FA
	2			2							1		DIFLUOROSILANE	A 629	FA
	2							1					WATER	9000	CA
	2							4		1			SULPHURIC ACID	6440	EA
	2									1			HYDROGEN SULPHIDE	G 9	BA
	2									1			HYDROGEN SULPHIDE	3942	EA
	2									2			HYDROGEN PERSULPHIDE	4926	EA
	3	1										GE	GERMANYL BROMIDE	A 443	FA
	3	1										GE	TRIDEUTERO GERMANYL BROMIDE	A 447	FA
	3		1									GE	GERMANYL CHLORIDE	A 442	FA
	3		1									GE	TRIDEUTERO GERMANYL CHLORIDE	A 446	FA
	3			1								GE	GERMANYL FLUORIDE	A 441	FA
	3			1								GE	TRIDEUTERO GERMANYL FLUORIDE	A 445	FA
	3				1							GE	GERMANYL IODIDE	A 444	FA
	3				1							GE	TRIDEUTERO GERMANYL IODIDE	A 448	FA
	3						1						AMMONIA	3911	EA
	3						1						AMMONIA	G 1	BA
	3						1						AMMONIA	P 2	BA
	3						1						AMMONIA	93	CA
	4						2						HYDRAZINE	3937	EA
	5	1				2							BROMO DIBORANE	834	GA
	5		1			2							CHLORODIBORANE	A 475	FA
	5		1								2		DISILANYL CHLORIDE	A 526	FA
	5			1							2		FLUORODISILANE	A 474	FA
	5								1		1		SILYL PHOSPHINE	6895	EA
	6											GE2	DIGERMANE	3454	EA
	6					2							DIBORANE	2180	FA
	6					2							DIBORANE	A 146	FA
	6					2							DIBORANE	826	GA
	6									1	2		DISILTHIANE	A 151	FA
	6									1	2		DISILYL SULPHIDE-D-6	7929	EA
	6									1	2		DISILYL SULPHIDE-D-6	6458	EA
	6									1	2		DISILYL SULPHIDE	6457	EA
	6									1	2		PERDEUTERODISILTHIANE (D6)	A 152	FA
	6										1	2	DISILYL SULPHIDE	7928	EA
	6										2	SE	DISILSELENANE	A 153	FA
	6										2	SE	PERDEUTERODISILSELENANE	A 154	FA
	9					5							PENTABORANE	2183	EA
	9						1				3		TRISILYLAMINE	6086	EA
	9						1				3		TRISILYLAMINE-D-9	6085	EA
	9						1				3		TRISILYL AMINE	A 21	FA
	10					4							TETRABORANE	2181	EA
	10					4							TETRABORANE	827	GA
	10					6							HEXABORANE	828	GA
	10										4		2-SILYLTRISILANE	A 776	FA
	10										4		TETRASILANE	A 775	FA
	11					5							DIHYDROPENTABORANE	2182	EA
	12						2				4		TETRASILYLHYDRAZINE (D12)	A 91	FA
	12						2				4		TETRASILYLHYDRAZINE	A 89	FA
		1		1				2		1			SULPHURYL BROMOFLUORIDE	6774	EA
		2		3					1				DIBROMOTRIFLUOROPHOSPHORANE	A 0939	FA
			1	1				2		1			SULPHURYL CHLORIDE FLUORIDE	A 829	FA
			1	1				2		1			SULPHURYL CHLORIDE FLUORIDE	A 701	FA
			1	3									CHLORINE TRIFLUORIDE	6442	EA
			1				1	1					NITROSYL CHLORIDE	3960	EA
			1				1	2					NITRYL CHLORIDE	A 0934	FA
			1				1	2					NITRYL CHLORIDE	2309	EA
			1				1	3					CHLORINE NITRATE	A 0932	FA
			2	1					1	1			PHOSPHOROTHIOIC DICHLORIDE FLUORIDE	A 0936	FA
			2	3					1				DICHLOROTRIFLUOROPHOSPHORANE	A 0938	FA
			2					1		1			THIONYL CHLORIDE	2302	EA
			2					1		1			THIONYL CHLORIDE	G 14	BA
			2					1		1			THIONYL CHLORIDE	1336	CA
			2					2		1			SULPHURYL CHLORIDE	1334	CA
			2					2		1			SULPHURYL CHLORIDE	2303	EA
			3						1	1			THIOPHOSPHORYL CHLORIDE	947	CA
			5						1				PHOSPHOROUS PENTACHLCRIDE	5114	FA
				1			1	1					NITROSYL FLUORIDE	529	EA

C H BR CL F I B N O P S SI M

C	H	BR	CL	F	I	B	N	O	P	S	SI	M	NAME	SPECTRUM NO.	LOCATION
				1			1	2					NITRYL FLUORIDE	A 0933	FA
				1			1	3					FLUORINE NITRATE	A 0931	FA
				2			2						TRANS-DIFLUCRODIAZEN	A 830	FA
				2				1		1			THIONYL FLUORIDE	A 639	FA
				2				2				SE	SELENIUM DIOXIDE DIFLUORIDE	A 828	FA
				2				2		1			SULPHURYL FLUORIDE	1031	EA
				2				2		1			SULPHURYL FLUORIDE	3686	EA
				2				2		1			SULPHURYL FLUORIDE	P 9	BA
				2				2		1			SULPHURYL DIFLUORIDE	A 88	FA
				2				3		1			FLUORINE FLUOROSULPHONATE	4446	EA
				2				3		1			FLUORINE FLUOROSULPHONATE	4526	EA
				2				6		2			PEROXYDISULPHURYLDIFLUORIDE	4525	EA
				2						2			DIFLUORODISULPHANE	A 636	FA
				2						2			THIOTHIONYL FLUORIDE	A 637	FA
				3		1							BORON TRIFLUORIDE	G 2	BA
				3			1						NITROGEN FLUORIDE	G 10	BA
				3			1						NITROGEN TRIFLUORIDE	A 769	FA
				3			1	2		1			DIFLUOROAMIDOSULPHURYL FLUORIDE	A 753	FA
				3			1	4		2			FLUCROIMIDODI(SULPHURYL FLUORIDE)	A 754	FA
				3			1			1			THIAZYL TRIFLUORIDE	G 13	BA
				3					1	1			THIOPHOSPHORYLTRIFLUORIDE	A 878	FA
				4								GE	GERMANIUM TETRAFLUORIDE	367	EA
				4				1		1			THIONYL TETRAFLUORIDE	4460	EA
				4						1			SULPHUR TETRAFLUORIDE	4132	EA
				4						1			SULPHUR TETRAFLUORIDE	A 638	FA
				4							1		SILICON TETRAFLUORIDE	A 640	FA
				4							1		SILICON TETRAFLUORIDE	G 12	BA
				5								V	VANADIUMPENTAFLUORIDE	A 784	FA
				5	1			1					IODOSYL PENTAFLUORIDE	A 693	FA
				6								MO	MOLYBDENUM HEXAFLUORIDE	530	EA
				6								U	URANIUM HEXAFLUORIDE	532	EA
				6								W	TUNGSTEN HEXAFLUORIDE	531	EA
				6			3		3				HEXAFLUOROCYCLOTRIPHOSPHAZENE	A 492	FA
				6			3		3				PHOSPHONITRILIC FLUORIDE TRIMER	A 361	FA
				6				1		1			PENTAFLUCRO SULPHUR HYPOFLUORITE	4461	EA
				6						1			SULPHUR HEXAFLUORIDE	3685	EA
				7								RH	RHENIUM HEPTAFLUORIDE	A 802	FA
				8			4		4				OCTAFLUOROCYCLOTETRAPHOSPHAZENE	A 493	FA
				8			4		4				PHOSPHONITRILIC FLUORIDE TETRAMER	A 362	FA
				10			5		5				PHOSPHONITRILIC FLUORIDE PENTAMER	A 364	FA
				10				1		2			BISPENTAFLUROOSULPHUR OXIDE	A 191	FA
				10				2		2			BISPENTAFLUOROSULPHUR PEROXIDE	A 192	FA
				10						2			DISULPHUR DECAFLUORIDE	3339	EA
				10						2			DISULPHUR DECAFLUORIDE	3340	EA
				10						2			DISULPHUR DECAFLUCRIDE	5301	EA
				10						2			DISULPHUR DECAFLUORIDE	426	EA
							1	1					NITRIC OXIDE	3953	EA
							1	1					NITRIC OXIDE	10559	CA
							1	1					NITRIC OXIDE	1000	GA
							1	2					NITROGEN DIOXIDE	3955	EA
							2	1					NITROUS OXIDE	3961	EA
							2	1					NITROUS OXIDE	6443	EA
							2	1					NITROUS OXIDE	G 11	BA
							2	1					NITROUS OXIDE	P 7	BA
							2	1					NITROUS OXIDE	1141	CA
							2	1					NITROUS OXIDE	A 3	FA
							2	4					DINITROGEN TETROXIDE	5112	EA
							4			4			NITROGEN TETRASULPHIDE	1812	EA
								2		1			SULPHUR DIOXIDE	841	CA
								2		1			SULPHUR DIOXIDE	3966	EA
								2		1			SULPHUR DIOXIDE	3488	EA
								2		1			SULPHUR DIOXIDE	A 2	FA
								3		1			SULPHUR TRIOXIDE	13443	CA

References

1 T. E. Jordan, *Vapour Pressure of Organic Compounds*, Interscience, London, 1954.
2 K. J. Sterling and R. J. Haines, *Anal. Chem.* **40** (8), 1395 (1968).
3 Research and Industrial Instrument Co. Ltd., Worsley Bridge Road, London, S.E.26.
4 Pye-Unicam Ltd., York Street, Cambridge, England.
5 Perkin-Elmer Ltd. (England), Beaconsfield, Bucks, England.
6 J. U. White, N. L. Alpert, W. M. Ward and W. S. Gallaway, *Anal. Chem.* **31,** 1267 (1959).
7 M. St. J. C. Flett and J. Hughes, *J. Chromatog.* **11,** 434 (1963).
8 Sir Howard Grubb Parsons and Co. Ltd., Walkergate, Newcastle-upon-Tyne, NE6 2YB.
9 A. M. Bartz and H. D. Ruhl, *Anal. Chem.* **36,** 1892 (1964).
10 P. A. Wilks and R. A. Brown, *Anal. Chem.* **36,** 1896 (1964).
11 Beckman Instrument Ltd., Fullerton, California, Bulletin 7073.
12 M. J. D. Low and S. K. Freeman, *Anal. Chem.* **39,** 194 (1967).
13 Carle Instruments, Anaheim, California, USA.
14 Wilks Scientific Corporation, South Norwalk, Connecticut, USA.
15 J. T. Ballinger, T. T. Bartels and J. H. Taylor, *J. Gas Chromatog.* **6,** 295 (1968).
16 M. J. D. Low and I. Coleman, *Spectrochim. Acta* **22,** 369 (1966).
17 Block Engineering Inc. Distributed by Dunn Analytical Instruments, Silver Spring, Maryland, USA.
18 M. J. D. Low and F. K. Claney, *Environ. Sci. Technol.* **1,** 73 (1967).
19 J. W. Robinson, C. Woodward and H. M. Barnes, *Anal. Chim. Acta* **43,** 119 (1968).
20 D. Welti and R. Stephany, *Appl. Spectry.* **22,** 678 (1968).
21 L. D. Smithson, L. B. Fall, F. D. Pitts and F. W. Bauer, *Tech. Doc. Rept.* No. RTD–TDR–4265. Research and Technology Division, Wright Patterson AFB, Ohio, USA, 1964.
22 R. A. Sparks, *Storage and Retrieval of Wyandotte/ASTM Infrared Spectral Data using an IBM* 1401 *Computer*, ASTM, Philadelphia Pa. USA, 1964.
23 T. A. Entzminger and E. A. Diephaus, *Storage and Retrieval of Wyandotte/ASTM Infrared Spectral Data using a Honeywell* 400 *Computer*, U.S. Public Health Service, Robert Taft Sanitary Engineering Center, Cincinnati, Ohio, 1964.
24 D. H. Anderson and G. L. Covert, *Anal. Chem.* **39,** 1288 (1967).
25 D. S. Erley, *Anal. Chem.* **40,** 894 (1968).
26 *Codes and Instructions for Wyandotte/ASTM Punched Cards indexing Spectral Absorption Data*, Philadelphia Pa., USA.

27 L. H. Cross, D. J. Shields, W. Proudlock and D. Welti, to be published.
28 L. H. Cross, J. Haw and D. J. Shields, *Symposium on Molecular Spectroscopy, Brighton*, 1968. Institute of Petroleum Hydrocarbon Research Group, London.
29 L. J. Bellamy, *Infrared Spectra of Complex Molecules*, 2nd ed., Methuen, London, 1958.
30 L. J. Bellamy, *Advances in Infrared Group Frequencies*, Methuen, London, 1968.
31 R. N. Jones and C. Sandorfy in *Technique of Organic Chemistry*, *Vol.* 9. (Ed. A. Weissberger), Interscience, New York, p. 247.
32 M. St. J. C. Flett, *Characteristic Frequencies of Chemical Groups in the Infrared*, Elsevier, Amsterdam.
33 G. H. Beaven, E. A. Johnson, R. G. J. Miller and H. A. Willis, *Molecular Spectroscopy*, 1st ed., Heywood and Sons, London, 1961.
34 *Irscot System*, English Edition (Eds. R. G. J. Miller and H. A. Willis), Heyden and Son, London, 1969.
35 A. V. Stuart and G. B. B. M. Sutherland, *J. Chem. Phys.* **24,** 559 (1956).
36 P. Tarte and R. Deponthiere, *J. Chem. Phys.* **26,** 962 (1957).
37 S. Krimm, C. J. Liang and G. B. B. M. Sutherland, *J. Chem. Phys.* **25,** 778 (1956).
38 Tien-Sung Lein and E. Fishman, *Spectrochim. Acta* **23A,** 491 (1967).
39 *Documentation of Molecular Spectroscopy*, Butterworths, London and Verlag-Chemie, Weinheim/Bergstr.
40 J. C. Evans, *Spectrochim. Acta* **16,** 1382 (1960).
41 M. Davies and R. L. Jones, *J. Chem. Soc.* 120 (1954).
42 R. K. Jakobsen, *Spectrochim. Acta* **21,** 433 (1965).
43 S. Bratoz, D. Hadzi and N. Sheppard, *Spectrochim. Acta* **8,** 249 (1956).
44 D. Hadzi and M. Pintar, *Spectrochim. Acta* **12,** 162 (1958).
45 R. Blinc and D. Hadzi, *Spectrochim. Acta* **15,** 82 (1959).
46 R. C. Millikan and K. S. Pitzer, *J. Chem. Phys.* **27,** 1305 (1957).
47 J. K. Wilmshurst, *J. Chem. Phys.* **25,** 478, 1171 (1956).
48 L. J. Bellamy, R. F. Lake and R. J. Pace, *Spectrochim. Acta* **19,** 443 (1963).
49 A. W. Pross and F. van Zeggerman, *Spectrochim. Acta* **16,** 563 (1960).
50 M. Haurie and A. Novak, *J. Chem. Phys.* **52,** 137, 146 (1965).
51 E. J. Hartwell, R. E. Richards and H. W. Thompson, *J. Chem. Soc.* 1436 (1948).
52 J. Bellanato and J. R. Barcelo, *Spectrochim. Acta* **16** 1333 (1960).
53 P. J. Corish and D. Chapman, *J. Chem. Soc.* 1746 (1957).
54 A. Finch, I. J. Hyams and D. Steele, *J. Mol. Spectry.* **16,** 103 (1965).
55 J. R. Barcelo and J. Bellanato, *Spectrochim. Acta* **8,** 27 (1956).
56 J. E. Stewart, *J. Chem. Phys.* **30,** 1259 (1959).
57 D. Hadzi and M. Skrbljak, *J. Chem. Soc.* 843 (1955).
58 J. C. Evans, *Spectrochim. Acta* **16,** 428 (1960).
59 R. Mecke and R. Mecke, *Chem. Ber.* **89,** 343 (1955).
60 J. C. Evans, *J. Chem. Phys.* **22,** 1228 (1954).
61 T. Miyazawa, T. Shimanouchi and S. Mizushima, *J. Chem. Phys.* **24,** 408 (1956).
62 R. L. Jones, *J. Mol. Spectry.* **2,** 581 (1958).
63 R. L. Jones, *J. Mol. Spectry.* **7,** 460 (1961).
64 R. L. Jones, *J. Mol. Spectry.* **11,** 411 (1963).

65 M. Davies, J. C. Evans and R. L. Jones, *Trans. Faraday Soc.* **51,** 761 (1955).
66 A. Gierer, *Z. Naturforsch.* **8B,** 644, 654 (1953).
67 J. Durig, *Spectrochim. Acta* **19,** 1225 (1963).
68 S. Pinchas, *Anal. Chem.* **27,** 2 (1955).
69 S. Pinchas, *Anal. Chem.* **29,** 334 (1957).
70 J. Overend and J. R. Scherer, *Spectrochim. Acta* **16,** 773 (1960).
71 W. J. Jones in *Infrared Spectroscopy and Molecular Structure* (Ed. M. Davies), Elsevier, Amsterdam, 1963, p. 111.
72 G. Herzberg, *Infrared and Raman Spectra,* Van Nostrand, New York, 1945.
73 A. R. Philpotts, *Proc. X^0 Colloq. Spect. Internat.* (Eds. E. R. Lippincott and M. Margoshes), Spartan Books, Washington, 1966, p. 577.
74 *Analytical Chemistry,* Annual Reviews.
75 J. G. Grasselli and M. K. Snavely in *Progress in Infrared Spectroscopy,* Vol. 3 (Ed. H. A. Szymanski), Plenum Press, New York, 1967, p. 55.
76 J. Haslam, A. R. Jeffs and H. A. Willis, *Analyst* **86,** 44 (1961).
77 H. W. Leggon, *Anal. Chem.* **33,** 1295 (1961).
78 M. D. D. Howlett and D. Welti, *Analyst* **91,** 291 (1966).
79 A. I. Fowlis and D. Welti, *Analyst* **92,** 639 (1967).
80 H. Copier and J. H. Van der Maas, *Spectrochim. Acta* **23A,** 2699 (1967).
81 A. S. Curry, J. F. Read, C. Brown and R. W. Jenkins, *J. Chromatog.* **38,** 200 (1968).
82 S. S. Chang, K. M. Brobst, C. E. Ireland and H. Tai, *Appl. Spectry.* **16,** 106 (1962).
83 R. Teranishi, J. W. Corse, J. C. Day and W. G. Jennings, *J. Chromatog.* **9,** 245 (1962).
84 P. Kratz, M. Jacobs and B. M. Mitzner, *Analyst* **84,** 671 (1959).
85 A. E. Thompson, *J. Chromatog.* **6,** 454 (1961).
86 B. M. Mitzner, *Chemist-Analyst* **51,** 20 (1962).
87 D. A. Shearer, B. C. Stone and W. A. McGugan, *Analyst* **88,** 147 (1963).
88 F. L. Snelson, *Chem. Ind.* 575 (1964).
89 M. Cartwright and A. Heywood, *Analyst* **91,** 33 (1966).
90 Gas chromatographic—multiple reflectance attachment, Wilks Scientific Corporation, South Norwalk, Connecticut, USA.
91 J. G. Grasselli and M. K. Snavely, *Appl. Spectry.* **16,** 190 (1962).
92 N. W. R. Daniels, *Chem. and Ind.* 1078 (1963).
93 G. J. Frisons, *Chemist-Analyst* **48,** 47 (1959).
94 B. M. Mitzner and M. H. Jacobs, *Chemist-Analyst* **48,** 104 (1959).
95 H. J. Gold, *Chemist-Analyst* **49,** 112 (1960).
96 A. Kreuchunas, *Chemist-Analyst* **49,** 82 (1960).
97 C. M. Drew, J. R. McNesby, S. R. Smith and A. S. Gordon, *Anal. Chem.* **28,** 979 (1956).
98 H. Szymanski, R. Povinelli, D. Stamires and G. Lynch, *Anal. Chem.* **31,** 2110 (1959).
99 W. J. Zubgh and A. Z. Connor, *Anal. Chem.* **32,** 912 (1960).
100 R. Self, *Nature* **189,** 223 (1961).
101 R. F. Kendall, *Appl. Spectry.* **21,** 31 (1967).
102 R. P. W. Scott, *3rd Wilkens Gas Chromatography Symposium—Amsterdam* 1964, Varian Aerograph AG.

103 D. M. W. Anderson, *Analyst* **84,** 50 (1959).

104 S. S. Chang, C. E. Ireland and H. Tai, *Anal. Chem.* **33,** 479 (1961).

105 J. T. Watson and K. Biemann, *Anal. Chem.* **36,** 1135 (1964).

106 R. Ryhage, *Anal. Chem.* **36,** 759 (1964).

107 P. M. Llewellyn and D. P. Littlejohn, *Pittsburg Conference on Analytical Chemistry and Applied Spectroscopy*, February, 1968.

108 S. R. Lipsky, W. J. McMurray and C. G. Horvath, *Gas Chromatography* 1966, (Ed. A. B. Littlewood), Institute of Petroleum, London, 1967, p. 299.

109 W. L. Senn, Jnr. and H. V. Drushel, *Anal. Chim. Acta* **25,** 328 (1961).

110 G. D. Propster, Wilks Scientific Corporation Reprint No. WSC 011–2.

111 H. Bober and K. Burner, *Z. Anal. Chem.* **238,** 1 (1968).

112 R. P. W. Scott, A. I. Fowlis, D. Welti and T. Wilkins, *Gas Chromatography* 1966, (Ed. A. B. Littlewood), Institute of Petroleum, London, 1967, p. 318.

113 M. J. D. Low, *Anal. Letters* **1,** 819 (1968).

114 The Warner and Swasey Co., 32–16 Downing Street, Flushing, New York, USA.

115 S. A. Dolin, H. A. Kruegle and B. Krakow, *Pittsburg Conference on Analytical Chemistry and Applied Spectroscopy*, Cleveland, Ohio, USA, March 1968.

116 S. A. Dolin, H. A. Kruegle and G. J. Penzias, *Appl. Optics* **6,** 267 (1967).

117 G. C. Pimentel, *Appl. Optics* **7,** 2155 (1968).

118 R. M. Hexter and C. W. Hand, *Appl. Optics* **7,** 2161 (1968).

119 H. J. Brabor and R. H. Towrin, *Appl. Optics* **7,** 2171 (1968).

120 *Table of Wavenumbers for the Calibration of Infrared Spectrometers, Pure and Applied Chemistry* **1,** 537 (1961).

121 Sadtler Research Laboratories, 3316 Spring Garden Street, Philadelphia, Pa. 19104, USA. Distributed in Europe by Heyden & Son Ltd.

122 Coblentz Society Spectra published by Sadtler Research Laboratories, 3316 Spring Garden Street, Philadelphia, Pa. 19104, USA. Distributed in Europe by Heyden & Son Ltd.

123 *Molecular Formula List of Compounds, Names and References to Published Infrared Spectra.* ASTM STP No. 331 and 331A. *Serial Number List of Compounds, Names and References to Published Infrared Spectra.* ASTM STP No. 358. Published by ASTM 1916 Race Street, Philadelphia 3, Pa., USA.
The revised Molecular Formula List (July 1969) is now referred to as publication AMD 31. The revised Serial Number List (July 1969) is now referred to as AMD 32. (Distributed in Europe by Heyman & Son Ltd.)

124 American Petroleum Institute, Research Project 44, Chemistry Dept., Agricultural and Mechanical College of Texas, College Station, Texas, USA.

125 NRC-NBS committee on *Spectral Absorption Data,* National Bureau of Standards, Washington 25, D.C., USA.

126 Manufacturing Chemists Assn. Research Project, Chemistry Dept., Agricultural and Mechanical College of Texas, College Station, Texas, USA.

127 Infrared Data Committee of Japan, Sanyo Shuppan Boeki Co., Hoyu Building, 8, 2-chome, Takara-Cho, Chuo-ku, Tokyo, Japan. Distributed in Europe by Heyden & Son Ltd.

128 *Infra-red Spectra of Gases and Vapors,* Vols. 7 and 2, D. S. Erley and B. H. Blake, The Dow Chemical Company, Midland, Michigan, USA.

Subject index